5 STEPS TO A 5™

AP Computer Science A

2019

AP Computer Science A

2019

Dean R. Johnson
Carol A. Paymer
Aaron P. Chamberlain

New York Chicago San Francisco Athens London Madrid
Mexico City Milan New Delhi Singapore Sydney Toronto

1 2 3 4 5 6 7 8 9 LHS 23 22 21 20 19 18

ISBN 978-1-260-12264-0
MHID 1-260-12264-6

e-ISBN 978-1-260-12265-7
e-MHID 1-260-12265-4

The series editor was Grace Freedson, and the project editor was Del Franz.

Series design by Jane Tenenbaum.

Contents

Preface

Way to go! By taking the AP Computer Science A Exam, you've decided to gain a deeper knowledge of the world around you. This knowledge will give you an advantage no matter what field you study in the future.

Computer science, ironically, is a much misunderstood subject. Even in this digital age, where people are dependent on their smartphones, they see their phone as a mysterious "black box." They receive messages and notifications on a daily basis, but, for some reason, they don't understand that someone had to write the software that makes their life richer. Software developers write the computer programs that people use every day. AP Computer Science A is a course that teaches you how to become a software developer.

The purpose of this book is to help you score a 5 on the AP Computer Science A Exam. Along the way, your Java programming skills and your ability to problem-solve should dramatically improve.

Here are some pieces of advice for you.

- **Be persistent.** Software developers hate it when they can't figure out how to make a program work and even when they succeed, they always try to think of better ways to solve the problem.
- **Try to see the big picture**. This is good advice not only when studying for this exam, but for life in general. The more you see problems from multiple perspectives, the more likely it is that you will understand how to solve them. Complex program designs require you to step back and view from a distance.
- **Be patient with yourself and others.** Everyone learns at a different rate. Some of the topics in AP Computer Science A are quite challenging, so you may have to reread concepts in this book until you understand them. If you are learning this with other students, help each other out. As a programmer, you should live by the mantra that "we are all in this together."
- **Too much of a good thing is a bad thing**. Software developers need to strike a balance between structure and creativity. That is, a good developer must understand how to do the technical things (structure) or else their program won't run. But programmers also need to be creative in order to solve problems in a clever way. Work on improving both of these two skills simultaneously so you can become the best programmer you can be.
- **Above all else, work toward original thought.** You are going to need to learn how to think for yourself in order to become a good programmer. Be curious and learn how things work so that you can apply skills and techniques to new situations. Also, learn how to explain things to other people, as this will develop you as a programmer and leader.

Acknowledgments

This book was made possible by the contributions of many people. First, I'd like to thank Grace Freedson and McGraw-Hill Education for giving me the opportunity to write this book. Thank you to my editors, Del Franz and Patricia Wallenburg, for their help in preparing the book for publication.

A huge thank you goes to my partner, Carol Paymer. This book would not have been possible without her tireless efforts of writing questions and explanations and proofreading. Carol's meticulous explanations of the solutions provide the student with the type of guided practice they need as they strive toward their goal of reaching a 5 on the AP Computer Science A Exam. Her superior editing skills make me look a lot better than I deserve; I couldn't have finished this project without her. Also, a big thank you to Aaron Chamberlain of Fort Atkinson High School for his contribution of many of the questions in this book.

A special thank you goes to Fort Atkinson High School computer science student Megan Charland, whose contributions in editing helped it to be seen from the student's point of view.

I want to thank many Wisconsin computer science colleagues for their role in reviewing this text. They include Mike King, John Quinn, Bob Juranitch, Dan Rhode, Donna Krasovich, Brittany Morgan, Theresa Breunig, John Jordan, and John Myers.

I'd also like to thank Patrick Monaghan of Blackthorne LLC, Whitewater, Wisconsin, for taking a leap of faith when hiring me as a developer. This real-world work experience enriched my classroom teaching methods and ultimately contributed to many of the technical understandings I've presented in this book.

Finally, to my wife, Sandy, it is difficult to quantify how much I love and appreciate you. You have unselfishly supported me, and as a result, I am a better teacher and person. You are truly the wind beneath my wings. This book is dedicated to you.

Dean R. Johnson

About the Authors

DEAN R. JOHNSON has taught mathematics and computer science at Fort Atkinson High School, Fort Atkinson, Wisconsin, since 1989. His students appreciate his passion and enthusiasm for teaching in his own unique style. He also works as a software consultant for a financial start-up, Blackthorne LLC, in Whitewater, Wisconsin. He is a certified Alice trainer and has contributed exercises for the *Learning to Program in Alice*, 3rd edition, textbook. He and his wife, Sandy, have four children, David, Jessica, Tommy, and Jordan, who are each amazing in their own way.

CAROL A. PAYMER teaches computer science at Campolindo High School in Moraga, California. She wrote her first computer program in high school in the early 1970s. She went on to earn an ScB from Brown University and an MS from Stanford before working at Bell Labs, Atari, and various start-ups that have vanished into the ether. Raising her three extraordinary children inspired her to explore teaching and she discovered her calling. She would like to dedicate this book to the many students who fill her days with joy and a sense of purpose and to Richard who runs beside her both uphill and down.

AARON P. CHAMBERLAIN began his career as a systems analyst at SCSI in Madison, Wisconsin. He also served in the United States Army Reserves as a Management Transportation Coordinator and was decorated with numerous medals and awards. Since 2016, he has taught mathematics and computer science at New Berlin Eisenhower High School in New Berlin, Wisconsin. He would like to dedicate his work on this book to his three children, Isabelle, Stefan, and Aria, who provide a continuous source of inspiration.

Introduction

Computer science provides limitless possibilities. It gives you power. It drives innovation. Computer science teaches skills that transfer to other disciplines: creativity, adaptability, structured processing, strategic problem solving, reflective analysis, relevant verification, and usability. Most importantly, it will define the future leaders of our world.

—Dr. Thomas Halbert, computer science teacher, Houston, Texas

Organization of the Book

This book was created to help you earn a 5 on the AP Computer Science A Exam. It is important that you understand how this book has been written specifically for you.

This book contains the following:

- An explanation of the highly successful Five-Step Program
- A suggested calendar based on how you want to prepare for the AP Computer Science A Exam
- A series of test-taking strategies
- A diagnostic test
- A thorough explanation of each topic on the AP Computer Science A Exam
- Two practice exams with explanations

Computer science is one of those fields that can be taught in many completely different ways; even the order in which things are taught can be different. Sure, there are some concepts that you have to understand before you can move on to others, but the reality is, there is a lot of flexibility when it comes to how to learn programming.

For instance, throw a dozen computer science teachers in a room and ask them if classes and objects should be taught first, or if the fundamentals of programming should be taught first. Then, grab a bag of popcorn and sit back and enjoy the show. Computer science teachers have been arguing which way is best for years. I have taught high school students both ways, and I believe there are valid arguments on both sides. This is one of those personal decisions that each teacher gets to make.

I have chosen to introduce the fundamentals of programming first. This way, when I introduce and explain classes and objects, I will reinforce these fundamental programming concepts. If your teacher teaches the concepts in a different order, don't worry, it will all work out in the end.

> The order that you learn the concepts in your computer science class may be different from the order that is presented in this book. And that's OK.

Advice from My Students

I asked my students for advice on writing this book. One thing they told me is to never explain a new topic using concepts that haven't been introduced yet. Students hate it when authors say, "Oh, I'm going to use something that I won't be explaining right now, but we'll get to it in a later chapter." I have tried really hard to design this book in a linear fashion, so I hope that I never do this to you.

No Forward Referencing
The concepts in this book move from easiest to hardest, and some of the concepts are even split into two sections (Basic and Advanced) so that the flow of the book is more natural. I will use earlier concepts to explain later concepts, but not the other way around.

One of my goals in writing this book was to strike the right balance between explaining topics in a way that *any* high school student can understand but still technical enough so students can earn a 5 on the exam. My students recommended that I start off by explaining each concept in a simple way so that anyone can understand it, and then go into greater detail. I hope you will appreciate my attempts at this. In other words, I've tried to make the book easy enough for beginning Java programmers to understand the basics so they are not completely lost, but then hard enough so that advanced students have the potential to earn a 5 on the AP exam.

Basic and Advanced Levels

It is for this reason that I have also chosen to create two levels of questions for each of the main concepts: Basic and Advanced. As a teacher, I have learned that students like to know that they know the basics of a concept before moving on to harder questions. If a book only has really hard practice questions, then most of the beginning students find them impossible to do, start feeling like failures, lose interest, and eventually drop out because they can't relate to anything the book is talking about.

I've also learned that it helps to try to find something that the student can relate to and explain the concept using that. I have tried to use examples that high school students can identify with.

The Concept Summary

Each concept begins with a concept summary. The summary tells what's covered in the concept, and it should give you a reason why you would want to learn the material covered in that concept. It also is a great tool for deciding if you need to look further at that topic.

The Rapid Review

Each concept ends with a "Rapid Review," which is a complete summary of the important ideas relating to the concept. The lists are full of wonderful one-liners that you should understand before taking the AP Computer Science A Exam. Make sure that you understand every line in the Rapid Review before taking the exam.

Learn the Rapid Review Content
Read every line in every Rapid Review and know what each means!

The Graphics Used in This Book

Powerful Tip: This graphic identifies a helpful tip for taking the AP Computer Science A Exam.

Fun Fact: This graphic identifies fun deviations from the subject matter.

Warning: This graphic points out a common error to avoid.

If you have any questions, suggestions, or would like to report an error, we'd love to hear from you. Email us at 5stepstoa5apcsa@gmail.com.

The Five-Step Program

This guide conducts you through the five steps necessary to prepare yourself for success on the exam. These steps will provide you with the skills and strategies that are vital to the exam as well as the practice that can lead you toward the perfect 5. Reading this guide will not guarantee you a 5 when you take the AP Computer Science A Exam in May. However, by understanding what's on the exam, using the test-taking strategies provided for each section of the test, and carefully reviewing the concepts covered on the exam, you will definitely be on your way!

Step One gives you the basic information you need to know about the exam and helps you determine which type of exam preparation you want to commit to. In Step One you'll find:

- The goals of the AP Computer Science A Exam
- Frequently asked questions about the exam
- An explanation of how you will be graded
- Calendars for three different preparation plan:
  - Full School Year Plan (September through May)
  - One-Semester Plan (January through May)
  - Six-Week Plan (six weeks prior to the exam)

Step Two introduces you to the types of questions you'll find on the actual test and, by identifying the content areas where you are weak, allows you to prioritize the areas you most need to review. In Step Two you'll find:

- A diagnostic exam in AP Computer Science A. This test is just like the real test except that it is only half its length
- Answers and explanations for all the exam questions. Read the explanations not just for the questions you miss, but also for all the questions and answer choices you didn't completely understand

Step Three helps you develop the strategies and techniques you need in order to do your best on the exam. In Step Three you'll learn:

- Strategies to tackle the multiple-choice questions efficiently and effectively
- Strategies to maximize your score on the free-response questions

Step Four develops the knowledge and skills necessary to do well on the exam. This is organized into 14 key concepts (Concepts 0–13). In Step Four, you'll review by using:

- Complete, but easy-to-follow, explanations of all of the concepts covered on the exam
- Review questions—both basic and advanced levels—for each concept, followed by step-by-step explanations so you'll understand anything you missed
- Rapid Reviews for each concept that you can use to make sure you have the knowledge and skills you'll need

Step Five will give you test-taking practice and develop your confidence for taking the exam. In Step Five you'll find:

- Two full-length practice exams that closely resemble the actual AP Computer Science A Exam. As well as checking your mastery of the concepts, you can use these exams to practice pacing and using the test-taking strategies given in Step Three
- Step-by-step explanations for all the exam questions so you can understand anything you missed and learn from your mistakes
- Scoring guidelines so you can assess your performance and estimate your score in order to see how if you have reached your goal

5 STEPS TO A 5™

AP Computer Science A

2019

Set Up Your Study Program

What You Need to Know About the AP Computer Science A Exam

IN THIS CHAPTER

Summary: This chapter provides you with background information on the AP Computer Science A exam. Learn about the AP exam, how exams are graded, what types of questions are asked, what topics are tested, and basic test-taking information.

Key Ideas

- The AP Computer Science A exam is about problem solving and the Java programming language.
- The exam has two parts: Multiple Choice and Free Response.
- Scoring a 4 or 5 will award you credit at nearly every college or university.

Background Information

A Brief History of the Exam

The first AP Computer Science exam was given in 1984, and the coding language used was Pascal. In 1999, C++ replaced Pascal. In 2004, Java replaced C++. As of 2015, the AP Computer Science A exam is the fastest-growing exam among all AP exams.

Goals of the Exam

The main goals of the AP Computer Science A exam are to test how proficient you are at problem solving and how well you know the Java programming language.

According to the College Board, here is a list of all of the goals of the test:

- Design, implement, and analyze solutions to problems.
- Use and implement commonly used algorithms.
- Develop and select appropriate algorithms and data structures to solve new problems.
- Write solutions fluently in an object-oriented paradigm.
- Write, run, test, and debug solutions in the Java programming language, utilizing standard Java library classes and interfaces from the AP Java subset.
- Read and understand programs consisting of several classes and interacting objects.
- Read and understand a description of the design and development process leading to such a program.
- Understand the ethical and social implications of computer use.

This list may seem rather ambiguous and open-ended; that's because it is. Just keep in mind that this test is all about learning how to solve different kinds of problems using Java. The best way to prepare for the exam is to follow the steps in this book and get plenty of practice designing and writing code.

Goal of the AP Computer Science A Exam

The main goals of the AP Computer Science A exam are to test how proficient you are at problem solving and how well you know the Java programming language.

Frequently Asked Questions About the Exam

Who Writes the AP Computer Science Exams? Who Grades Them?

The AP Computer Science A exam is designed by a committee of college professors and high school AP Computer Science teachers. The process takes years to ensure that the exam questions reflect the high quality and fairness that is expected of the College Board.

Similarly, the free-response questions are scored by hundreds of college professors and high school AP Computer Science teachers. The AP exam readers are thoroughly trained so that all exams are graded consistently.

Why Take the AP Computer Science A Exam?

There are several benefits of taking the AP Computer Science A exam. First is the college credit. If you score a 4 or 5 on the exam, you will earn credit from nearly every college or university. This saves you money. Second, you will start off at a higher level in your college coursework so that you can advance faster.

Finally, and my favorite, is that learning computer science will improve your ability to think for yourself and problem solve. Even if you don't plan on becoming a software developer, or you fear that you will not pass the exam, the skills and thought process that you learn in computer science will benefit you. Your world will be filled with rainbows and flying unicorns. Well, maybe not, but I'm just seeing if you are reading this.

Computer Science Makes You Think
Learning how to program improves your ability to think. You learn how things work.

When Is the Exam and How Do I Register for It?

The College Board publishes an exam schedule each year. The schedule for the various AP exams is two weeks long and starts on the first Monday of May. In the past, the AP Computer Science A exam has been given during the first week.

If you are taking an AP Computer Science class, you will probably get information from someone called an AP coordinator at your school. Normally, he or she likes to have everyone sign up online before March 1. If you are not taking an AP Computer Science class and just taking this exam on your own, contact the AP coordinator for your local school district by March 1. That person will help you locate the testing site and help you register. You can take as many AP exams as you want (even if you didn't take a class!), but they cost around $100 per exam.

What Is the Format of the AP Computer Science A Exam?

The exam is three hours long and there are two sections: multiple choice and free response.

Section I	Multiple Choice	40 Questions	90 Minutes	40 points (1 point per problem)	50% of Exam Score
Section II	Free Response	4 Questions	90 Minutes	36 points (9 points per problem)	50% of Exam Score

How Is My Final Score Calculated?

Each question in the multiple-choice section is worth 1 point (total of 40 points). Each of the free-response questions is worth 9 points (total of 36 points). Since there are an unequal number of points for the multiple-choice and the free-response sections, the total from the free-response section is multiplied by 1.1111. This makes each section worth 50 percent.

The total from the two sections is fed into a chart similar to the one below and your final grade is decided. This chart is intended to serve as a guide, and the ranges will vary slightly from year to year.

Composite Score Range	AP Score	Interpretation
62–80	5	Extremely Well Qualified
44–61	4	Well Qualified
31–43	3	Qualified
25–30	2	Possibly Qualified
0–24	1	No Recommendation

Example: Computing a Score on the AP Computer Science A Exam
You score 31 on Section I (Multiple Choice).
You score 28 on Section II (Free Response).
Multiply 28 by 1.1111. This equals 31.1108.
Your total is 31 + 31.1108 = 62.1108, which is rounded to 62.
Your composite score is 62. Your AP score is a 5. Congratulations!

What Concepts Are Included in the Multiple-Choice Section?

The table below gives a rough idea of what topics are covered on the multiple-choice section of the test. The percentages change from year to year, but the emphasis stays about the same. Keep in mind that many of the questions belong to more than one category, which is why the percentages don't add up to 100 percent.

Concept	Percent of Multiple-Choice Questions
Programming Fundamentals	55–75%
Algorithms & Problem Solving	25–45%
Data Structures	20–40%
Object-Oriented Programming	15–25%
Logic	5–15%
Recursion	5–15%
Software Engineering	2–10%

What Is the Lab Requirement for the Exam?

Starting with the 2015 exam, a lab requirement replaced the GridWorld case study as a way for the student to get hands-on experience with a complex package with lots of class hierarchy. Your teacher should provide you with a chance to work with a big, complex set of related classes that helps you understand concepts like inheritance, polymorphism, abstract classes, and interfaces. Your teacher has the choice of either using the labs that are provided by the College Board or using something completely different.

The three labs that are provided by the College Board are:

1. Magpie Chatbot Lab—An exploration of some of the basics of Natural Language Processing.
2. Elevens Lab—A simple game of solitaire.
3. Picture Lab—An exploration in modifying digital pictures.

I See Materials on the GridWorld Case Study Online. What's with That?

That is old news. The GridWorld case study is no longer tested on the AP Computer Science A exam. It was dropped after the 2014 exam. If you want to check it out, there are some pretty interesting things you can do with it, but the GridWorld case study won't be on your exam.

What Is the Difference Between the AP Computer Science A and the AP Computer Science Principles Exams?

Starting with the 2016–2017 school year, a new AP Computer Science course called AP Computer Science Principles is being offered. The AP Computer Science Principles course introduces students to computational thinking skills and promotes understanding of the impact of computers in our world. The biggest difference between these two courses is that AP Computer Science A focuses mainly on programming, whereas only a small part of the AP Computer Science Principles course is devoted to actual programming.

CHAPTER 2

How to Plan Your Time

IN THIS CHAPTER

Summary: The right preparation plan for you depends on your study habits, your own strengths and weaknesses, and the amount of time you have to prepare for the test. This chapter recommends some possible plans to get you started.

Key Ideas

- It helps to have a plan—and stick with it!
- You should select the study plan that best suits your situation and adapt it to fit your needs.

Three Approaches to Preparing for the AP Computer Science A Exam

It's up to you to decide how you want to use this book to study for the AP Computer Science A exam. In this chapter you'll find three plans, each of which provides a different schedule for your review effort. Choose one or combine them if you want. Adapt the plan to your strengths and weaknesses and the way you like to study. If you are taking an AP Computer Science A course at your school, you will have more flexibility than someone learning the material independently.

The Full School-Year Plan

Choose this plan if you like taking your time going through the material. Following this path will allow you to practice your skills and develop your confidence gradually. This is a good choice if you want to use this book as a resource while taking an AP Computer Science A course.

The One-Semester Plan

Choose this plan if you are OK with learning a lot of material in a fairly short amount of time. You'll need to be a pretty good student who can grasp concepts quickly. This plan is also a good choice if you are currently taking an AP Computer Science A course.

The Six-Week Plan

This option is available if any one of these sounds like you:

- You are enrolled in an AP Computer Science A course and want to do a final review before the exam.
- You are enrolled in an AP Computer Science A course and want to use this book as a reference to refresh your memory.
- You are not currently enrolled in an AP Computer Science A course, but you are a fluent Java programmer and want to know what is tested on the exam.

When to Take the Practice Exams

You should take the practice exams *prior* to May. The AP Computer Science A exam is usually given during the first week of May. If you wait until May to take the practice exams, you won't have enough time to review the concepts that you don't fully understand.

The Three Plans Compared

The chart summarizes and compares the three study plans.

Month	Full School-Year Plan	One-Semester Plan	Six-Week Plan
September	Diagnostic Exam Concepts 0–1	—	—
October	Concepts 2–4	—	—
November	Concepts 5–6	—	—
December	Concepts 7–8	—	—
January	Concept 9	Diagnostic Exam Concepts 0–6	—
February	Concepts 10–11	Concepts 7–9	—
March	Concepts 12–13 Practice Exam 1 Review all concepts	Concepts 10–13	Diagnostic Exam Concepts 0–13
April	Practice Exam 2 Review all concepts	Practice Exam 1 Review all concepts Practice Exam 2	Practice Exam 1 Review all concepts Practice Exam 2
May	Review all concepts	Review all concepts	Review all concepts

Calendars for Preparing for Each of the Plans

The Full School-Year Plan

SEPTEMBER
- Read Concepts 0–1.
- Learn how the book is put together.
- Determine your approach.
- Skim the practice exams.
- Choose an IDE, the *integrated development environment* in which you will write your code.
- Take the diagnostic exam.
- Learn all of the primitive variables.
- Learn the basic conditional structures.
- Learn the basic looping structures.

OCTOBER
- Read Concepts 2–4.
- Learn how to create your own class.
- Learn how to create your own methods.
- Learn the String class.
- Learn the methods from the Math class.

NOVEMBER
- Read Concepts 5–6.
- Learn the array.
- Learn the 2-D array.
- Learn the ArrayList.
- Learn all the basic algorithms.

DECEMBER
- Read Concepts 7–8.
- Learn the advanced features of classes.
- Learn how to extend classes.
- Learn inheritance.
- Learn polymorphism.

JANUARY
- Read Concept 9.
- Learn about abstract classes.
- Learn about interfaces.

FEBRUARY
- Read Concepts 10–11.
- Learn how to read recursive code.
- Learn all the sorting algorithms.

MARCH
- Read Concepts 12–13.
- Learn all the advanced algorithms.
- Learn how to design large-scale object-oriented applications.

APRIL
- Take Practice Exam 1.
- Review the material on the exam.
- Take Practice Exam 2.
- Review the material on the exam.

MAY
- Do any last minute reviewing you need to.
- Good luck!

The One-Semester Plan

JANUARY
- Read Concepts 0–6.
- Learn how the book is put together.
- Determine your approach.
- Skim the practice exams.
- Take the diagnostic exam.
- Choose an IDE, the *integrated development environment* in which you will write your code.
- Be familiar with all variable types.
- Learn the basic conditional structures.
- Learn the basic looping structures.
- Learn how to create your own class.
- Learn the String class and Math class.
- Learn the array, 2-D array, and ArrayList.

FEBRUARY
- Read Concepts 7–9.
- Learn the advanced features of classes.
- Learn inheritance and polymorphism.
- Learn abstract classes and interfaces.

MARCH
- Read Concepts 10–13.
- Learn how to read recursive code.
- Learn the sorting algorithms.
- Learn all the advanced algorithms.
- Learn how to design large-scale object-oriented applications.

APRIL
- Take Practice Exam 1.
- Review the material on the exam.
- Take Practice Exam 2.
- Review the material on the exam.

MAY
- Do any last minute reviewing you need to.
- Good luck!

The Six-Week Plan

MARCH

- Take the diagnostic exam.
- Read all Concepts: 0–13.
- Learn
  - all variable types
  - all conditional statements
  - all looping structures
  - all complex data structures
  - how to create a full class
  - inheritance and polymorphism
  - how to read recursive methods
  - all sorting algorithms
  - all the advanced algorithms
  - how to design and implement large-scale object-oriented applications

APRIL

- Take Practice Exam 1.
- Review the material on the exam.
- Take Practice Exam 2.
- Review the material on the exam.

MAY

- Do any last-minute reviewing you need to.
- Good luck!

Determine Your Test Readiness

Take a Diagnostic Exam

IN THIS CHAPTER

Summary: This step contains a diagnostic exam that is exactly one-half the length of the actual AP Computer Science A Exam. However, in all other ways, it closely matches what you can expect to find on the real test. Use this test to familiarize yourself with the AP Computer Science A Exam and to assess your strengths and weaknesses as you begin your review for the test.

Key Ideas

- Familiarize yourself with the types of questions and the level of difficulty of the actual Computer Science A Exam early in your test preparation process.
- Use the diagnostic exam to identify the content areas on which you need to focus your test preparation efforts.

Using the Diagnostic Exam

The purpose of the diagnostic exam is to give you a feel for what the actual AP Computer Science A Exam will be like and to identify content areas that you most need to review. This diagnostic exam is one-half the length of the real exam.

When to Take the Diagnostic Exam

Since one purpose of the diagnostic exam is to give you a preview of what to expect on the AP Computer Science A Exam, you should take the exam earlier rather than later. However, if you are attempting the diagnostic exam without having studied Java yet, you may feel overwhelmed and confused. If you are starting this book in September, you may want to read a sampling of questions from the diagnostic exam, but then save the test until later when you know some Java. Taking the diagnostic exam when you begin to review will help you identify what content you already know and what content you need to revisit; then you can alter your test prep plan accordingly.

Take the diagnostic exam in this step when you begin your review, but save both of the full-length practice exams at the end of this book until after you have covered all of the material and are ready to test your abilities.

How to Administer the Exam

When you take the diagnostic exam, try to reproduce the actual testing environment as closely as possible. Find a quiet place where you will not be interrupted. Do not listen to music or watch a movie while taking the exam! You will not be able to do this on the real exam. Set a timer and stop working when the 45 minutes are up for each section. Note how far you have gotten so you can learn to pace yourself, but take some extra time to complete all the questions so you can find your areas of weakness. Use the answer sheet provided and fill in the correct ovals with a #2 pencil. Although this is a computer science exam, the AP exam is a paper-and-pencil test.

Part I of the exam contains multiple-choice questions. On the actual exam you'll have 90 minutes to complete 40 multiple-choice questions. For this diagnostic exam, give yourself 45 minutes to complete the 20 multiple-choice questions. Note that there is no penalty for guessing on the exam, so if you're not sure of an answer, eliminate obviously wrong answer choices and guess. You'll find more helpful strategies to use in answering the multiple-choice questions in Step 3.

Part II of the exam consists of free-response questions. On the AP Computer Science A Exam, you'll have 90 minutes to complete four questions, each containing multiple parts. On this diagnostic exam, give yourself 45 minutes to complete the two free-response questions. Strategies you can use to approach the free-response question efficiently and effectively can be found in Step 3. Read those strategies after you've tried the diagnostic exam and become familiar with the types of questions on the exam.

After Taking the Diagnostic Exam

Following the exam, you'll find not only the answers to the test questions, but also complete explanations for each answer. Don't just read the explanations for the questions you missed; you also need to understand the explanations for the questions you got right but weren't sure of. In fact, it's a good idea to work through the explanations for *all* the questions. Working through the step-by-step explanations is one of the most effective review tools in this book.

If you missed a lot of questions on the diagnostic exam or are just starting out learning how to program in Java, don't stress out! Step 4 of this book explains all of the concepts that will appear on the AP Computer Science A Exam. Read Concepts 0–13 in Step 4, do the practice questions for each concept, and read the explanations so you understand what the correct answer is and why it is correct. Then you'll be well prepared for the AP Computer Science A Exam.

On the other hand, if you did well on some of the questions on the diagnostic exam and understand the explanations for these questions, you may be able to skip some of the concepts in Step 4. Look at the summaries and key ideas that begin each of the concepts in Step 4. If you're reasonably sure you understand a concept already, you may want to skip to that concept's review questions or the "Rapid Review" and then move on to the next concept. A good test prep plan focuses on the areas you most need to review.

Now let's get started.

Diagnostic Exam

Multiple-Choice Questions

ANSWER SHEET

1	Ⓐ	Ⓑ	Ⓒ	Ⓓ	Ⓔ	11	Ⓐ	Ⓑ	Ⓒ	Ⓓ	Ⓔ
2	Ⓐ	Ⓑ	Ⓒ	Ⓓ	Ⓔ	12	Ⓐ	Ⓑ	Ⓒ	Ⓓ	Ⓔ
3	Ⓐ	Ⓑ	Ⓒ	Ⓓ	Ⓔ	13	Ⓐ	Ⓑ	Ⓒ	Ⓓ	Ⓔ
4	Ⓐ	Ⓑ	Ⓒ	Ⓓ	Ⓔ	14	Ⓐ	Ⓑ	Ⓒ	Ⓓ	Ⓔ
5	Ⓐ	Ⓑ	Ⓒ	Ⓓ	Ⓔ	15	Ⓐ	Ⓑ	Ⓒ	Ⓓ	Ⓔ
6	Ⓐ	Ⓑ	Ⓒ	Ⓓ	Ⓔ	16	Ⓐ	Ⓑ	Ⓒ	Ⓓ	Ⓔ
7	Ⓐ	Ⓑ	Ⓒ	Ⓓ	Ⓔ	17	Ⓐ	Ⓑ	Ⓒ	Ⓓ	Ⓔ
8	Ⓐ	Ⓑ	Ⓒ	Ⓓ	Ⓔ	18	Ⓐ	Ⓑ	Ⓒ	Ⓓ	Ⓔ
9	Ⓐ	Ⓑ	Ⓒ	Ⓓ	Ⓔ	19	Ⓐ	Ⓑ	Ⓒ	Ⓓ	Ⓔ
10	Ⓐ	Ⓑ	Ⓒ	Ⓓ	Ⓔ	20	Ⓐ	Ⓑ	Ⓒ	Ⓓ	Ⓔ

AP Computer Science A Diagnostic Exam

Part I
Multiple Choice
Time: 45 minutes
Number of questions: 20
Percent of total score: 50

Directions: Choose the best answer for each problem. Some problems take longer than others. Consider how much time you have left before spending too much time on any one problem.

Notes:

- The diagnostic exam is exactly one-half the length of the actual AP Computer Science A Exam.
- You may assume that all import statements have been included where they are needed.
- You may assume that the parameters in method calls are not null.
- You may assume that declarations of variables and methods appear within the context of an enclosing class.

1. Consider the following method.

```
public int someMethod(int val)
{
    for (int i = 2; i < 7; i++)
    {
        if((val + i) % 2 == 0)
        {
            val += 3;
        }
    }
    return val;
}
```

What value is returned by the call `someMethod(13)`?

(A) `17`
(B) `25`
(C) `28`
(D) `31`
(E) Nothing is returned. There is a compile-time error.

2. Consider the following code segment.

```
int num1 = 2;
int num2 = 13;
int result = 4;

if ((num1 < 5) && (num2 < 5))
    result = num1 - num2;
else if ((num1 == 2) && (num2 < 2))
    result = num2 - num1;
else
    result = num1 + num2;
System.out.println(result);
```

What is printed as a result of executing the code segment?

(A) `-11`
(B) `4`
(C) `11`
(D) `13`
(E) `15`

3. Assume `list` is an `ArrayList<Integer>` that has been correctly constructed and populated with the following items.

```
[13, 7, 0, 5, 12, 6, 10]
```

Consider the following method.

```
public int calculate(ArrayList<Integer> numbers)
{
    int sum = 0;

    for (Integer n : numbers)
    {
        if (n - 8 > 0)
        {
            sum = sum + n;
        }
    }
    return sum;
}
```

What value is returned by the call `calculate(list)`?

(A) `10`
(B) `11`
(C) `13`
(D) `35`
(E) `45`

4. Consider the following class declarations.

```
public class Planet
{
    private String name;
    private double mass;
    private int position;

    public Planet()
    {  /* implementation not shown  */  }

    public Planet(String name)
    {  /* implementation not shown  */  }

    public Planet(String name, int position)
    {  /* implementation not shown  */  }

}

public class DwarfPlanet extends Planet
{
    private double distance;
    public DwarfPlanet(String name)
    {  /* implementation not shown  */  }
}
```

Which of the following declarations compiles without error?

I. `Planet mars = new Planet();`
II. `Planet pluto = new DwarfPlanet("Pluto");`
III. `Planet ceres = new DwarfPlanet();`

(A) I only
(B) II only
(C) I and II only
(D) I and III only
(E) I, II, and III

5. Consider the following code segment.

```
List<String> supernatural = new ArrayList<String>();

supernatural.add("Vampire");
supernatural.add("Werewolf");
supernatural.add("Ghost");
supernatural.set(0, "Zombie");
supernatural.add(2, "Mummy");
supernatural.add("Witch");
supernatural.remove(3);
System.out.println(supernatural);
```

What is printed as a result of executing the code segment?

(A) `[Zombie, Werewolf, Mummy, Witch]`
(B) `[Zombie, Werewolf, Ghost, Witch]`
(C) `[Zombie, Werewolf, Mummy, Ghost]`
(D) `[Zombie, Vampire, Werewolf, Mummy, Ghost]`
(E) `[Zombie, Vampire, Werewolf, Mummy, Witch]`

6. Consider the following interface and class declarations.

```
public interface Polygon
{
    /**
     * @return whether the polygon contains the point defined by (x, y)
     */
    boolean contains(int x, int y);
}

public class ClickableZones
{
    private ArrayList<Polygon> zones;

    public int getNumberOfZones(int x, int y)
    {
        int total = 0;
        for (Polygon p : zones)
        {
            /*  missing code  */
        }
        return total;
    }

    /* Additional implementation not shown */
}
```

Which of the following could replace `/* missing code */` so that the method `getNumberOfZones` returns the number of `Polygons` that contain the coordinate point defined by `x` and `y`?

(A)
```
if (polygon contains x, y)
    return true;
```

(B)
```
if (p.contains(x, y))
    return true;
```

(C)
```
if (!p.contains(x, y))
    return false;
```

(D)
```
if (Polygon.contains(x, y))
    total++;
```

(E)
```
if (p.contains(x, y))
    total++;
```

7. Consider the following partial class declaration.

```
public class Park
{
    private String name;
    private boolean playground;
    private int acres;

    public Park(String myName, boolean myPlayground, int myAcres)
    {
        name = myName;
        playground = myPlayground;
        acres = myAcres;
    }

    public String getName()
    {  return name;  }

    public boolean hasPlayground()
    {  return playground;  }

    public int getAcres()
    {  return acres;  }

    /* Additional implementation not shown */
}
```

Assume that the following declaration has been made in the main method of another class.

```
Park park = new Park("Central", true, 300);
```

Which of the following statements compiles without error?

(A) `int num = park.acres;`
(B) `String name = central.getName();`
(C) `boolean play = park.hasPlayground();`
(D) `int num = park.getAcres(acres);`
(E) `park.hasPlayground = true;`

8. Consider the following code segment.

```
String idk = "mnomnomno";
for (int i = 0; i < idk.length(); i++)
{
    if (idk.substring(i, i + 1).equals("m"))
    {
        idk = idk.substring(0, i) + idk.substring(i + 1, idk.length());
    }
}
System.out.println(idk);
```

What is printed as a result of executing the code segment?

(A) mmm

(B) nonono

(C) mnomno

(D) nomnono

(E) mnomnomno

9. Consider the following method.

```
public int loopy(int n)
{
    if (n % 7 == 0)
    {
        return n;
    }
    return loopy(n + 3) + 2;
}
```

What value is returned by the call `loopy(12)`?

(A) 12

(B) 21

(C) 23

(D) 27

(E) 29

10. Consider the following class declarations.

```
public class Letter
{
    private String letter = "letter";

    public String toString()
    {   return letter;   }

    /* Additional implementation not shown */
}

public class ALetter extends Letter
{
    private String letter = "a";

    public String toString()
    {   return letter;   }

    /* Additional implementation not shown */
}

public class BLetter extends Letter
{
    private String letter = "b";

    public String toString()
    {   return letter;   }

    /* Additional implementation not shown */
}

public class CapALetter extends ALetter
{
    private String letter = "A";

    /* Additional implementation not shown */
}
```

Consider the following code segment.

```
Letter x = new ALetter();
Letter y = new BLetter();
ALetter z = new CapALetter();

System.out.print(x);
System.out.print(y);
System.out.print(z);
```

What is printed as a result of executing the code segment?

(A) `abA`
(B) `aba`
(C) `letterlettera`
(D) `letterletterletter`
(E) Nothing is printed. There is a compile-time error.

11. Consider the following method.

```
public String lengthen(String word)
{
    int index = 0;
    while (index < word.length())
    {
        word = word + word.substring(index, index + 1);
        index += 2;
    }
    return word;
}
```

What is returned by the call `lengthen("APCS")`?

(A) `"APCS"`
(B) `"APCSACAA"`
(C) `"APCSAPCS"`
(D) Nothing is returned. Run-time error: `StringIndexOutOfBoundsException`
(E) Nothing is returned. The call will result in an infinite loop.

12. Consider the following code segment.

```
int[] array = {-3, 0, 2, 4, 5, 9, 13, 1, 5};
for (int n = 1; n < array.length - 1; n++)
{
    if (array[n] - array[n - 1] <= array[n] - array[n + 1])
        System.out.print(array[n] + "   ");
}
```

What is printed as a result of executing the code segment?

(A) `1`
(B) `13`
(C) `13 1`
(D) `2 4 5`
(E) `2 4 5 9`

13. Assume that `k`, `m`, and `n` have been declared and correctly initialized with `int` values. Consider the following statement.

```
boolean b1 = (n >= 4) || ((m == 5 || k < 2) && (n > 12));
```

For which statement below does `b2 = !b1` for all values of `k`, `m`, and `n`?

(A) `boolean b2 = (n >= 4) && ((m == 5 && k < 2) || (n > 12));`
(B) `boolean b2 = (n < 4) || ((m != 5 || k >= 2) && (n <= 12));`
(C) `boolean b2 = (n < 4) && (m != 5) && (k >= 2) || (n <= 12);`
(D) `boolean b2 = (m == 5 || k < 2) && (n > 12);`
(E) `boolean b2 = (n < 4);`

14. Consider the following code segment.

```
int[] ray = new int[11];

for (int i = 0; i < ray.length; i++)
{
    ray[i] = i * 2;
}

for (int m = 0; m < 5; m++)
{
    for (int n = 0; n < 7; n += 2)
    {
        if (m + n > 8)
        {
            System.out.print(ray[m + n]);
        }
    }
}
```

What is printed as a result of executing the code segment?

(A) Nothing is printed. Runtime error: `ArrayIndexOutOfBounds`
(B) `18`
(C) `181620`
(D) `68`
(E) `1820`

15. Consider the following method.

```
public int mystery(String code, int index)
{
    if (code.indexOf("c") == index)
    {
        return index;
    }
    return mystery(code.substring(2), index + 1);
}
```

Assume that the string `codeword` has been declared and initialized as follows.

```
String codeword = "advanced placement";
```

What value is returned by the call `mystery(codeword, 9)`?

(A) `5`
(B) `6`
(C) `7`
(D) Nothing is returned. Infinite recursion causes a stack overflow error.
(E) Nothing is returned. Run-time error: `StringIndexOutOfBoundsException`

16. Consider the following method.

```
public void switcheroo(int num, int index, int[] nums)
{
    int temp = num;
    num = nums[index];
    nums[index] = temp;
    index++;
}
```

Consider the following code segment.

```
int[] val = {5, 7, 4, -2, 8, 12};
int num = 10;
int index = 3;
switcheroo(num, index, val);

System.out.println("num = " + num + "  val[" + index + "] = " + val[index]);
```

What is printed as a result of executing the code segment?

(A) `num = 10  val[3] = 10`
(B) `num = 10  val[3] = -2`
(C) `num = 10  val[4] = 8`
(D) `num = -2  val[3] = 10`
(E) `num = -2  val[4] = 8`

17. Consider the following code segment.

```
for (int h = 2; h <= 6; h += 2)
{
    for (int k = 30; k > 0; k -= 10)
    {
        System.out.print(h + k + "    ");
    }
}
```

Consider these additional code segments.

I.
```
int num = 32;
int count = 0;
for (int i = 0; i < 9; i++)
{
    System.out.print(num + "    ");
    num += 2;
    if (count % 3 == 0)
    {
        count = 0;
        num -= 14;
    }
}
```

II.
```
int num = 32;
while (num < 38)
{
    System.out.print(num + "    ");
    num -= 10;
    if (num < 10)
    {
        num += 32;
    }
}
```

III.
```
for (int h = 0; h <= 3; h++)
{
    for (int k = 30; k > 0; k -= 10)
    {
        System.out.print(k + h + "    ");
    }
}
```

Which of the code segments produce the same output as the original code segment?

(A) I only
(B) II only
(C) III only
(D) II and III only
(E) I, II, and III

18. Consider the following method.

```
public void mystery (int[] array)
{
    for (int i = 1; i < array.length; i++)
    {
        int j;
        int key = array[i];
        for (j = i - 1; j >= 0 && array[j] > key; j--)
        {
            array[j + 1] = array[j];
        }
        array[j + 1] = key;
    }
}
```

The method above could be best described as an implementation of which of the following?

(A) Insertion Sort
(B) Binary Search
(C) Selection Sort
(D) Merge Sort
(E) Sequential Sort

19. Consider the following statement.

```
int number = (int)(Math.random() * 21 + 13);
```

After executing the statement, what are the possible values for the variable `number`?

(A) All integers from 13 to 21 (inclusive).
(B) All real numbers from 13 to 34 (not including 34).
(C) All integers from 13 to 34 (inclusive).
(D) All integers from 13 to 33 (inclusive).
(E) All real numbers from 0 to 21 (not including 21).

20. Consider the following class declaration.

```
public class City
{
    private String name;
    private int population;

    public City(String myName, int myPop)
    {
        name = myName;
        population = myPop;
    }

    public String getName()
    {   return name;   }

    public int getPopulation()
    {   return population;   }

    /*  Additional implementation not shown */
}
```

Assume `ArrayList<City> cities` has been properly instantiated and populated with `City` objects.

Consider the following code segment.

```
int maxPop = Integer.MIN_VALUE;
for (int i = 0; i < cities.size(); i++)
{
    /* missing code */
}
```

Which of the following should replace `/* missing code */` so that, after execution is complete, `maxPop` will contain the largest population that exists in the `ArrayList`?

(A)
```
City temp = cities[i];
if (temp.getPopulation() > maxPop)
{
    maxPop = temp.getPopulation();
}
```
(B)
```
City temp = cities.get(i);
if (temp.population > maxPop)
{
    maxPop = temp.population;
}
```
(C)
```
if (cities.get(i + 1).getPopulation() > cities.get(i).getPopulation())
{
    maxPop = cities.get(i + 1).getPopulation();
}
```
(D)
```
if (cities.get(i).getPopulation() > maxPop)
{
    maxPop = cities.get(i).getPopulation();
}
```
(E) `maxPop` should have been set to `Integer.MAX_VALUE`. This cannot work as written.

STOP. End of Part I.

AP Computer Science A Diagnostic Exam

Part II
Free Response
Time: 45 minutes
Number of questions: 2
Percent of total score: 50%

Directions: Write all of your code in Java. Show all your work.

Notes:

- The diagnostic exam is exactly one-half the length of the actual AP Computer Science A Exam.
- You may assume all imports have been made for you.
- You may assume that all preconditions are met when making calls to methods.
- You may assume that all parameters within method calls are not null.
- Be aware that you should, when possible, use methods that are defined in the classes provided as opposed to duplicating them by writing your own code.

1. Complex Numbers

In Mathematics, a complex number is a number that is composed of both a real component and an imaginary component. Complex numbers can be expressed in the form $a + bi$ where a and b are real numbers and i is the imaginary number $\sqrt{-1}$ (which means that $i^2 = -1$). In complex expressions, a is considered the *real part* and b is considered the *imaginary part.*

You can add, subtract, multiply and divide complex numbers. Addition with complex numbers involves adding the *real* parts and the *imaginary* parts as two separate sums and expressing the answer as a new complex number. Here are some examples of the addition of complex numbers.

Example 1: $(3 + 6i) + (4 + 2i)$
$= (3 + 4) + (6i + 2i)$
$= 7 + 8i$

Example 2: $(10 - 3i) + (-4 + 2i)$
$= (10 - 4) + (-3i + 2i)$
$= 6 - i$

Multiplication of complex numbers involves using the distributive property, just like multiplying polynomials. The first term is always a real number (no i), the middle two terms have an i and can be combined, and the last term has an i^2. Remember that $i^2 = -1$, so, after simplifying i^2 to -1 and multiplying, the last term is a real number (no i) and can be added to the first term.

$$(2 + 3i)(4 + 2i)$$

$2 * 4 = 8$ $2 * 2i = 4i$ $3i * 4 = 12i$ $3i * 2i = 6i^2 = -6$

Combine like terms: $8 + (-6) = 2$ and $4i + 12i = 16i$.

Write in complex number format: $2 + 16i$.

Here are some additional examples of multiplication of complex numbers.

Example 1: $(3 + 6i)(-4 + 2i)$
$= (3)(-4) + (3)(2i) + (6i)(-4) + (6i)(2i)$
$= -12 + 6i - 24i - 12$
$= -24 - 18i$

Example 2: $(2 + 4i)(3 - 5i)$
$= (2)(3) + (2)(-5i) + (4i)(3) + (4i)(-5i)$
$= 6 - 10i + 12i + 20$
$= 26 + 2i$

Write the complete `ComplexNumber class`, including:

- any necessary instance variables
- a constructor that takes two `int` variables, `a` and `b`, representing the components of the complex number expressed $a + bi$
- the `add` method that takes another `ComplexNumber` object as a parameter, adds to `this ComplexNumber` object as described above, and returns the solution as a `ComplexNumber` object
- the `multiply` method that takes another `ComplexNumber` object as a parameter, multiplies with `this ComplexNumber` object as described above, and returns the solution as a `ComplexNumber` object
- accessor methods `getA()` and `getB()` for the components of the complex number

An outline of the class including method declarations follows.

```
public class ComplexNumber
{
   // instance variables

   //constructor
   public ComplexNumber(int a, int b)
   {  /* to be completed */  }

   // add
   public ComplexNumber add(ComplexNumber cNum)
   {  /* to be completed */  }

   //multiply
   public ComplexNumber multiply(ComplexNumber cNum)
   {  /* to be completed */  }

   //accessors
   public int getA()
   {  /* to be completed */  }

   public int getB()
   {  /* to be completed */  }
}
```

2. Coin Collector

The High School Coin Collection Club needs new software to help organize its coin collections. Each coin in the collection is represented by an object of the `Coin` class. The `Coin` class maintains three pieces of information for each coin: its country of origin, the year it was minted, and the type of coin it is. Because coin denominations vary from country to country, the club has decided to assign a coin type of 1 to the coin of lowest denomination, 2 to the next lowest, and so on. For American coins, `coinType` is assigned like this:

Coin Name	Value	`coinType`
Penny	$0.01	1
Nickel	$0.05	2
Dime	$0.10	3
Quarter	$0.25	4
Half-Dollar	$0.50	5
Dollar	$1.00	6

```
public class Coin
{
    String country;
    int year;
    int coinType;

    public Coin(String cCountry, int cYear, int cType)
    {
        country = cCountry;
        year = cYear;
        coinType = cType;
    }

    public String getCountry()
    {  return country;  }

    public int getYear()
    {  return year;  }

    public int getCoinType()
    {  return coinType;  }
}
```

The Coin Club currently keeps track of its coins by maintaining an `ArrayList` of `Coin` objects for each country. The coins in the `ArrayList` are in order by year, oldest to newest. If two or more coins were minted in the same year, those coins appear in a random order with respect to the other coins from the same year.

The Coin Club has acquired some new collection boxes of various sizes to store their coins. The boxes are rectangular and contain many small compartments in a grid of rows and columns. The club will store coins from different countries in different boxes.

The `CoinCollectionTools` class below assists the coin club in organizing and maintaining their collection.

```
public class CoinCollectionTools
{
   private Coin[][] coinBox;

   /** Constructor instantiates coinBox and fills it with
    *  default Coin objects
    *
    *  @param country the country of origin of the coins in this box
    *  @param rows the number of rows in the coin box
    *  @param columns the number of columns in the coin box
    */
   public CoinCollectionTools(String country, int rows, int columns)
   {
      /* to be completed in part (a) */
   }

   /** Creates and returns a completed coinBox grid by adding the
    *  Coin objects from the parameter ArrayList to the coinBox
    *  in column-major order.
    *
    *  @param myCoins the list of Coin objects to be added
    *                 to the coinBox
    *  Precondition:  myCoins is in order by year
    *  Precondition:  the size of myCoins is not greater than the
    *                 the total number of compartments available
    *  @return coinBox the completed 2-D array of Coin objects
    */
   public Coin[][] fillCoinBox(ArrayList<Coin> myCoins)
   {
      /*  to be implemented in part (b)  */
   }

  /** Returns an ArrayList of Coin objects sorted by coinType
   *
   *  @return ArrayList of Coin objects
   *  Precondition: all cells in coinBox contain a valid Coin object
   *  Precondition: coinBox is ordered by year in column-major
   *                order followed by default Coin objects
   *  Postcondition: Coins in ArrayList are in order grouped by coinType
   */
  public ArrayList<Coin> fillCoinTypeList()
  {
     /*  to be implemented in part (c)  */
  }

  /* Additional implementation not shown */
}
```

(a) The `CoinCollectionTools` class constructor initializes the instance variable `coinBox` as a two-dimensional array of `Coin` objects with dimensions specified by the parameters. It then instantiates each of the `Coin` objects in the array as a default `Coin` object with `country` equal to the country name passed as a parameter, `year` equal to 0, and `coinType` equal to 0.

Complete the `CoinCollectionTools` class constructor.

```
/** Constructor instantiates coinBox and fills it with
 * default Coin objects
 *
 * @param country the country of origin of the coins in this box
 * @param rows the number of rows in the coin box
 * @param columns the number of columns in the coin box
 */
public CoinCollectionTools(String country, int rows, int columns)
```

(b) The Coin Club intends to fill the collection boxes from their list of coins, starting in the upper left corner and moving down the columns in order until all `Coin` objects have been placed in a compartment.

The `fillCoinBox` method takes as a parameter an `ArrayList` of `Coin` objects in order by year minted and returns a chart showing their position in the box, filled in column-major order.

You may assume that `coinBox` is initialized as intended, regardless of what you wrote in part (a).

Complete the method `fillCoinBox`.

```
/** Creates and returns a completed coinBox chart by adding the
 *  Coin objects from the parameter ArrayList to the coinBox
 *  in column-major order.
 *
 *  @param myCoins the list of Coin objects to be added
 *                 to the coinBox
 *  Precondition:  myCoins is in order by year
 *  Precondition:  the size of myCoins is not greater than the
 *                 the total number of compartments available
 *  @return coinBox the completed two-dimensional array of Coin objects
 */
public Coin[][] fillCoinBox(ArrayList<Coin> myCoins)
```

(c) Sometimes the Coin Club would prefer to see a list of its coins organized by coin type.

The `fillCoinTypeList` method uses the values in `coinBox` to create and return an `ArrayList` of `Coin` objects filled first with all the `Coin` objects of type 1, then type 2, and so on through type 6. You may assume that no country has more than 6 coin types. Note that the number of coins from any specific type may be 0.

Since the original `coinBox` was filled in column-major order, `Coin` objects should be retrieved from the `coinBox` in column-major order. This will maintain ordering by year within each coin type.

Remember that the `CoinCollectionTools` class constructor filled the `coinBox` with default `Coin` objects with a `coinType` of 0, so no entry in the `coinBox` is `null`.

You may assume that `coinBox` is initialized and filled as intended, regardless of what you wrote in parts (a) and (b).

Complete the method `fillCoinTypeList`.

```
/** Returns an ArrayList of Coin objects sorted by coinType
 *
 *  @return ArrayList of Coin objects
 *  Precondition: all cells in coinBox contain a valid Coin object
 *  Precondition: coinBox is ordered by year in column-major
 *                order followed by default Coin objects
 *  Postcondition: Coins in ArrayList are in order grouped by coinType
 */
public ArrayList<Coin> fillCoinTypeList()
```

STOP. End of Part II.

Diagnostic Exam Answers and Explanations

Part I (Multiple-Choice) Answers and Explanations

Bullets mark each step in the process of arriving at the correct solution.

1. The answer is B.
- When we first enter the loop, val = 13 and i = 2. The if condition is looking for even numbers. Since val + i is odd, we skip the if clause. Increment i to 3. i < 7 so we continue.
- val = 13, i = 3. This time val + i is even, so add 3 to val. val = 16, increment i to 4, i < 7 so we continue.
- val = 16, i = 4. val + i is even, add 3 to val. val = 19, increment i to 5, i < 7, continue.
- val = 19, i = 5. val + i is even, add 3 to val. val = 22, increment i to 6, i < 7, continue.
- val = 22, i = 6. val + is even, add 3 to val. val = 25, increment i to 7. This time, when we check the loop condition, i is too big, so we exit the loop.
- val = 25 and that is what is returned.

2. The answer is E.
- The first if condition evaluates to (2 < 5 && 13 < 5), which is false, so we skip to the else clause.
- The else clause has its own if statement. The condition evaluates to (2 == 2 && 13 < 2), which is false, so we skip to the else clause.
- result = 2 + 13 = 15, which is what is printed.

3. The answer is D.
- The for-each loop can be read like this: for each Integer in numbers, which I am going to call n.
- The if clause is executed only when the element is > 8, so the loop adds all the elements greater than 8 to the sum variable.
- The elements greater than 8 are 13, 12, and 10, so sum = 35, and that is what the method returns.

4. The answer is C.
- Option I is correct. The reference variable type Planet matches the Planet object being instantiated, and the Planet class contains a no-argument constructor.
- Option II is correct. The reference variable type Planet is a superclass of the DwarfPlanet object being instantiated, and the DwarfPlanet class contains a constructor that takes one String parameter.
- Option III is not correct. Although the reference variable type Planet is a superclass of the DwarfPlanet object being instantiated, the DwarfPlanet class does not contain a no-argument constructor. Unlike a method, the constructor of the parent class is not inherited.

5. The answer is A.

- Let's picture the contents of our ArrayList in a table. After the 3 adds, we have:

0	1	2
Vampire	Werewolf	Ghost

Setting 0 to Zombie gives us:

0	1	2
Zombie	Werewolf	Ghost

Adding Mummy at position 2 pushes Ghost over one position:

0	1	2	3
Zombie	Werewolf	Mummy	Ghost

Witch gets added at the end:

0	1	2	3	4
Zombie	Werewolf	Mummy	Ghost	Witch

After the remove at index 3, Ghost goes away and Witch shifts over one:

0	1	2	3
Zombie	Werewolf	Mummy	Witch

6. The answer is E.

- Options A, B, and C can be eliminated because they return a boolean. If we look at the method declaration, we can see that this method must return an int.
- The difference between D and E is a matter of syntax. We are trying to say if the element contains (x, y) increment total. The for-each loop can be read like this: for each Polygon in zones, which I am going to call p… and there's the answer. I'm going to call each Polygon p, so I want p.contains(x, y).

7. The answer is C.

- Option A will not compile, because acres is a private instance variable.
- Option B will not compile, because the name of the object is park, not central.
- Option C is correct. The method is called properly and it returns a boolean.
- Option D will not compile, because getAcres does not take a parameter.
- Option E will not compile, because hasPlayground is not a public boolean variable.

8. The answer is B.

- The for loop goes through the entire string, looking at each character individually. If the character is an m, the substrings add the section of the string *before* the m to the section of the string *after* the m, leaving out the m.
- The result is that all the m's are removed from the string, and the rest of the string is untouched.

9. The answer is D.

- This is a recursive method. Let's trace the calls. The parts in *italics* were filled in on the way back up. That is, the calls in the plain type were written top to bottom until the base case returned a value. Then the answer were filled in *bottom to top*.

  loopy (12) = loopy(15) + 2 = *25 + 2 = **27**,* which gives us our final answer.
  loopy(15) = loopy(18) + 2 = *23 + 2 = 25*
  loopy(18) = loopy(21) + 2 = *21 + 2 = 23*
  loopy(21) Base Case! return 21

10. The answer is B.

- The compiler looks at the left side of the equals sign and checks to be sure that whatever methods are called are available to variables of that type. Since Letter and ALetter both contain toString methods, the compiler is fine with the code. Option E is incorrect.
- The run-time environment looks at the right side of the equals sign and calls the version of the method that is appropriate for that type.
- System.out.print(x) results in an implicit call to the ALetter toString method, and prints "a".
- System.out.print(y) results in an implicit call to the BLetter toString method and prints "b".
- System.out.print(z) tries to make an implicit call to a CapALetter toString method, but there isn't one, so it moves up the hierarchy and calls the toString method of the ALetter class and prints "a".

11. The correct answer is B.

- On entry, word = "APCS" and index = 0.
  word = word + its substring from 0 to 1, or "A". word = "APCSA".
- Add 2 to index, index = 2. word.length() is 5, 2 < 5 so continue.
  word = word + its substring from 2 to 3, or "C". word = "APCSAC".
- Add 2 to index, index = 4. word.length() is 6, 4 < 6, so continue.
  word = word + its substring from 4 to 5, or "A". word = "APCSACA".
- Add 2 to index, index = 6. word.length() is 7, 6 < 7, so continue.
  word = word + its substring from 6 to 7, or "A". word = "APCSACAA".
- Add 2 to index, index = 8. wordlength() is 8, 8 is not < 8 so the loop exits.
- Notice that since word is altered in the loop, word.length() is different each time we evaluate it at the top of the loop. You can't just replace it with 4, the length of word at the beginning of the method. You must re-evaluate it each time through the loop.

12. The answer is C.

- Each time through the loop:
  - We are considering element n.
  - We calculate array[n] – array[n – 1] and compare it to array[n] – array[n + 1]. In other words subtract the element *before* from the nth element, then subtract the element *after* from the nth element. If the *before* subtraction <= the *after* subtraction, print n.
  - Let's make a table. As a reminder, here is our array:

```
{-3, 0, 2, 4, 5, 9, 13, 1, 5}
```

n	array[n] – array[n – 1]	array[n] – array[n + 1]	<= ?
1	0 – (-3) = 3	0 – 2 = -2	no
2	2 – 0 = 2	2 – 4 = -2	no
3	4 – 2 = 2	4 – 5 = -1	no
4	5 – 4 = 1	5 – 9 = -4	no
5	9 – 5 = 4	9 – 13 = -4	no
6	13 – 9 = 4	13 – 1 = 12	YES
7	1 – 13 = -12	1 – 5 = -4	YES

- Printing array[n] in the YES cases gives us 13 1. Remember that we are printing the *element* not the *index*.

13. The answer is C.

- We want to negate our expression, so we are trying to solve this (note the added ! at the beginning).

```
!((n >= 4) || ((m == 5 || k < 2) && (n > 12)))
```

- DeMorgan's theorem tells us that we can distribute the !, but we must change AND to OR and OR to AND when we do that. Let's take it step by step.
- Distribute the ! to the 2 expressions around the || (which will change to &&).

```
!(n >= 4) && !((m == 5 || k < 2) && (n > 12)))
```

- !(n >=4) is the same as (n < 4).

```
(n < 4) && !((m == 5 || k < 2) && (n > 12)))
```

- Now let's distribute the ! to the expression around the second && (which becomes ||).

```
(n < 4) && !(m == 5 || k < 2) || !(n > 12)
```

- Fix the !(n > 12) because that's easy.

```
(n < 4) && !(m == 5 || k < 2) || (n >= 12)
```

- One to go! Distribute the ! around the || in parentheses (which becomes &&).

```
(n < 4) && !(m == 5) && !(k < 2) || (n >= 12)
```

- Simplify those last two simple expressions.

```
(n < 4) && (m != 5) && (k >= 2) || (n >= 12)
```

- A good way to double-check your solution is to assign values to m, n, and k and plug them in. If you don't think you can simplify the expression correctly, assigning values and plugging them in is another way to find the answer, though you may need to check several sets of values to be sure you've found the expression that works every time.

14. The answer is E.

- Consider Option A. Can an index be out of bounds? The largest values m and n will reach are 4 and 6, respectively. 4 + 6 = 10, and ray[10] is the 11th element in the array (because we start counting at 0). The array has a length of 11, so we will not index out of bounds.
- Look at the first for loop. The array is being filled with elements that are equal to twice their indices, so the contents of the array look like { 0, 2, 4, 6, 8, … }.
- Look at the nested for loop. The outer loop will start at m = 0 and continue through m = 4. For each value of m, n will loop through 0, 2, 4, 6, because n is being incremented by 2. (It's easy to assume all loops use n++. Look!)
- We are looking for m + n > 8. Since the largest value for n is 6, that will not happen when m is 0, 1, or 2.
- The first time m + n is greater than 8 is when m = 3 and n = 6.

  ray[3 + 6] = ray[9] = 2 * 9 = 18. Print 18.
- The next time m + n is greater than 8 is when m = 4 and n = 6.

  ray[4 + 6] = ray[10] = 2 * 10 = 20. Print 20.
- Notice that we aren't printing any spaces, so the 1820 are printed right next to each other.

15. The answer is E.

- This is a recursive method. Let's trace the calls.
  mystery("advanced placement", 9) = mystery("vanced placement", 10)
  mystery("vanced placement", 10) = mystery("nced placement", 11)
  mystery("nced placement", 11) = mystery("ed placement", 12)
  mystery("ed placement", 12) = mystery ("placement", 13)
- By this point we should have noticed that we are getting farther and farther from the base case. What will happen when we run out of characters? Let's keep going and see.
  mystery("placement", 13) = mystery("lacement", 14)
  mystery("lacement", 14 = mystery("cement", 15)
  mystery("cement", 15) = mystery("ment", 16)
  mystery ("ment", 16)= mystery("nt", 17)
- "nt".substring(2) is, somewhat surprisingly, a valid expression. You are allowed to begin a substring just past the end of a string. This call will return an empty string.
  mystery ("nt", 17) = mystery("", 18)
- This time code.substring(2) fails: `StringIndexOutOfBoundsException`.

16. The answer is A.

- This problem requires you to understand that primitives are passed by value and objects are passed by reference.
- When a primitive argument (or actual parameter) is passed to a method, its value is copied into the formal parameter. Changing the formal parameter inside the method will have no effect on the value of the variable passed in; num and index will not be changed by the method.
- When an object is passed to a method, its reference is copied into the formal parameter. The actual and formal parameters become aliases of each other; that is, they both point to the same object. Therefore, when the object is changed inside the method, those changes will be seen outside of the method. Changes to array nums inside the method will be seen in array val outside of the method.
- num and index remain equal to 10 and 3, respectively, but val[3] has been changed to 10.

17. The answer is B.

- The original code segment is a nested loop. The outer loop has an index, which will take on the values 2, 4, 6. For each of those values, the inner loop has an index, which will take on the values 30, 20, 10. The code segment will print:
  (2+30) (2+20) (2+10) (4+30) (4+20) (4+10) (6+30) (6+20) (6+10)
  = 32 22 12 34 24 14 36 26 16
  We need to see which of I, II, III also produce that output.
- Option I is a for loop. On entry, num = 32, count = 0, i = 0.
  - 32 is printed, num = 34, count % 3 = 0, so we execute the if clause and set count = 0 and num = 20.
  - Next time through the loop, 20 is printed . . . oops, no good! Eliminate option I.
- Option II is a while loop. On entry, num = 32.
  - 32 is printed, num = 22, if condition is false.
  - Next time through the loop, 22 is printed, num = 12, if condition is false.
  - 12 is printed, num = 2, if condition is true, num = 34.
  - 34 is printed, num = 24, if condition is false.
  - 24 is printed, num = 14, if condition is false.

- 14 is printed num = 4, if condition is true, num = 36.
- 36 is printed, num = 26, if condition is false.
- 26 is printed, num = 16, if condition is false.
- 16 is printed, num = 6, if condition is true, num = 38.
- Loop terminates – looks good! Option II is correct.

- Option III is a nested for loop.
  - The first time through the loops, h = 0, k = 30,
  - h + k = 30, so 30 is printed. That's incorrect. Eliminate option III.

18. The answer is A.

- This is one way of implementing the Insertion Sort algorithm. The interesting thing about this implementation is the use of a compound condition in the for loop. This condition says "Continue until you reach the beginning of the array OR until the element we are looking at is bigger than the key." When the for loop exits, the "key" is put into the open slot at position j. That's an Insertion Sort algorithm.
- Looking at this problem a different way:
  - It can't be B. If this were a search algorithm, there would have to be a parameter to tell us what element we are looking for, and that isn't the case.
  - It can't be D because Merge Sort is recursive and there's no recursion in this code segment.
  - It can't be E, because we know Sequential *Search* but there isn't a Sequential Sort.
  - That just leaves Selection Sort. Selection Sort has nested for loops, but the loops have no conditional exit. They always go to the end of the array. And on exit, elements are swapped. There's no swap code here.

19. The answer is D.

- The general form for generating a random number between *high* and *low* is

```
(int)(Math.random * (high - low + 1) + low
```

- high – low + 1 = 21, low = 13, so high = 33.
- The correct answer is integers between 13 and 33 inclusive.

20. The answer is D.

- Option A is incorrect. It uses array syntax rather than ArrayList syntax to retrieve the element.
- Option B is incorrect. It attempts to access the instance variable population directly, but population is private (as it should be).
- Option C is incorrect. First of all, the algorithm is wrong. It is only comparing the population in consecutive elements of the ArrayList, not in the ArrayList overall. In addition, it will end with an IndexOutOfBoundsException, because it uses (i + 1) as an index.
- Option D works correctly. It accesses the population using the getter method, and it compares the accessed population to the previous max.
- Option E is incorrect. If we began by setting maxPop = Integer.MAX_VALUE, then that is the value maxPop will have when the code segment completes.

Part II (Free-Response) Solutions

Please keep in mind that there are multiple ways to write the solution to a free-response question, but the general and refined statements of the problem should be pretty much the same for everyone. Look at the algorithms and coded solutions, and determine if yours accomplishes the same task.

General penalties (assessed only once per problem):
-1 using a local variable without first declaring it
-1 returning a value from a void method or constructor
-1 accessing an array or ArrayList incorrectly
-1 overwriting information passed as a parameter
-1 including unnecessary code that causes a side effect like a compile error or console output

1. Complex Numbers

General Problem: Write a ComplexNumber class that will represent a complex number and allow addition and multiplication of two ComplexNumber objects.

Refined Problem: Write a ComplexNumber class that includes:

- instance variables representing the real and imaginary components of the complex number,
- a valid constructor that takes 2 parameters,
- methods for multiplication and addition of ComplexNumbers that take the second ComplexNumber object as a parameter and return the answer in a ComplexNumber object, and
- accessors (getters) for the instance variables.

Algorithm:

- Declare two instance variables, a and b.
- Write a constructor that assigns passed values to the instance variables.
- Write an add method that adds the a and b components of the passed object to the a and b components of this object, places the result in a new ComplexNumber object, and returns it.
- Write a multiply method that does the 4 required multiplications using the a and b components of the passed object and the a and b components of this object, simplifies the result remembering to multiply the last term by -1, places the result in a new ComplexNumber object, and returns it.
- Write the accessors (getters) for a and b.

Java Code:

```
public class ComplexNumber
{
    // instance variables
    private int a;
    private int b;

    // constructor
    public ComplexNumber(int aIn, int bIn)
    {
        a = aIn;
        b = bIn;
    }

    // add
    public ComplexNumber add(ComplexNumber cNum)
    {
        int newA = a + cNum.getA();
        int newB = b + cNum.getB();
        ComplexNumber ans = new ComplexNumber (newA, newB);
        return ans;
    }

    // multiply
    public ComplexNumber multiply(ComplexNumber cNum)
    {
        int newA = a * cNum.getA() + (-1) * b * cNum.getB();
        int newB = a * cNum.getB() + b * cNum.getA();
        ComplexNumber ans = new ComplexNumber (newA, newB);
        return ans;
    }

    // accessors
    public int getA()
    {
        return a;
    }
    public int getB()
    {
        return b;
    }
}
```

Common Errors:

- Remember to instantiate a new ComplexNumbers object in the add and multiply methods.
- Check the arithmetic carefully in the multiply method, and don't forget the -1.

Java Alternate Solution:

- The add and multiply methods can be written without the extra variables, but the statements become awfully long and complex.

```
// add
public ComplexNumber add(ComplexNumber cNum)
{
    return new ComplexNumber (a + cNum.getA(), b + cNum.getB());
}

// multiply
public ComplexNumber multiply(ComplexNumber cNum)
{
    return new ComplexNumber
            (a * cNum.getA() + (-1) * b * cNum.getB(),
             a * cNum.getB() + b * cNum.getA());
}
```

Scoring Guidelines:

+1 Instantiates 2 private instance variables and initializes them to passed parameters in the constructor
+3 Implements the add method
- **+1** Accesses the 2 components of the parameter ComplexNumber
- **+1** Adds the components of the parameter to the components of this object
- **+1** Instantiates and returns a new ComplexNumber object, containing the correct sum

+5 Implements the multiply method
- **+1** Multiplies this real component by other real component and multiplies this imaginary component by other imaginary component by -1
- **+1** Multiplies this real component by other imaginary component and this imaginary component by other real component
- **+1** Adds the four results to find the real and imaginary components of the product
- **+1** Instantiates and returns a new ComplexNumber object, containing the correct product
- **+1** Correctly implements accessors for both instance variables

2. Coin Collector

(a) General Problem: Complete the CoinCollectionTools class constructor.

Refined Problem: Instantiate the instance variable coinBox as a new array of the size specified by the parameters. Traverse the array filling every cell with a Coin object instantiated with country = the country parameter, year = 0, and coinType = 0.

Algorithm:

- Instantiate coinBox as a new array of Coin objects with dimensions [rows][columns].
- Outer loop: traverse the rows of the array.
  Inner loop: traverse the columns of the array.
    - Instantiate a new Coin object, passing parameters (country, 0, 0).

Java Code:

```
public CoinCollectionTools(String country, int rows, int columns)
{
    coinBox = new Coin[rows][columns];
    for (int row = 0; row < rows; row++)
        for (int col = 0; col < columns; col++)
            coinBox[row][col] = new Coin(country,0,0);
}
```

Common Errors:

- If you look above the constructor in the code for the class, you will see that the coinBox has been declared as an instance variable. If you write:

  ```
  int[][] coinBox = new int[rows][columns];
  ```

  then you are declaring a different array named coinBox that exists only within the constructor. The instance variable coinBox has not been instantiated.

Java Code Alternate Solution:

If you read the whole problem before starting to code, you might have noticed that the other two parts work in column-major order. You can write this in column-major order also. Since you are filling every cell with the same information, it doesn't make any difference.

```
public CoinCollectionTools(String country, int rows, int columns)
{
    coinBox = new Coin[rows][columns];
    for (int col = 0; col < columns; col++)
        for (int row = 0; row < rows; row++)
            coinBox[row][col] = new Coin(country,0,0);
}
```

(b) General Problem: Complete the fillCoinBox method.

Refined Problem: The parameter myCoins is an ArrayList of Coin objects in order by year minted. Assign them to the coinBox grid in column-major order.

Algorithm:

- Create a count variable.
- Loop through all of the Coin objects in myCoins using variable count as the loop counter.
  - Update the row and column variables based on count.
  - Get the next Coin object from the ArrayList, and place it in the coinBox location specified by the row and column variables.
  - Increment the count variable.
- Return the completed coinBox.

Java Code:

```
public Coin[][] fillCoinBox(ArrayList<Coin> myCoins)
{
    int count = 0;
    int row;
    int column;
    while (count < myCoins.size())
    {
        row = count % coinBox.length;
        column = count / coinBox.length;
        coinBox[row][column] = myCoins.get(count);
        count++;
    }
    return coinBox;
}
```

Java Code Alternate Solution #1:
You may not have thought of using % and / to keep track of column and row. Here's another way to do it.

```
public Coin[][] fillCoinBox(ArrayList<Coin> myCoins)
{
    int count = 0;
    int row = 0;
    int column = 0;
    while (count < myCoins.size())
    {
        coinBox[row][column] = myCoins.get(count);
        if (row < coinBox.length - 1)
        {
            row++;
        }
        else
        {
            row = 0;
            column++;
        }
        count++;
    }
    return coinBox;
}
```

Java Code Alternate Solution #2:
This solution bases its loops on the grid, rather than the ArrayList.

```
public Coin[][] fillCoinBox(ArrayList<Coin> myCoins)
{
    int row = 0;
    int column = 0;
    int count = 0;

    // The first condition in the outer loop is not actually
    // necessary, because myCoins will run out before coinBox
    // goes out of bounds.
    while (column < coinBox[0].length && myCoins.size() > count)
    {
        while (row < coinBox.length && myCoins.size() > count)
        {
            coinBox[row][column] = myCoins.get(count);
            count++;
            row++;
        }
        row = 0;
        column++;
    }
    return coinBox;
}
```

Common Errors:

- Do not count on the fact that you can fill the entire grid. It is tempting to write nested for loops that traverse the whole grid, but if there are fewer Coin objects in the ArrayList than elements in the grid, your program will terminate with an IndexOutOfBoundsException.
- It is common to write the row and column loops in the wrong order. In column-major order, columns vary slower than rows (we do all the rows before we change columns), so the column loop is the outer loop. In row-major order, the row loop is the outer loop.
- Even though we are filling our array in column-major order, the syntax for specifying the element we want to fill is coinBox[row][column], not the other way around.
- If you used remove instead of get when accessing the Coin objects in the ArrayList, you modified the list and that's not allowed. It's called *destruction of persistent data* and may be penalized.
- In the solutions that loop through the grid (Alternates #2 and #3), be careful not to use incorrect notation for the end conditions. In general, the number of rows is arrayName.length and the number of columns is arrayName[row].length. When processing in column-major order, the loop that varies the column is the outer loop. We cannot use the loop variable row as the array index when finding the length of a column, because it does not exist outside of the inner loop. Since this is not a ragged array, it is safe to use [0] as our index.

(c) General Problem: Complete the fillCoinTypeList method.

Refined Problem: Given a coinBox as created by part (a) and filled in part (b), create a list that contains Coins in order by coin type (1–6). If Coins are retrieved from the coinBox in column-major order, they will already be in order by year.

Algorithm:

- Loop 1: Complete the inner loops 6 times, once for each coin type 1–6.
- Loop 2: Loop through the columns.
- Loop 3: Loop through the rows.
- If the Coin object at the row-column location specified by the loop counters of loops 2 and 3 matches the coinType specified from the loop counter of loop 1:
  - Add the Coin object to the ArrayList.
- Return the ArrayList of Coin objects.

Java Code:

```
public ArrayList<Coin> fillCoinTypeList()
{
   ArrayList<Coin> myCoins = new ArrayList<Coin>();
   for (int type = 1; type <= 6; type++)
      for (int col = 0; col < coinBox[0].length; col++)
         for (int row = 0; row < coinBox.length; row++)
            if (coinBox[row][col].getCoinType() == type)
               myCoins.add(coinBox[row][col]);
   return myCoins;
}
```

Common Errors:

- You should not create a new Coin object to add to the myCoins list. The Coin object already exists in the array.
- Do not worry about the default Coins added in the constructor. Since they have a coinType of 0, they will be ignored.

Scoring Guidelines:

Part (a):	**CoinCollectionTools constructor**	**2 points**
+1	Traverses the array using a nested loop. No bounds errors, no missed elements	
+1	Instantiates new Coin objects for each cell in the array	

Part (b):	**fillCoinBox**	**4 points**
+1	Accesses every element in myCoins. No bounds errors, no missed elements	
+1	Accesses all appropriate elements of coinBox. No bounds errors, no missed elements	
+1	Accesses elements of coinBox in column-major order	
+1	Returns correctly filled coinBox	

Part (c):	**fillCoinTypeList**	**3 points**
+1	Loops through coinBox at least once in column major order. No bounds errors, no missed elements	
+1	Locates Coins of type 1, followed by types 2–6 in the correct order	
+1	Instantiates, fills, and returns a correct ArrayList<Coin>	

Sample Driver:

There are many ways to write these methods. Maybe yours is a bit different from our sample solutions and you are not sure if it works. Here is a sample driver program. Running it will let you see if your code works, and will help you debug it if it does not.

Copy CoinCollectionToolsDriver into your IDE along with the complete Coin and CoinCollectionTools classes (including your solutions). You will also need to add this import statement as the first line in your CoinCollectionTools class: `import java.util.ArrayList;`

```
import java.util.ArrayList;

public class CoinCollectionToolsDriver {

   public static void main(String[] args) {
      CoinCollectionTools tools = new CoinCollectionTools("USA", 3, 4);
      int[] years = { 1920, 1930, 1940, 1940, 1950, 1950,
            1950, 1960, 1970, 1980, 1990 };
      int[] types = { 1, 2, 4, 1, 2, 3, 3, 4, 4, 2, 4 };
      String[] typeNames = { "", "penny", "nickel", "dime", "quarter" };
      Coin[][] coinBox = new Coin[3][4];
      ArrayList<Coin> coins = new ArrayList<Coin>();

      for (int i = 0; i < years.length; i++)
         coins.add(new Coin("USA", years[i], types[i]));

      System.out.println("fillCoinBox test\nExpecting:");
      System.out.println("penny 1920\tpenny 1940\tdime 1950\tnickel 1980"
            + "\nnickel 1930\tnickel 1950\tquarter 1960\tquarter 1990"
            + "\nquarter 1940\tdime 1950\tquarter 1970\t0");

      System.out.println("\nYour answer:");
      coinBox = tools.fillCoinBox(coins);

      for (int row = 0; row < coinBox.length; row++) {
         for (int col = 0; col < coinBox[row].length; col++)
            System.out.print(typeNames[coinBox[row][col].getCoinType()] + " "
                  + coinBox[row][col].getYear() + "\t");
         System.out.println();
      }

      System.out.println("\nfillCoinTypeList test\nExpecting:");
      System.out.println("penny 1920, penny 1940, nickel 1930, "
            + "nickel 1950, nickel 1980, dime 1950, "
            + "\ndime 1950, quarter 1940, quarter 1960, "
            + "quarter 1970, quarter 1990,");
      System.out.println("\nYour answer:");
      coins = tools.fillCoinTypeList();
      for (int i = 0; i < coins.size(); i++) {
         System.out.print(typeNames[coins.get(i).getCoinType()]
            + " " + coins.get(i).getYear()+ ", ");
         if (i == coins.size() / 2)
            System.out.println();
      }
   }
}
```

Develop Strategies for Success

Strategies to Help You Do Your Best on the Exam

IN THIS CHAPTER

Summary: This chapter supplies you with strategies for taking the multiple-choice and the free-response sections of the AP Computer Science A Exam. The strategies will help you get the best score you are capable of.

Key Ideas

- Many questions require you to read and process code. Practice by hand-tracing hundreds of pieces of code.
- When you don't know an answer, eliminate choices you know are incorrect and guess. There is no penalty for guessing.
- You need to understand how points are awarded on the free-response section and when deductions are taken. Also learn what errors are not considered important enough to result in a deduction.
- You should make an attempt to write some kind of response to every free-response question, even when you feel you don't understand the question or know the answer. You may earn a point or two even if you don't answer the whole question or get it all right.
- On the multiple-choice section, you must clearly understand the difference between the System.out.print() and System.out.println() statements and be able to read code that uses each of them. However, on the free-response questions, you are never allowed to use either of these; they are considered undesirable side effects.

Strategies for the Multiple-Choice Section

The multiple-choice section of the exam determines how well you can read someone else's code. There are many problems involving the fundamentals of programming such as if and if-else statements, looping structures, complex data structures, and boolean logic. Some of the questions have to do with class hierarchy, inheritance, abstract classes, and interfaces.

Hand-Tracing Code

Hand-tracing code is the act of pretending you are the computer in order to predict the output of some code or the value of a variable. A large number of the questions on the multiple-choice portion require you to read and process code. You should hand-trace hundreds of sample pieces of code to prepare for the exam.

Don't Do Too Much in Your Head!
It's very easy to make mental mistakes when hand-tracing code. Write out the current values of all the variables in the white space that is provided on the exam.

When You Don't Know, Guess

There is no penalty for guessing on the multiple-choice questions. Make sure you eliminate the choices that don't make sense before guessing. If you are running out of time, quickly make guesses for the questions you weren't able to get to.

Read All of the Options Before Selecting Your Answer

Before you fill in the bubble, be sure to look at every one of the options. It's possible that you may have misunderstood the question. Looking at each of the choices could help point you in the right direction. However, some people disagree and say it's a waste of time to read all answer choices. So, if you prefer, only take this advice when you feel it will help.

Input from the User

You will never have to write code that asks for input from the user. Instead, the question will make a generic statement in the form of a comment that shows that they are requesting input from the user as in the example below.

Generic Input from the User
The AP Computer Science Exam will use a generic reference to show input from the user. When you see this, just assume that the variable receives an appropriate value from a nameless, faceless user.

```
int myValue =  /* call to a method that reads an integer number */
```

or

```
double myValue =  /* read user input */
```

Strategies for the Free-Response Section

The free-response section of the exam determines how well you can write your own code. Although there are an infinite number of possible types of questions that could be asked, it is common for you to be asked to:

- Write an entire method body that processes data in a specific way.
- Write an entire class given the specifications.
- Write an interface or abstract class given the specifications.
- Write an overloaded constructor for a class given the specifications.

How the Free-Response Questions Are Graded

The free-response question (FRQ) section is graded by an expert group of high school and college computer science teachers and professors. A real person who knows computer science will be reading your code to determine if you have solved the problem correctly. Your written code must be readable by another human being.

The Intent of the Question

Every year, the *intent* of each question is explained to the people who are grading the test as part of their training. This is normally a one-line statement that describes the purpose of the question. You should try to predict what the intent of the question is for each of the free-response questions when you are taking the exam. Thinking about the intent of the question will help you see the big picture and then zero in on writing the code that solves the problem. If you are oblivious to the intent of the question, then you are likely to go off course and solve a problem that is not the point of the question.

Scoring Guidelines

In order to help the exam graders be consistent when grading tens of thousands exams, a scoring rubric is used. This rubric dictates how many points are awarded for writing valid code that solves specific components of the problem. The problem is broken down into its most important components and each of these is assigned a point value. The total of all of these components from all sections (a, b, c, etc.) is 9 points.

Scoring for the Free-Response Questions
You start off with zero points for each free-response question. As the reader finds valid code that is part of the rubric, you earn points. Each component of the rubric is worth one point. Your goal is to earn 9 points, the maximum awarded for each question.

1-Point Penalty

After your response is graded against the rubric and your maximum score is calculated, the grader decides if points should be deducted from your total. If you break a major rule, then points are subtracted from your total. The College Board 1-point penalties include:

- **Array/collection access confusion**
  You must use the correct syntax when accessing elements in arrays and ArrayLists.

- **Extraneous code that causes side effects**
  This is a fancy way of saying your code doesn't compile or you have written a System.out.print() or System.out.println() statement. Don't ever print anything to the console!

- **Local variables used but none declared**
  Be sure to declare every variable before using it.

- **Destruction of persistent data**
  Anything that is passed to a method or constructor should be left alone. Do not change the argument that is passed. Reassign it to a different variable and only modify that one.

- **Void method or constructor that returns a value**
  Don't ever have a return statement in a void method or constructor.

You Can't Get a Negative Score

- The graders won't deduct points if you haven't earned any points.
- Penalty points can only be deducted in a part of the question that has earned credit so they won't take away points from a different section.
- A penalty can be assessed only once for a question, even if it occurs multiple times or in multiple parts of that question.

Non-Penalized Errors

I feel a little guilty telling you this and as a teacher it goes against my philosophy, but I'm going to tell you anyway. The readers of the exam are told to look the other way when seeing certain kinds of errors for which they would normally deduct points in their classroom. A full list is found in the Appendix, but here are a few of them:

- Using = instead of = = and vice versa
- Confusing length and size when using an array, ArrayList, or String
- Using a keyword as a variable name
- Confusing [] with () or even <>

Now listen up! I'm not telling you to be careless when you write your code or to make those mistakes on purpose. I'm just telling you that, on the AP Computer Science A exam, the readers will ignore minor syntax mistakes like these.

What Is Valid Code?

Any code that works and satisfies the conditions of the problem will be given credit as valid code. Please remember the graders are humans; so don't go writing crazy, Rube-Goldberg-style code to solve a really easy problem. When in doubt, go at the problem in the most straightforward, logical way and stick to the AP computer science Java subset.

Fun Fact: *Rube Goldberg was a popular cartoonist who depicted gadgets that performed simple tasks in indirect, convoluted ways. He is the inspiration behind the many Rube Goldberg Machine Contests that are held worldwide.*

Brute Force Versus Efficiency and Elegance

There are many ways to solve problems in our world. As a software developer, your goal is to find the most *elegant* way to solve any problem. Sometimes that comes easy and other times, not so much.

On the AP Computer Science A exam, elegance and efficiency are not considered (unless the problem specifically asks for it). When you are asked to write code to solve a problem, try to go at the problem in the most reasonable way first. If you can't think of an efficient way to solve the problem, then use **brute force**. In other words, if the solution doesn't come to you, crank out the code in a way that works even if it is a really slow way to solve the problem.

Answer the Question
Above all else, answer the question. Or at least, make an attempt at it. Don't be afraid to write something down even if you don't have a full solution. Remember that the rubric for every question has components that are pieces of the solution. The one thing that you put down just may be a component that will get you a point or two.

Comments Are Not Graded

Don't spend time writing comments for your Java code. The readers will not read anything that is part of a Java comment.

Cross Out, Don't Erase

If you think of a new solution, write it out first and then cross out the old one. Don't waste time by erasing the old solution. The graders are trained to ignore anything that is crossed out.

Output to the Console

On the AP Computer Science A exam, the use of output statements is a little strange. On the multiple-choice section, you must clearly understand the difference between the System.out.print() and System.out.println() statements and be able to read code that uses each of them.

However, on the free-response questions, you are *never* allowed to write a System.out.print() or System.out.println() instruction! You will lose points if the reader reads any kind of output statement in your response to a free-response question.

Output to the Console!
In general, on the Free Response section, it is highly unlikely that you will be asked to write a System.out.println statement. So, if they don't ask for it, don't write one. You will get penalized if you write one and they don't ask you to.

[illegible]

[illegible]

Cross Out, Don't Erase

[illegible]

Output to the Console

[illegible]

Review the Knowledge You Need to Score High

Background on Software Development

IN THIS CONCEPT

Summary: This concept will give you background on Java and what a software developer is. You will also learn the purpose of software, and the tools you will use to prepare for the exam.

Key Ideas

- Java is the programming language of the AP Computer Science A Exam.
- Java is an object-oriented programming language.
- Software developers design and implement code for countless applications.
- An integrated development environment (IDE) is the tool that you will use when you write your Java code.

What Is Java?

So, you want to learn how to write software for other people. Well, you could write all your programs on paper using a pencil. Then, the users could just read your code and pretend that they're running your program. There's just one minor problem with that: to write software, you need software.

Java is a general-purpose, object-oriented computer programming language that was developed by James Gosling in the mid-1990s when he worked at Sun Microsystems (Sun was later acquired by Oracle). Java is different from other languages in several ways, but the major difference is that it was designed to be **machine-independent**. This means that a Java program could be written on any platform (Windows, Mac, Linux, etc.) and then could be run on any platform. Prior to Java, software was **machine-dependent** which meant that you could only run your program on the same type of machine that you wrote it on.

Here is a simplistic explanation of how Java accomplishes its machine-independent process. Suppose you write a program in Java. The computer can't run the program as you've written it, because it doesn't understand Java. It can only run a program whose instructions are written in **machine code**. A **compiler** is a program that converts your Java program into code called **bytecode**. The compiler checks your program for **syntax errors**, and when all of these errors are eliminated, the compiler generates a new program that is readable by a **Java virtual machine (JVM)**. The JVM is where the Java program actually runs since it translates the bytecode into machine code. The JVM is a part of the **Java Runtime Environment (JRE)** that is provided by Oracle. Oracle writes a JVM for numerous platforms like Windows, which is how a Java program can run on any machine.

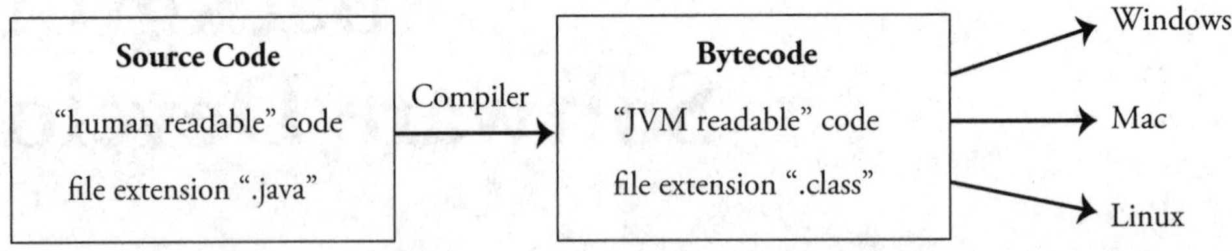

> **It's a Big Java World Out There**
> Java has thousands of built-in classes that all Java programmers can use. Since the focus of this book is to help you score a 5 on the AP Computer Science A Exam, only the features that are tested on the exam are included in this book.

What Is a Software Developer?

Software developers are the wizards of our technology-based world. They are the people who design and create the programs that run on smartphones, tablets, smart TVs, computers, cars, and so on. It is stunning how many people the software developer affects.

Software developers may work individually, but most of the time, they work as part of a team of developers. It is essential that they write clean, readable code so that others can read and modify it, if needed. Since other developers will often read your code, it is important to insert comments into your code. In the real world of programming, most programmers despise commenting their code, but it is a necessary evil, and every developer appreciates reading code that has been well documented.

One of the most important skills that a programmer needs to have is figuring out how to design and subsequently implement code for a project. The software developer's job can be summarized with this simple graphic:

This is the exciting, yet challenging part of the developer's role in project development. Computer programmers get a high level of satisfaction from solving problems, and this is why it is a high priority on the AP Computer Science A Exam.

In many classrooms, students work independently on their programming assignments. This is fine when you are learning how to code; however, few software projects in the work world are like that. Projects can have anywhere from a small handful of programmers to hundreds of programmers. There is a hierarchy among the developers that separates the types of work to be done.

What Is OOP (Object-Oriented Programming)?

Object-oriented programming is an approach to programming that uses the concept of classes and objects. It's a brilliant approach since *we live in an object-oriented world.* Everywhere you look, there are objects. These objects have attributes that make them different from other objects. Many of these objects can perform some kind of action, or you can get information from them. In OOP, classes define how an object will be constructed. Then, objects are created from these classes, and you manipulate them in your program. Kind of makes you feel like you rule the world, doesn't it?

Viewing the World Through the Eyes of a Software Developer

Something I challenge all my computer science students to do is to begin viewing the world through the eyes of a software developer. As you learn new concepts, try to relate them to things in your daily life. If you are a gamer, start to analyze the games that you play and see if you can figure out how the developers made the game. My students tell me that their minds "open up" after doing this, and it helps them to be better programmers since they are constantly making connections between the computer science topics and the world that they live in.

For the Good of All Humankind

Let's be honest, in our world, there are good software developers and there are evil software developers. The hackers who live on the dark side will always exist, and our job as noble software developers is to make the world a better place. Integrity is vital to our side and is a badge to be worn with pride. Therefore, we must vow to only write software that is for the good of all humankind and stand up against all evil software developers.

As developers, we strive to:

- Make the best product possible
- Keep our clients' information safe and private
- Honor the intellectual property of others
- Make ethically moral software

Choosing Your IDE

An Integrated Development Environment, or IDE, is a piece of software that allows you to write software more easily. Professional IDEs, such as Eclipse or IntelliJ, have really awesome features that make creating programs easier. Other IDEs are great for introducing a newbie to programming.

There are many excellent IDEs for Java. Your computer science teacher will probably choose the IDE for you. While it is important to learn how to write a program with an IDE, it is also important to remember that *you will not be able to use a computer* on the AP Computer Science A Exam. So, whatever program you use to write your Java code, you *must* learn how to do it without the use of a computer.

No Computer!
Computers are *not used* on the AP Computer Science A Exam.

HelloWorld

In whatever IDE you choose to write your programs, type in the following code for the HelloWorld program. Compile it and run it. This will get you to the point where you can begin this book.

PROGRAM

```
public class HelloWorld
{
   public static void main(String[] args)
   {
      System.out.println("Hello World.");
   }
}
```

OUTPUT (This is what is displayed on the console screen.)

```
Hello World.
```

Spacing and Indenting in Your Program
Programmers have the freedom to put spacing and indentions in their program in any way they want. Typically, programmers follow spacing guidelines to make it easier to read. The `HelloWorld` example shown above demonstrates a classic form for spacing and indenting code.

You may also see a `HelloWord` program that looks like the next example. This program will run in exactly the same way as the previous example even though the spacing is different.

```
public class HelloWorld {
   public static void main(String[] args){
      System.out.println("Hello World.");
   }
}
```

❯ Rapid Review

- Java is a general-purpose, object-oriented computer programming language.
- Java source code is translated into bytecode by the compiler.
- Java has thousands of built-in classes that all Java programmers can use.
- A software developer is a person who writes computer programs.
- Software developers must be creative but also pay attention to detail.
- An object-oriented language is a programming model that uses classes and objects.
- To be a better programmer you should start to view the world through the eyes of a software developer.
- Software developers must keep their clients' information safe, honor the intellectual properties of others, and make ethically moral software.
- An Integrated Development Environment (IDE) is a piece of software that allows you to write software more easily.
- A computer is not used on the AP Computer Science A Exam.
- HelloWorld is traditionally the first computer program that a beginning Java programmer writes.

Fundamentals of Programming

IN THIS CONCEPT

Summary: You will learn the fundamental building blocks that all programmers need to write software. The syntax, which is the technical way of writing the code, will be written in Java; however, all of the structures that you will learn in this concept will apply to any language that you come across in the future. To pass the AP Computer Science A Exam, and to be a developer, you must master all of these fundamental concepts.

Key Ideas

- Variables allow you to store information that is used by the program.
- Conditional statements allow you to branch in different directions in your program.
- Looping structures allow you to repeat instructions many times.
- The computer follows the order of operations when performing mathematical calculations.
- Software developers write comments to document their code.
- The console screen is where simple input and output are displayed.
- Good software developers can debug their code and the code of others.

Introduction

Do you want to keep track of a score in a game that you want to write? Do you want your program to make choices based on whatever the user wants? Will you be doing any math to get answers for the user? Do you want an easy way to do something a million times? In this concept you will learn the fundamental building blocks that programmers use to write software to do all these things.

Syntax

Syntax is not a fine you have to pay for doing something you're not supposed to do. The syntax of a programming language describes the correct way to type the code so the program will run. An example of a syntax error is forgetting to put a semicolon after an instruction. Another example is not putting a pair of parentheses in the correct place. You have to fix all syntax errors before your program will run. If your program has any syntax errors, the **compiler** will respond with a **compile-time** error. This type of error prevents the compiler from doing its job of turning your Java code into bytecode. As soon as you have eliminated all syntax errors from your program, your program can be compiled and then **run (executed)**. A list of common syntax errors can be found in the Appendix.

Inline Comment

An inline comment is a way for a programmer to tell someone who is reading the code a secret message. The way to make an inline comment is to make two forward slashes, like this: //.

```
// This is an inline comment.
// The computer ignores everything after the two forward slashes on the same line.
// I will use inline comments as a way to give you secret messages throughout this book.
```

The Console Screen

The **console** screen is the simplest way to input and output information when running a program. The two most common instructions to display information to the screen are **System.out.println()** and **System.out.print()**.

Summary of Console Output

CASE 1: To display some text and move the cursor onto the next line:

```
System.out.println("text goes here");
```

CASE 2: To display some text and NOT move the cursor onto the next line:

```
System.out.print("text goes here");
```

CASE 3: To only move the cursor onto the next line:

```
System.out.println();
```

Example

The difference between the print() and the println() statements:

Predict the Output of This Code:

```
line 1: System.out.print("Players gonna play, ");
line 2: System.out.println("play, play, play, play");
line 3: System.out.print("Haters gonna hate, hate, hate, ");
line 4: System.out.println("hate, hate");
line 5: System.out.println();
line 6: System.out.println("I shake it off, I shake it off");
```

Output on the Console Screen:

```
Players gonna play, play, play, play, play
Haters gonna hate, hate, hate, hate, hate

I shake it off, I shake it off
```

The Semicolon

The semicolon is a special character that signifies the end of an instruction. Every distinct line of code should end with a semicolon. It tells the compiler that an instruction ends here.

```
System.out.println("some text");  // the semicolon separates instructions
```

Primitive Variables

Numbers can be stored and retrieved while a program is running if they are given a home. The way that integers and decimal numbers are stored in the computer is by declaring a **variable**. When you declare a variable, the computer sets aside a space for it in the working memory of the computer (RAM) while the program is running. Once that space is declared, the programmer can assign a value to the variable, change the value, or retrieve the value at any time in the program. In other words, if you want to keep track of something in your program, you have to create a home for it, and declaring a variable does just that.

In Java, the kind of number that the programmer wants to store must be decided ahead of time and is called the **data type**. For the AP Computer Science A Exam, you are required to know two primitive data types that store numbers: the **int** and the **double**.

Variable

A variable is a simple data structure that allows the programmer to store a value in the computer. The programmer can retrieve or modify the variable at any time in the program.

The `int` and `double` Data Types

Numbers that contain decimals can be stored only in the data type called the **double**. Integers can be stored in either the data type called the **int** or the **double**. Remember that this book is designed for the AP Computer Science A Exam. There are several other data types in Java that can store numbers; however, they are not tested on the AP Computer Science A Exam.

When it is time for you to create a variable, you need to know its data type. It's also a common practice to give the variable a starting value if you know what it should be. This process is called *declaring a variable and initializing it.*

The General Form for Declaring a Variable in Java

```
datatype variableName;
```

The General Form for Declaring a Variable and Then Initializing It

```
datatype variableName = value;
```

Examples

Declaring int and double primitive variables in Java:

```
int numberOfPokemon = 25;          // numberOfPokemon gets a value of 25
double health = 112.7;             // health gets the value of 112.7
double gpa = 3;                    // a double can receive an int value
double a = 2.4, b = 5.6, c = 4;    // multiple variables on same line
```

The following graphic is intended to give you a visual of how **primitive variables** are stored in the computer's **RAM (random access memory)**. The value of each variable is stored in the computer's memory at a specific **memory address**. The name of the variable and its value are stored together. Whenever you declare a new primitive variable, think of this picture to imagine what is going on inside the computer. The way this process works inside of a computer is bit more complicated than this, but I hope the diagram helps you imagine how variables are stored in memory.

Memory Address: 0000 numberOfPokemon = 25	*Memory Address: 0001* health = 112.7	*Memory Address: 0002* gpa = 3.0
Memory Address: 0003 a = 2.4	*Memory Address: 0004* b = 5.6	*Memory Address: 0005* c = 4.0
Memory Address: 0006 *<empty>*	*Memory Address: 0007* *<empty>*	*Memory Address: 0008* *<empty>*
Memory Address: 0009 *<empty>*	*Memory Address: 000A* *<empty>*	*Memory Address: 000B* *<empty>*

Naming Convention and Camel Case

The manner in which you choose a variable name should follow the rules of a **naming convention**. A naming convention makes it easy for programmers to recognize and recall variable names.

Obviously, you are familiar with uppercase and lowercase, but let me introduce you to **camel case**, the technique used to create **identifiers** in Java. All primitive variable names start with a lowercase letter; then for each new word in the variable name, the first letter of the new word is assigned an uppercase letter.

All variable names should be descriptive, yet concise. The name should describe exactly what the variable holds and be short enough to not become a pain to write every time you use it. Single letter variable names should be avoided, as they are not descriptive.

Examples

Mistakes when declaring int and double primitive variables in Java:

```
double 5dollars = 900.0;        // Error: name cannot begin with a number
int #tickets = 5;               // Error: name cannot start with a # symbol
int theNumberOfPokemonThatAshCaughtDuringHisLifetime = 151;
// the name is valid, but is a terrible choice because it is way too long
```

Summary of Variable Names in Java

- Variable names should be meaningful, descriptive, and concise.
- Variable names cannot begin with a number or symbol (the dollar sign and underscore are exceptions to this rule).
- By naming convention, variable names use camel case.

The `boolean` Data Type

If you want to store a value that is not a number, but rather a **true** or **false** value, then you should choose a **boolean** variable. The boolean data type is stored in the computer memory in the same way as ints and doubles.

Examples

Declaring boolean variables in Java:

```
boolean userHasWon = false;       // userHasWon is false
boolean unicornIsVisible = true;  // unicornIsVisible is true
boolean bananaCount = 5;          // Error: boolean cannot store a number
```

One-Letter Variable Names

On the free-response section of the AP Computer Science A Exam, you should name your variables in a meaningful way so the reader thinks you know what you're doing. Declaring all your variable names with single letters that don't explain what is stored is considered bad programming practice.

Keywords in Java

A **keyword** is a reserved word in Java. They are words that the compiler looks for when translating the Java code into bytecode. Examples of keywords are **int**, **double**, **public**, and **boolean**. Keywords can never be used as the names of variables, as it confuses the compiler and therefore causes a syntax error. A full list of Java keywords appears in the Appendix.

Example

The mistake of using a keyword as a variable name:

```
int public = 6;          // Syntax Error: public is a Java keyword
```

Mathematical Operations

Order of Operations

Java handles all mathematical calculations using the **order of operations** that you learned in math class (your teacher may have called it PEMDAS or GEMA). The order of operations does exactly that: it tells you the order to evaluate any expression that may contain parentheses, multiplication, division, addition, subtraction, and so on. The computer will figure out the answers in exactly the same way that you learned in math class.

Operator	Priority
Parentheses	Top priority
Multiplication, Division, Modulo	Done in order from left to right (equal precedence)
Addition, Subtraction	Done in order from left to right (equal precedence)

Modulo

The **modulus** (or **mod**) operator appears often on the AP Computer Science A Exam. While it is not deliberately taught in most math classes, it is something that you are familiar with.

The mod operator uses the percent symbol, **%**, and produces the remainder after doing a division. Back in grade school, you may have been taught that the answer to 14 divided by 3 was 4 R 2, where the *R* stood for *remainder*. If this is how you were taught, then mod will be simple for you. The result of 14 % 3 is 2 because there are 2 left over after dividing 14 by 3.

Modulo Example	Answer	Explanation
20 % 7	6	20 divided by 7 is 2 with a remainder of 6
100 % 20	0	100 divided by 20 is 5 with a remainder of 0
41 % 12	5	41 divided by 12 is 3 with a remainder of 5
0 % 2	0	0 divided by 2 is 0 with a remainder of 0

Using Modulo to Find Even or Odd Numbers

The mod operator is great for determining if a number is even or odd. If someNumber % 2 = 0, then someNumber is even. Or, if someNumber % 2 = 1, then someNumber is odd.

Furthermore, the mod operator can be used to find a multiple of *any* number. For example, if someNumber % 13 = 0, then someNumber is a multiple of 13.

Division with Integers

When you divide an int by an int in Java, the result is an int. This is referred to as **integer division**. The decimal portion of the answer is ignored. The technical way to say it is that the result is **truncated**, which means that the decimal portion of the answer is dropped.

Arithmetic Operator	Java Symbol	Example	Result	Justification
Addition	+	7 + 3	10	Simple addition
Subtraction	–	7 – 3	4	Simple subtraction
Multiplication	*	7 * 3	21	Simple multiplication
Division	/	7 / 3	2	7 divided by 3 is 2 (remainder is dropped)
Modulo	%	7 % 3	1	7 divided by 3 is 2 with a remainder of 1

Examples

The examples show how division works with int and double values in Java. If either number is a double, you get a double result.

- A double divided by a double is a double. // example: 10.0 / 4.0 = 2.5
- A double divided by an int is a double. // example: 10.0 / 4 = 2.5
- An int divided by a double is a double. // example: 10 / 4.0 = 2.5
- An int divided by an int is an int. // example: 10 / 4 = 2

`ArithmeticException`

Any attempt to divide by zero or perform someNumber % 0 will produce a **run-time error** called an **ArithmeticException.** The program crashes when it tries to do this math operation.

Examples

Getting an ArithmeticException error by doing incorrect math:

```
int num1 = 5, num2 = 0;
int num3 = num1 / num2;          // ArithmeticException: / by zero
int num4 = num1 % num2;          // ArithmeticException: / by zero
```

`ArithmeticException`

Dividing by zero or performing a mod by zero will produce an error called an **ArithmeticException**.

Modifying Number Variables

The Assignment Operator

The equal sign, =, is called an **assignment operator** because you use it when you want to assign a value to a variable. It acts differently from the equal sign in mathematics since the equal sign in math shows that the two sides have the same value. The assignment operator gives the left side the value of whatever is on the right side. The right side of the equal sign is computed first and then the answer is assigned to the variable on the left side.

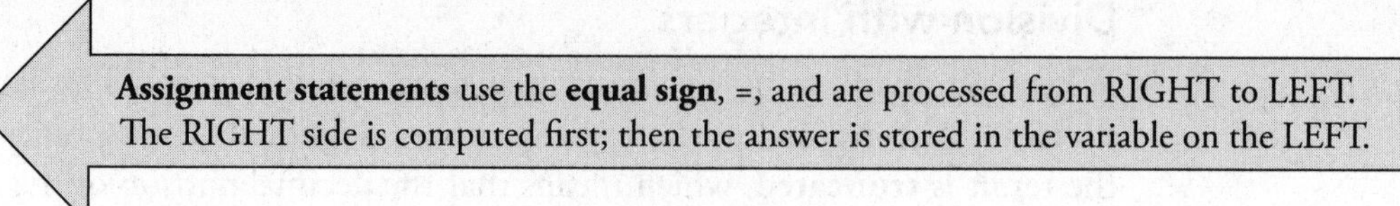

Assignment statements use the **equal sign**, =, and are processed from RIGHT to LEFT. The RIGHT side is computed first; then the answer is stored in the variable on the LEFT.

Accumulating

I'm pretty sure that if you were writing some kind of game, you would want to keep track of a score. You may also want a timer. In order to do this, you will need to know how to count and accumulate. When giving a variable a value, the right side of the assignment statement is computed first, and the result is given to the variable on the left side of the statement (even if it is the same variable!). The left side of the assignment statement is replaced by the right side. So, to modify a number variable, you must change it and then store it back on itself.

Example

How to accumulate in Java using the assignment operator:

```
int score = 0;          // create a score variable and initialize it to zero
score = score + 100; //  add 100 to the variable score
```

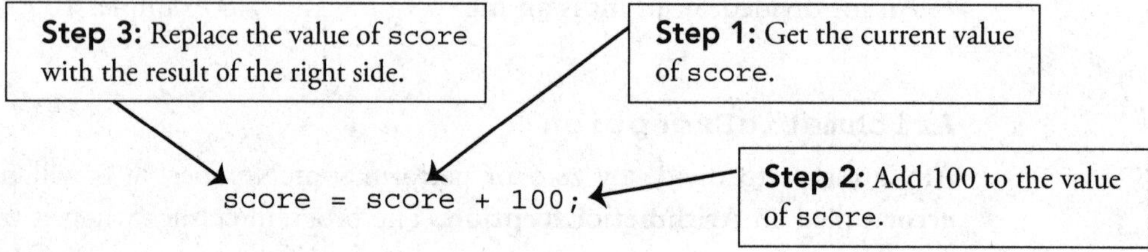

Short-Cut Operators

Java provides another way to modify the value of a variable called a **short-cut operator**. There is a short-cut operator for addition, subtraction, multiplication, division, and modulo. It's called a short-cut operator because you only have to write the variable name one time rather than twice like in the previous explanation.

Examples

How to modify a variable using short-cut operators:

```
int score = 90;
score += 10;            // adds 10 to score: score is now 100
score -= 25;            // subtracts 25 from score: score is now 75
score *= 10;            // multiplies score by 10: score is now 750
score /= 5;             // divides score by 5: score is now 150
score %= 3;             // score mod 3: score is now 0
```

Incrementing and Decrementing a Variable

To **increment** a variable means to *add* to the value stored in the variable. This is how we can *count up* in Java. There are three common ways to write code to add *one* to a variable. You may see any of these techniques for incrementing by one on the AP Computer Science A Exam.

```
i++;                    // increments i by one
i = i + 1;              // increments i by one
i+=1;                   // increments i by one
```

To **decrement** a variable means to *subtract* from its value. This is how we can *count down* in Java. There are three common ways to subtract *one* from a variable in Java.

```
i--;                    // decrements i by one
i = i - 1;              // decrements i by one
i-=1;                   // decrements i by one
```

Casting a Variable

On the AP Computer Science A Exam, you will be tested on the correct way to **cast** variables. Casting is a way to tell a variable to temporarily become a different data type for the sake of performing some action, such as division. The next example demonstrates why casting is important using division with integers. Observe how the cast uses parentheses.

The Most Common Type of Cast

The most common types of casts are from an int to a double or a double to an int.

Example 1

Here is the *correct way* to cast an int to a double: Make the variable pretend that it is a double *before* doing the division.

Hey hits, I know you are an int, but could you just temporarily pretend that you are a double so that I can give this softball player her correct batting average? If you don't, her average will be 0.0.

```
int atBats = 5;
int hits = 2;
double battingAverage = (double)hits / atBats;      // battingAverage is 0.4
```

Example 2

Here is the *wrong way* to cast an int to a double. This example does not produce the result we want, because we are casting the *result* of the division between the integers. Remember that 2 divided by 5 is 0 (using integer division).

```
int atBats = 5;
int hits = 2;
double battingAverage = (double)(hits / atBats); // battingAverage is 0.0
```

Manually Rounding a double

On the AP Computer Science A Exam, you will have to know how to round a double to its nearest whole number. This requires a cast from a double to an int.

Example 1

Manually rounding a positive decimal number to the nearest integer:

```
double roundMe = 53.6;
int result = (int)(roundMe + 0.5);                  // result is 54
```

Example 2

Manually rounding a negative decimal number to the nearest integer:

```
double roundMe = -53.6;
int result = (int)(roundMe - 0.5);                    // result is -54
```

Relational Operators

Double Equals Versus Single Equals

Java compares numbers the same way that we do in math class, using **relational operators** (like greater than or less than). A **condition** is a comparison of two values using a relational operator. To decide if two numbers are equal in Java, we compare them using the **double equals**, ==. This is different from the **single equals**, =, which is called an **assignment operator** and is used for assigning values to a variable.

The not Operator: !

To compare if two numbers are **not equal** in Java, we use the **not operator**. It uses the exclamation point, **!**, and represents the opposite of something. It is also referred to as a **negation** operator.

Table of Relational Operators

The following table shows the symbols for comparing numbers using relational operators.

Comparison	Java Symbol
Greater than	>
Less than	<
Greater than or equal to	>=
Less than or equal to	<=
Equal to	==
Not equal to	!=

Examples

Demonstrating relational operators within conditional statements:

```
int a = 6;
System.out.println(a == 10);   // false is printed;  6 is not equal to 10
System.out.println(a < 10);    // true is printed;  6 is less than 10
System.out.println(a >= 10);   // false is printed;  6 is less than 10
System.out.println(a != 10);   // true is printed;  6 is not equal to 10
```

The ! as a Toggle Switch

The not operator, !, can be used in a clever way to reverse the value of a boolean.

Example

Using the not operator to toggle player turns for a two-player game:

```
boolean playerOneTurn = true;          // playerOneTurn is true
playerOneTurn = !playerOneTurn;        // playerOneTurn is false
playerOneTurn = !playerOneTurn;        // playerOneTurn is true
```

Logical Operations

Compound Conditionals: Using AND and OR

AND and **OR** are **logical operators**. In Java, the AND operator is typed using two ampersands (**&&**) while the OR operator uses two vertical bars (||). These two logical operators are used to form **compound conditional statements.**

Logical Operator	Java Symbol	Compound Conditional	Result with Explanation
AND	`&&`	*`condition1 && condition2`*	True only if both conditions are true. False if either condition is false.
OR	`||`	*`condition1 || condition2`*	True if either condition is true. False only if both conditions are false.

The Truth Table

A **truth table** describes how AND and OR work.

First Operand (A)	Second Operand (B)	`A && B`	`A || B`
True	True	True	True
True	False	False	True
False	True	False	True
False	False	False	False

Example 1

Suppose you are writing a game that allows the user to play as long as their score is less than the winning score and they still have time left on the clock. You would want to allow them to play the game as long as *both* of these two conditions are true.

```
boolean continuePlaying = (score < winningScore && timeLeft > 0);
```

Example 2

Computing the results of compound conditionals:

```
int a = 3, b = 4, c = 5;
System.out.println(a <= 3 && b != 4);                 // false is printed
System.out.println(b % 2 == 1 || c / 2 == 1);         // false is printed
System.out.println(a > 2 && (b > 5 || c < 6));        // true is printed
```

Explanation: `(3 <= 3) && (4 != 4)` → `true && false` → `false`
Explanation: `(4 % 2 == 1) || (5 / 2 == 1)` → `false || false` → `false`
Explanation: `(3 > 2) && (4 > 5 || 5 < 6)` → `true && (false || true)` → `true && true` → `true`

Fun Fact: *The boolean variable is named in honor of George Boole who founded the field of algebraic logic. His work is at the heart of all computing.*

Short-Circuit Evaluation

Java uses **short-circuit evaluation** to speed up the evaluation of compound conditionals. As soon as the result of the final evaluation is known, then the result is returned and the remaining evaluations are not even performed.

Find the result of: `(1 < 3 || (a >= c - b % a + (b + a / 7) - (a % b +c)))`

In a *split second,* you can determine that the result is true *because 1 is less than 3.* End of story. The rest is never even evaluated. Short-circuit evaluation is useful when the second half of a compound conditional might cause an error. This is tested on the AP Computer Science A Exam.

Example

Demonstrate how short-circuit evaluation can prevent a division by zero error.

If the count is equal to zero, the second half of the compound conditional is never evaluated, thereby avoiding a division by zero error.

```
int count = 0;
int total = 0;
boolean result = (count != 0 && total/count > 0);   // result is false
```

DeMorgan's Law

On the AP Computer Science A Exam, you must be able to evaluate complex compound conditionals. **DeMorgan's Law** can help you decipher ones that fit certain criteria.

Compound Conditional	Applying DeMorgan's Law
!(a && b)	!a || !b
!(a || b)	!a && !b

An easy way to remember how to apply DeMorgan's Law is to think of the distributive property from algebra, but with a twist. Distribute the ! to both of the conditionals and also change the logical operator to its opposite. Note the use of the parentheses and remember that the law can be applied both forward and backward.

Example

Computing the results of complex logical expressions using DeMorgan's Law:

```
int a = 2, b = 3;
boolean result1 = !(b == 3 && a < 1);       //result1 is true
boolean result2 = !(a != 2 || b <= 4);      // result2 is false
```

Explanation for `result1`: DeMorgan's Law says that you should *distribute* the negation operator. Therefore, the negation of b == 3 is b != 3 and the negation of a < 1 is a >= 1:

```
!(b == 3 && a < 1) ➔ (b != 3) || (a >= 1) ➔ (3 != 3) || (2 >= 1)
                                          ➔ false || true ➔ true
```

Explanation for `result2`: The negation of a != 2 is a == 2 and the negation of b <= 4 is b > 4:

```
!(a != 2 || b <= 4) ➔ (a == 2) && (b > 4) ➔ (2 == 2) && (3 > 4)
                                          ➔ true && false ➔ false
```

> **The Negation of a Relational Operator**
>
> The negation of *greater than* (>) is *less than or equal to* (<=) and vice versa.
> The negation of *less than* (<) is *greater than or equal to* (>=) and vice versa.
> The negation of *equal to* (==) is *not equal to* (!=) and vice versa.

Conditional Statements

If you want your program to branch out into different directions based on the input from the user or the value of a variable, then you will want to have a **conditional** statement in your program.

The `if` Statement

The **if** statement is a conditional statement. In its simplest form, it allows the program to execute a specific set of instructions if some certain condition is met. If the condition is not met, then the program skips over those instructions.

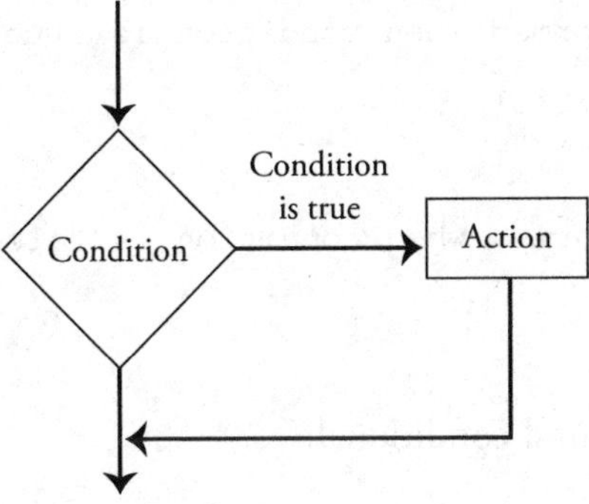

The `if-else` Statement

The **if-else** statement allows the program to execute a specific set of instructions if some certain condition is met and a different set of instructions if the condition is not met. I will sometimes refer to the code that is executed when the condition is true as the *if clause* and the code that follows the else as the *else clause.*

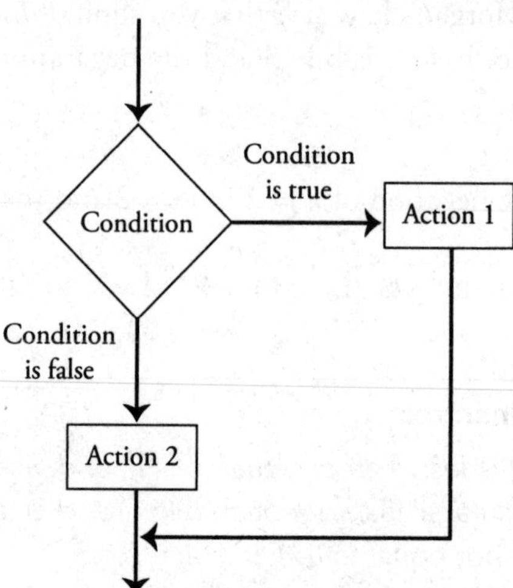

The syntax for if and if-else statements can be tricky. Pay attention to the use of the curly braces.

Example 1

The if statement for one or more lines of code:

```
if (condition)
{
   // one or more instructions to be performed when condition is true
}
```

Example 2

The if-else statement for one or more lines of code for each result:

```
if (condition)
{
   // instructions to be performed when condition is true
}
else
{
   // instructions to be performed when condition is false
}
```

Example 3

The if-else statement using a compound conditional:

```
if (condition1 && condition2)
{
   // condition1 and condition2 are both true
}
else
{
   // either condition1 or condition2 is false (or both are false)
}
```

Example 4

Nested if-else statements for one or more lines of code (watch the curly braces!):

```
if (condition1)
{
    // condition1 is true
    if (condition2)
    {
        // condition1 is true and condition2 is true
    }
    else
    {
        // condition1 is true and condition2 is false
    }
}
else
{
    // condition1 is false
    if (condition3)
    {
        // condition1 is false and condition3 is true
    }
    else
    {
        // condition1 is false and condition3 is false
    }
}
```

Example 5

When you want to execute only one line of code when something is true, you don't actually need the curly braces. Java will execute the first line of code after the if statement if the result is true. This is not recommended but it is possible that you will see this on the AP Computer Science A Exam.

```
if (condition)
    // single instruction to be performed when condition is true
```

Example 6

The same goes for an if-else statement. If you want to execute only one line of code for each value of the condition, then no curly braces are required.

```
if (condition)
    // single instruction to be performed when condition is true
else
    // single instruction to be performed when condition is false
```

Can You Spot the Error in This Program?

```
if (condition);
{
    // instructions to be performed when condition is true
}
```

Answer: Never put a semicolon on the same line as the condition. It ends the if-statement right there. In general, *never put a semicolon before a curly brace.*

The Dangling `else`

If you don't use curly braces properly within if-else statements, you may get a **dangling else**. A dangling else attaches itself to the nearest if statement and not necessarily to the one that you may have intended. Use curly braces to avoid a dangling else and control which instructions you want executed. The dangling else does not cause a compile-time error, but rather it causes a **logic error**. The program doesn't execute in the way that you expected.

Example

The second else statement attaches itself to the nearest preceding if statement:

```
if (condition1)
   // instruction performed when condition1 is true
   if (condition2)
      // instruction performed when condition1 is false and condition2 is true
   else
      // instruction performed when condition1 is false and condition2 is false
```

Scope of a Variable

The word **scope** refers to the code that knows that a variable exists. Local variables, like the ones we have been declaring, are only known within the block of code in which they are defined. This can be confusing for beginning programmers. Another way to say it is: a variable that is declared inside a pair of curly braces is only known inside that set of curly braces.

Curly Braces and Blocks of Code

Curly braces come in pairs and the code they surround is called a **block of code**.

```
{
   // A group of instructions inside a pair of curly braces
   // is called a "block of code". Variables declared inside a block
   // of code are only known inside that block of code.
}
```

Looping Statements

The `for` Loop

Suppose your very strict music teacher catches you chewing gum in class and tells you to write out "I will not chew gum while playing the flute" 100 times. You decide to use your computer to do the work for you. How can you write a program to repeat something as many times as you want? The answer is a **for loop**.

General Form for Creating a `for` Loop

```
for (initialize loop control variable(LCV);condition using LCV; modify LCV)
{
   // instructions to be repeated
}
```

The for loop uses a **loop control variable** to repeat its instructions. This loop control variable is typically an int and is allowed to have a one-letter name like i, j, k, and so on. This breaks the rule of having meaningful variable names. Yes, the loop control variable is a rebel.

Explanation 1: The `for` loop for those who like to read paragraphs . . .
When a for loop starts, its loop control variable is declared and initialized. Then, a comparison is made using the loop control variable. If the result of the comparison is true, the instructions that are to be repeated are executed one time. The loop control variable is then modified in some way (usually incremented or decremented, but not always). Next, the loop control variable is compared again using the same comparison as before. If the result of this comparison is true, the loop performs another **iteration**. The loop control variable is again modified (in the same way that it was before), and this process continues until the comparison of the loop control variable is false. When this happens, the loop ends and we **exit (or terminate) the loop**. The total number of times that the loop repeats the instructions is called the number of iterations.

Explanation 2: The `for` loop for those who like a list of steps . . .
Step 1: Declare and initialize the loop control variable.
Step 2: Compare the loop control variable in some way.
Step 3: If the result of the comparison is true, then execute the instructions one time. If the result of the comparison is false, then skip to Step 6.
Step 4: Modify the loop control variable in some way (usually increment or decrement, but not always).
Step 5: Go to Step 2.
Step 6: Exit the loop and move on to the next line of code in the program.

Explanation 3: The `for` loop for those who like pictures . . .

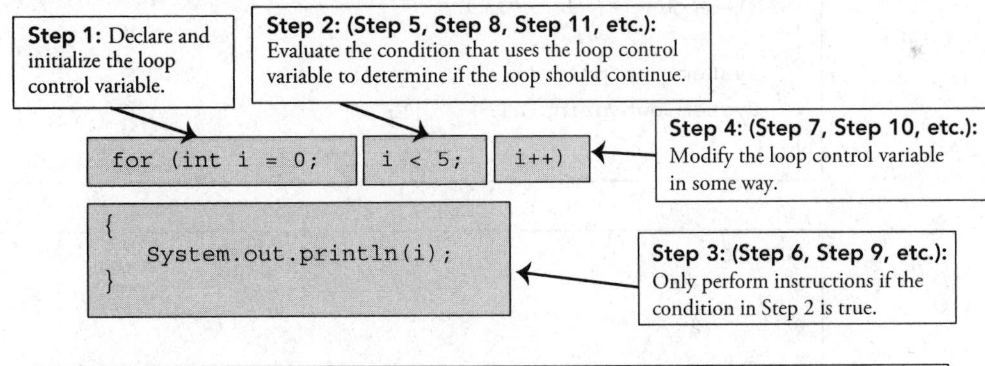
Step 1: Declare and initialize the loop control variable.
Step 2: (Step 5, Step 8, Step 11, etc.): Evaluate the condition that uses the loop control variable to determine if the loop should continue.
for (int i = 0;
i < 5;
i++)
Step 4: (Step 7, Step 10, etc.): Modify the loop control variable in some way.
{
System.out.println(i);
}
Step 3: (Step 6, Step 9, etc.): Only perform instructions if the condition in Step 2 is true.

Last step: Exit the loop when the condition in Step 2 becomes false. Move on to the next instruction in the program.

The Loop Control Variable

The most common approach when using a for loop is to declare the loop control variable in the for loop. If this is the case, then the loop control variable is only known inside the for loop. Any attempt to use the loop control variable in an instruction after the loop will get a compile-time error (cannot be resolved to a variable).

However, it is possible to declare a loop control variable prior to the for loop code. If this is the case, then the loop control variable is known before the loop, during the loop, and after the loop.

Example 1

A simple example of a for loop:

```
for (int i = 0; i < 3; i++)
{
    System.out.println("hello");
}
```

```
OUTPUT
hello
hello
hello
```

Example 2

The loop control variable is often used inside the loop as shown here:

```
for (int j = 9; j > 6; j--)
{
    System.out.print(j + "        ");
    System.out.println(10 - j);
}
```

```
OUTPUT
9     1
8     2
7     3
```

Example 3

In this example, an error occurs when an attempt is made to print the loop control variable after the loop has finished. The variable was declared inside the loop and is not known after the loop exits:

```
for (int k = 1; k <= 4; k++)
{
    System.out.println(k);
}
System.out.println(k*10);
```

```
Compile-time error: k cannot be resolved to a variable
```

Example 4

In this example, the loop control variable is declared in a separate instruction before the loop. This allows the variable to be known before, during, and after the loop:

```
int i;
for (i = 1; i < 5; i++)
{
    System.out.println(i);
}
System.out.println(i);
```

```
OUTPUT
1
2
3
4
5
```

Fun Fact: *In an early programming language called Fortran, the letters i, j, and k were reserved for storing integers. To this day programmers still use these letters for integer variables even though they can be used to store other data types.*

The Nested `for` Loop

A for loop that is inside of another for loop is called a **nested for loop**. The **outer loop control variable** (OLCV) is used to control the outside loop. The **inner loop control variable** (ILCV) is used to control the inside loop. There is no limit to the number of times you can nest a series of for loops.

General Form for a Nested for Loop

```
for (initialize OLCV; condition using OLCV; modify OLCV)
{
    for (initialize ILCV; condition using ILCV; modify ILCV)
    {
        // instructions to be repeated
    }
}
```

Execution Count Within a Nested for Loop

When a nested for loop is executed, the **inner loop** is performed in its entirety for every iteration of the **outer loop**. That means that the number of times that the instructions inside the inner loop are performed is equal to the product of the number of times the outer loop is performed and the number of times the inner loop is performed.

Example

To demonstrate execution count for a nested for loop: The outer loop repeats a total of three times (once for i = 0, then for i = 1, and then for i = 2). The inner loop repeats four times (once for j = 1, then for j = 2, then for j = 3, and then for j = 4). The inner loop repeats (starts over) for every iteration of the outer loop. Therefore, the System.out.println statement executes a total of 12 times:

```
for (int i = 0; i < 3; i++)                    // the outer loop repeats 3 times
{
    for (int j = 1; j < 5; j++)                // the inner loop repeats 4 times
    {
        System.out.println(i + "    " + j);    // total of 12 iterations (3 * 4)
    }
}
```

```
OUTPUT
0    1
0    2
0    3
0    4
1    1
1    2
1    3
1    4
2    1
2    2
2    3
2    4
```

Execution Count for a Nested `for` Loop

To find the total number of times that the code inside the inner for loop is executed, multiply the number of iterations of the outer loop by the number of iterations of the inner loop.

```
for (int x = 4; x > 0; x--)            // the outer loop repeats 4 times
{
    for (int y = 9; y <= 15; y++)      // the inner loop repeats 7 times
    {
        // some instruction            // total of 28 iterations (4 * 7)
    }
}
```

The `while` Loop

Consider the software that is used to check out customers at a store. The clerk drags each item over the scanner until there aren't any more items to be scanned. *Viewing this through the eyes of a programmer*, I can suggest that a **while loop** is used to determine the final cost of all the items. The process repeats until there aren't any more items. Now, in contrast, if the clerk asked you how many items you had in your cart before he started scanning, then I would suggest that a for loop is being used to determine the final cost of all the items.

General Form for a `while` Loop

Recommended form: Repeat instructions (using curly braces):

```
while (condition)
{
    // one or more instructions to be repeated
}
```

Legal (but not recommended) form: Repeat an instruction (without using curly braces):

```
while (condition)
    // single instruction to be repeated
```

The while loop repeats instructions just like the for loop, but it does it differently. Instead of having a loop control variable that is a number like the for loop, the while loop can use any kind of data type to control when it is to perform another iteration of the loop. I like to use the phrase *as long as* as a replacement for the word *while* when reading a while loop. In other words: *As long as the condition is true, I will continue to repeat the instructions.*

Explanation 1: The `while` loop for those who like to read paragraphs . . .

Declare and initialize some variable of any data type *before* the loop begins. Then compare this variable at the start of the loop. If the result of the comparison is true, the instructions inside the loop get executed one time. Then, modify the variable in some way. Next, compare the variable again in the same way as before. If the result is true, the instructions get executed one more time. This process continues until the result of the comparison is false. At this point, the loop ends and we exit the loop. You must make certain that the variable that is being compared is changed inside of the loop. If you forget to modify this variable, you may end up with a loop that never ends!

Explanation 2: The `while` loop for those who like a list of steps . . .

Step 1: Declare and initialize some variable of any data type.
Step 2: Compare this variable in some way.
Step 3: If the result of the comparison is true, then execute the instructions one time. If the result of the comparison is false, then skip to Step 6.
Step 4: Modify the variable that is to be compared.
Step 5: Go to Step 2.
Step 6: Exit the loop and move on to the next line of code in the program.

Example 1

A simple example of a while loop:

```
int count = 0;
while (count < 5)
{
    System.out.println(count);
    count++;
}
```

```
OUTPUT
0
1
2
3
4
```

Example 2

Use a while loop to require a user to enter the secret passcode for a vault that contains a million dollars. Do not allow the user to exit the loop until they get it right. If they get it right the first time, do not enter the loop (do not execute any of the instructions inside the loop).

```
System.out.print("Enter the secret passcode: ");
int secretPasscode = /* int provided by user */
while (secretPasscode != 1234)
{
    System.out.println("Sorry that is incorrect. Try again: ");
    secretPasscode = /* int provided by user */
}
System.out.println("Congratulations. You have access to the vault.");
```

```
OUTPUT
Enter the secret passcode: 9999
Sorry that is incorrect. Try again: 8888
Sorry that is incorrect. Try again:  1234
Congratulations. You have access to the vault.
```

A Flag in Programming

A **flag** in programming is a virtual way to signal that something has happened. Think of it as, "I'm waving a flag" to tell you that something important just occurred. Normally, boolean variables are chosen for flags. They are originally set to false, then when something exciting happens, they are set to true.

Example 3
Use a while loop to continue playing a game as long as the game is not over:

```
boolean gameOver = false;
while (!gameOver)
{
    // play the game
    // When a player wins, set gameOver = true
}
```

Strength of the `while` Loop

A strength of the while loop is that you don't have to know how many times you will execute the instructions before you begin looping. You just have to know when you want it to end.

The Infinite Loop

If you make a mistake in a looping statement such that the loop never ends, the result is an **infinite loop**.

Example 1
Accidentally getting an infinite loop when using a for loop:

```
for (int i = 0; i < 10; i--)
{
    i++;
    // i will always revert back to 0 causing an infinite loop
}
```

Example 2
Accidentally getting an infinite loop when using a while loop:

```
int i = 0;
while (i < 10)
{
    // do something forever because you forgot to increment i
}
```

Boundary Testing

A very common error when using a for loop or a while loop is called a **boundary error**. It means that you accidently went one number past or one number short of the right amount. *The loop didn't perform the precise number of iterations.* This concept is tested many times on the AP Computer Science A Exam. The only real way to prevent it is to hand-trace your code very carefully to make sure that your loop is executing the correct number of times.

Goldilocks and the Three Looping Structures

Always test that your for loops, nested for loops and while loops are executing the correct number of times—not too many, but not too few. Just the right number of times. Hint: If you use:

```
for (int i = 0; i < maximum; i++)
```

then the loop will execute maximum number of times.

Bases Other Than Decimal

Binary, Octal, and Hexadecimal

As humans, we are familiar with base 10, or **decimal notation**. However, computers don't use base 10, they use **base 2**, **base 8**, or **base 16**. The AP Computer Science A Exam requires that you understand how to convert between these different number base systems and also understand the underpinnings of **place value** systems other than decimal.

System	Base	Digits	Number of Digits
Decimal	10	0,1,2,3,4,5,6,7,8,9	10
Binary	2	0,1	2
Octal	8	0,1,2,3,4,5,6,7	8
Hexadecimal	16	0,1,2,3,4,5,6,7,8,9,A,B,C,D,E,F	16

Each of these number systems uses place value and is controlled by its base. Think back to when you were in grade school learning about really big numbers. The number 2375 in decimal means two thousands, three hundreds, seven tens, and five ones. This is because each of the digits resides in a place that represents a power of 10. This system is called *place value*, because, yes, you guessed it, each place has a value.

Notice that we ran out of digits in hexadecimal and started to use the first six letters of the English alphabet. This is the secret code of the hexadecimal. Just note that A = 10, B = 11, C = 12, D = 13, E = 14, and F = 15.

Fun Fact: *The word* **bit** *comes from the words* binary digit. *A bit is represented by either a 1 or a 0, which represents on or off. The symbol for the power button is actually a combination of a 1 and a 0.*

Decimal Numbers (Base 10)

This graphic explains what a number in base 10 means. Now, let's apply this same idea to convert numbers from binary, octal, and hexadecimal to base 10.

2 3 7 5 dec $= \mathbf{2} \times 10^3 = 2000$

$\mathbf{3} \times 10^2 = 300$

$\mathbf{7} \times 10^1 = 70$

$+\mathbf{5} \times 10^0 = 5$

10^3 10^2 10^1 10^0 2375dec

Example 1

Converting a binary (base 2) number to decimal:

$$\begin{aligned} \mathbf{1\ 0\ 0\ 1}\ \text{bin} &= \mathbf{1} \times 2^3 = 8 \\ &\quad \mathbf{0} \times 2^2 = 0 \\ &\quad \mathbf{0} \times 2^1 = 0 \\ &\ +\mathbf{1} \times 2^0 = 1 \\ \hline &\qquad\qquad 9\text{dec} \end{aligned}$$

$2^3\ \ 2^2\ \ 2^1\ \ 2^0$

Therefore, $1001_{\text{bin}} = 9_{\text{dec}}$

Example 2

Converting an octal (base 8) number to decimal:

$$\begin{aligned} \mathbf{2\ 3\ 7\ 5}\ \text{oct} &= \mathbf{2} \times 8^3 = 1024 \\ &\quad \mathbf{3} \times 8^2 = 192 \\ &\quad \mathbf{7} \times 8^1 = 56 \\ &\ +\mathbf{5} \times 8^0 = 5 \\ \hline &\qquad\qquad 1277\text{dec} \end{aligned}$$

$8^3\ \ 8^2\ \ 8^1\ \ 8^0$

Therefore, $2375_{\text{oct}} = 1277_{\text{dec}}$

Example 3

Converting a hexadecimal (base 16) number to decimal (recall: A = 10 and F = 15):

$$\begin{aligned} \mathbf{2\ A\ F\ 5}\ \text{hex} &= \mathbf{2} \times 16^3 = 8192 \\ &\quad \mathbf{10} \times 16^2 = 2560 \\ &\quad \mathbf{15} \times 16^1 = 240 \\ &\ +\ \mathbf{5} \times 16^0 = 5 \\ \hline &\qquad\qquad 10997\text{dec} \end{aligned}$$

$16^3\ \ 16^2\ \ 16^1\ \ 16^0$

Therefore, $2AF5_{\text{hex}} = 10997_{\text{dec}}$

Question: Why do computer scientists confuse Halloween with Christmas?

Answer: Because $31_{\text{OCT}} = 25_{\text{DEC}}$

Commenting Your Code

Inline Comments

I've been using inline comments throughout this concept and will continue to use them in the rest of the book. They begin with two forward slashes, //. All text that comes after the slashes (on the same line) is ignored by the compiler.

```
// This is an example of an inline comment.
```

Multiple-Line Comments

When your comment takes longer than one line, you will want to use a multiple-line comment. These comments begin with /* and end with */. The compiler ignores everything in between these two special character sequences.

```
/*
  This is how to write a multiple-line comment. Everything typed here is
  ignored by the compiler; even mi mysteaks is speling, so be particularly
  careful of your spelling when you are typing multiple-line comments!
*/
```

Javadoc Comments

You will see **Javadoc comments** (also called **documentation comments**) on the AP Computer Science A Exam, especially in the FRQ section. These comments begin with /** and end with */. The compiler ignores everything in between the two character sequences.

```
/**
 *  This is an example of a Javadoc comment.
 *  You will see these "documentation" comments on the AP CS exam.
 */
```

Types of Errors

Compile-Time Errors

When you don't use the correct **syntax** in Java, the compiler yells at you. Well, it doesn't actually yell at you, it just won't compile your program and gives you a **compile-time error**. A list of the most common compile-time errors is located in the Appendix.

A brief list of compile-time errors:

- Forgetting a semicolon at the end of an instruction
- Forgetting to put a data type for a variable
- Using a keyword as a variable name
- Forgetting to initialize a variable
- Forgetting a curly brace (a curly brace doesn't have a partner)

Run-Time Exceptions

If your program crashes while it is running, then you have a **run-time error**. The compiler will display the error and it may have the word **exception** in it.

These are the run-time errors that you are required to understand on the AP Computer Science A Exam:

- ArithmeticException (explained in this Concept)
- NullPointerException (explained in Concept 3: The String Class)
- IndexOutOfBoundsException (explained in Concept 5: Data Structures)
- ArrayIndexOutOfBoundsException (explained in Concept 5: Data Structures)
- IllegalArgumentException (explained in Concept 7: Classes and Objects)

Logic Errors

When your program compiles and runs without crashing, but it doesn't do what you expected it to do, then you have a **logic error**. A logic error is the most challenging type of error to fix because you have to figure out where the problem is in your program. Is your math correct? Are your if-statements comparing correctly? Does your loop actually do what it's supposed to do? Are any statements out of order or are you missing something? Logic errors require you to read your code very carefully to determine the source of the error. My advice is to help other people fix their errors so you can ask for help from them when you need it.

Logic Errors on the AP Computer Science A Exam

You will have to analyze code in the multiple-choice section of the exam and find hidden logic errors.

Debugging

The process of removing the errors in your program (compile-time, run-time, and logic) is called **debugging** your program. On the AP Computer Science A Exam, you will be asked to find errors in code.

Fun Fact: *Grace Murray Hopper documented the first actual computer bug on September 9, 1947. It was a moth that got caught in Relay #70 in Panel F of the Harvard Mark II computer.*

`System.out.println` as a Debugging Device

A common way to debug a computer program is to peek inside the computer while it is running and display the current values of variables on the console screen. We aren't actually opening up the computer. We are just displaying the current values of the important variables. By printing the values of the variables at precise moments, you can determine what is going on during the running of the program and hopefully figure out the error.

Example

Figure out what this loop is doing by writing an output statement to the console screen. Printing the values of a, b, and i can help you figure out what the program is doing.

```
int a = 1;
for (int i = 1; i < 10; i++)
{
   a += i;
   int b = a % i;
   System.out.println(i + "   " + a + "   " + b);      // Debugging trick
}
```

```
OUTPUT
1  2  0
2  4  0
3  7  1
4  11  3
5  16  1
6  22  4
7  29  1
8  37  5
9  46  1
```

› Rapid Review

Variables

- Variables store data and must be declared with a data type.
- The int, double, and boolean types are called primitive data types.
- Variables may be initialized with a value or be assigned one later in the program.
- Variable names may begin with a letter, dollar sign, or underscore. They cannot begin with a number.
- Camel case is used for all variable names starting with a lowercase letter.
- Choose meaningful names for variables.
- A keyword is a word that has special meaning to the compiler such as *while* or *public.*
- A variable name cannot be the same as a keyword.
- The int data type is used to store integer data.
- The double data type is used to store decimal data.
- To cast a variable means to temporarily change its data type.
- The most common type of cast is to cast an int to a double.
- The boolean data type is used to store either a true or false value.

Math and Logic

- The arithmetic operators are +, −, *, /, and %.
- The modulo operator, %, returns the remainder of a division between two integers.
- Java evaluates all mathematical expressions using the order of operations.
- The precedence order for all mathematical calculations is parentheses first, then *, / and % equally from the left to right, then + and − equally from left to right.
- Java uses integer division when dividing an int by an int. The result is a truncated int.
- "Truncating" means dropping (not rounding) the decimal portion of a number.
- The equal sign, =, is called the assignment operator.
- To accumulate means to add (or subtract) a value from a variable.
- Short-cuts for performing mathematical operations are +=, −=, *=, /=, and %=.
- The relational operators in Java are >, >=, <, <=, ==, and !=.
- "To increment" means to add one to the value of a number variable.
- "To decrement" means to subtract one from the value of a number variable.
- A condition is an expression that evaluates to either true or false.
- The logical operator AND is coded using two ampersands, &&.
- The && is true only when both conditions are true.
- The logical operator OR is coded using two vertical bars, ||.

- The || is true when either condition or both are true.
- Software developers use logical operators to write compound conditionals.
- The NOT operator (negation operator) is coded using the exclamation point, !.
- The ! can be used to flip-flop a boolean.
- DeMorgan's Law states: !(A && B) = !A || !B and also !(A || B) = !A && !B.
- When evaluating conditionals, the computer uses short-circuit evaluation.

Programming Statements

- The if and the if-else are called conditional statements since the flow of the program changes based upon the evaluation of a condition.
- Curly braces come in pairs and the code they contain is called a block of code.
- Curly braces need to be used when more than one instruction is to be executed for if, if-else, for, and while statements.
- The scope of a variable refers to the code that knows that the variable exists.
- Variables declared within a conditional are only known within that conditional.
- The for loop is used to repeat one or more instructions a specific number of times.
- The for loop is a good choice when you know exactly how many times you want to repeat a set of instructions.
- The for loop uses a numeric loop control variable that is compared and also modified during the execution of the loop. When the comparison that includes this variable evaluates to false, the loop exits.
- Variables declared within a for loop are only known within that for loop.
- A nested for loop is a for loop inside of a for loop.
- The number of iterations of a nested for loop is the product of the number of iterations for each loop.
- The while loop does not require a numeric loop control variable.
- The while statement must contain a conditional that compares a variable that is modified inside the loop.
- Choose a while loop when you don't know how many times you need to repeat a set of instructions but know that you want to stop when some condition is met.
- An infinite loop is a loop that never ends.
- Forgetting to modify the loop control variable or having a condition that will never be satisfied are typical ways to cause infinite loops.
- Variables declared within a while loop are only known within the while loop.
- A boundary error occurs when you don't execute a loop the correct number of times.

Miscellaneous

- The System.out.println() statement displays information to the console and moves the cursor to the next line.
- The System.out.print() statement displays information to the console and does not move the cursor to the next line.
- Computer scientists use bases other than decimal such as binary, octal, and hexadecimal.
- Programs are documented using comments. There are three different types of comments: inline, multiple line, and Javadoc.
- There are three main types of errors: compile-time, run-time, and logic.
- Compile-time errors are caused by syntax errors.
- *Exception* is another name for error.
- There are many types of run-time exceptions and they are based on the type of error.

- Logic errors are difficult to fix because you have to read the code very carefully to find out what is going wrong.
- Debugging is the process of removing errors from a program.
- Printing variables to the console screen can help you debug your program.

› Review Questions

Basic Level

1. Consider the following code segment.

```
int var = 12;
var = var % 7;
var--;
System.out.println(var);
```

What is printed as a result of executing the code segment?

(A) 0
(B) 1
(C) 2
(D) 4
(E) 5

2. Consider the following code segment.

```
int count = 1;
int value = 31;
while (value >= 10)
{
    value = value - count;
    count = count + 3;
}
System.out.println(value);
```

What is printed as a result of executing the code segment?

(A) 4
(B) 9
(C) 10
(D) 13
(E) 31

3. Consider the following code segment.

```
int count = 5;
double multiplier = 2.5;
int answer = (int)(count * multiplier);
answer = (answer * count) % 10;
System.out.println(answer);
```

What is printed as a result of executing the code segment?

(A) 0
(B) 2.5
(C) 6
(D) 12.5
(E) 60

4. Consider the following code segment.

```
int num1 = 9;
int num2 = 5;
if (num1 > num2)
{
   System.out.print((num1 + num2) % num2);
}
else
{
   System.out.print((num1 - num2) % num2);
}
```

What is printed as a result of executing the code segment?

(A) `0`
(B) `2`
(C) `4`
(D) `9`
(E) `14`

5. Consider the following code segment.

```
int count = 5;
for (int i = 3; i < 7; i = i + 2)
{
   count += i;
}
System.out.println(count);
```

What is printed as a result of executing the code segment?

(A) `5`
(B) `7`
(C) `9`
(D) `13`
(E) `20`

6. Consider the following code segment.

```
int value = initValue;
if (value > 10)
   if (value > 15)
      value = 0;
else
   value = 1;
System.out.println("value = " + value);
```

Under which of the conditions below will this code segment print `value = 1`?

(A) `initValue = 8;`
(B) `initValue = 12;`
(C) `initValue = 20;`
(D) Never. `value = 0` will always be printed.
(E) Never. Code will not compile.

7. Consider the following code segment.

```
int incr = 1;
for (int i = 0; i < 10; i+= incr)
{
    System.out.print(i - incr + "    ");
    incr++;
}
```

What is printed as a result of executing the code segment?

(A) `-1   0   2   5`
(B) `-1   0   3   7`
(C) `-1   1   4   8`
(D) `0   2   5   9`
(E) `0   1   3   7`

8. Which of the following code segments will print exactly eight `"$"` symbols?

I.
```
for (int i = 0; i < 8; i++)
{
    System.out.print("$");
}
```

II.
```
int i = 0;
while (i < 8)
{
   System.out.print("$");
   i++;
}
```

III.
```
for (int i = 7; i <= 30 ; i+=3)
{
   System.out.print("$");
}
```

(A) I only
(B) II only
(C) I and II only
(D) I and III only
(E) I, II, and III

9. Assume that `a`, `b` and `c` have been declared and initialized with `int` values. The expression

```
!(a > b || b <= c)
```

is equivalent to which of the following?

(A) `a > b && b <= c`
(B) `a <= b || b > c`
(C) `a <= b && b > c`
(D) `a < b || b >= c`
(E) `a < b && b >= c`

Advanced Level

10. Assume that x, y and z have been declared as follows:

```
boolean x = true;
boolean y = false;
boolean z = true;
```

Which of the following expressions evaluates to true?

(A) `x && y && z`
(B) `x && y || z && y`
(C) `!(x && y) || z`
(D) `!(x || y) && z`
(E) `x && y || !z`

11. Consider the following code segment.

```
for (int j = 0; j < 10; j++)
{
    for (int k = 10; k > j; k--)
    {
        System.out.print("*");
    }
}
```

How many "*" symbols are printed as a result of executing the code segment?

(A) 5
(B) 10
(C) 45
(D) 55
(E) 100

12. Consider the following code segment.

```
boolean b1 = true;
boolean b2 = true;
int x = 7;
while (b1 || b2)
{
    if (x > 4)
    {
        b2 = !b2;
    }
    else
    {
        b1 = !b1;
    }
    x--;
}
System.out.print(x);
```

What is printed as a result of executing the code segment?

(A) 3
(B) 4
(C) 6
(D) 7
(E) Nothing will be printed. Infinite loop.

13. Which of the following three code segments will produce the same output?

I.
```
int i = 1;
while (i < 12)
{
    System.out.print(i + " ");
    i *= 2;
}
```

II.
```
for (int i = 4; i > 0; i--)
{
    System.out.print((4 - i) * 2 + 1 + " ");
}
```

III.
```
for (int i = 0; i < 4; i++)
{
    System.out.print((int)Math.pow(2, i) + " ");
}
```

(A) I and II only
(B) II and III only
(C) I and III only
(D) I, II, and III
(E) All three outputs are different.

14. Consider the following code segment.

```
double count = 6.0;
for (int num = 0; num < 5; num++)
{
    if (count != 0 && num / count > 0)
    {
        count -= num;
    }
}
System.out.println(count);
```

What is printed as a result of executing the code segment?

(A) `0.0`
(B) `-4.0`
(C) `5`
(D) The program will end successfully, but nothing will be printed.
(E) Nothing will be printed. Run-time error: `ArithmeticException`.

15. Consider the following code segment.

```
int n = 0;
while (n < 20)
{
    System.out.print(n % 4 + " ");
    if (n % 5 == 2)
    {
        n += 4;
    }
    else
    {
        n += 3;
    }
}
```

What is printed as a result of executing the code segment?

(A) `0 3 2 1 0 0 3`
(B) `0 0 3 2 2 1`
(C) `0 3 2 1 0 0`
(D) `0 0 3 2 2 1 0`
(E) Many numbers will be printed. Infinite loop.

16. Convert 93 from base 10 to base 16.

(A) A9
(B) 93
(C) 5D
(D) 139
(E) 147

17. Convert AF from base 16 to base 10.

(A) 115
(B) 175
(C) 250
(D) 400
(E) 1015

18. Which of these number representations is equivalent to 108 base 10?

I. 1101100 base 2
II. 154 base 8
III. 6C base 16

(A) I only
(B) II only
(C) I and II only
(D) I and III only
(E) I, II, and III

› Answers and Explanations

Bullets mark each step in the process of arriving at the correct solution.

1. The answer is D.

- The value of var begins at 12.
- The operation var % 7 finds the remainder after var is divided by 7. Since 12 / 7 is 1 remainder 5, the value of var is now 5.
- var-- means subtract one from var, so the value of var is now 4, and that's what is printed.

2. The answer is B.

- The first time we evaluate the while condition, value = 31, which is >= 10, so the loop executes.
  - In the loop, we subtract count from value which becomes 30,
  - and then we add 3 to count, which becomes 4.
- Then we go back to the top of the while loop and re-evaluate the condition. 30 >= 10, so we execute the loop again.
  - This time we subtract 4 from value, since count is now 4. value = 26
  - and count = 7.
- Back up to the top, 26 >= 10, so on we go.
  - value = 27 – 7 = 19 and count = 10.
- Since 10 >= 10, we will execute the loop one more time.
  - value = 19 – 10 = 9 and count = 13.
- This time when we evaluate the condition, it is no longer true, so we exit the loop and print value, which equals 9.

3. The answer is A.

- count * multiplier = 12.5
- When we cast 12.5 to an int by putting (int) in front of it, we *truncate* (or cut off) the decimals, so the value we assign to answer is 12.
- 12 * 5 = 60. 60 % 10 = 0 because 60 / 10 has no remainder.
- So we print 0.

4. The answer is C.

- Since 9 > 5, we execute the statements after the if and skip the statements after the else.
- 9 + 5 = 14. 14 % 5 = 4 (since 15 / 5 = 2 remainder 4), so we print 4.

5. The answer is D.

- When we first enter the for loop count = 5 and i = 3.
- The first time through the loop, we execute count += i, which makes count = 8, and we increment i by 2, which makes i = 5.
- i < 7, so we execute the loop again; count = 8 + 5 = 13, and i = i + 2 = 7.
- This time the condition i < 7 fails and we exit the loop. 13 is printed.

6. The answer is B.

   - This is a confusing one. We count on indenting to be correct and we use it to evaluate code, but in this case it is deceiving us. Remember that indenting is for humans; Java doesn't care how we indent our code.
   - An else clause always binds to the *closest* available if. That means the else belongs to if (value > 15) not to if (value > 10). The code should be formatted like this:

```
if (value > 10)
    if (value > 15)
        value = 0;
    else
        value = 1;
System.out.println("value = " + value);
```

   - This makes it clearer that value = 1 will be printed if the first condition is true (so value has to be greater than 10) and the second condition is false (so value has to be less than or equal to 15). The only value in the answers that fits those criteria is 12, so initValue =12 is the correct answer.

7. The answer is A.

   - The first time through the loop, incr = 1 and i = 0. 0 – 1 = -1 which is printed; then incr is increased to 2. The increment portion of this for loop is a little unusual. Instead of i++ or i = i + 2, it's i = i + incr, so, since incr = 2 and i = 0, i becomes 2.
   - 2 < 10 so we continue. 2 – 2 = 0, which is printed, incr = 3 and i = 2 + 3 = 5.
   - 5 < 10 so we continue. 5 – 3 = 2, which is printed, incr = 4 and i = 5 + 4 = 9.
   - 9 < 10 so we continue. 9 – 4 = 5, which is printed, incr = 5 and i = 9 + 5 = 14.
   - This time we fail the condition, so our loop ends having printed -1 0 2 5

8. The answer is E.

   - The first option is the classic way to write a for loop. If you always start at i = 0 and go until i < n, you know the loop will execute n times without having to think about it (using i++ of course). So this first example prints eight "$".
   - The second example is the exact same loop as the first example, except written as a while loop instead of a for loop. You can see the i = 0, the i < 8, and the i++. So this example also prints eight "$".
   - The third example shows why we always stick to the convention used in the first example. Here we have to reason it out: i will equal 7, 10, 13, 16, 19, 22, 25, 28 before finally failing at 31. That's eight different values, which means eight times through the loop. So this example also prints eight "$".

9. The answer is C.

   - We start out by using DeMorgan's Law (notice that || becomes && when we distribute the !).

```
!(a > b || b <= c) ➔ !(a > b) && !(b <= c)
```

   - !> is the same as <= (don't forget the =), and !<= is the same as >, so

```
!(a > b) && !(b <= c) ➔ a <= b && b > c
```

10. The answer is C.

- Rewrite the expressions, filling in true or false for x, y, and z. Remember order of operations. && has precedence over ||.
- true && false && true
  - true && false = false, and since false && anything is false, this one is false.
- true && false || true && false
  - Order of operations is && before || so this simplifies to false || false = false.
- !(true && false) || true
  - We don't need to think about the first half because as soon as we see || true, we know this one is true.

We've found our answer, but let's keep going so we can explain them all.

- !(true || false) && true
  - true || false is true, !true is false. false && anything is false, so this one is false.
- true && false || !true
  - true and false is false, !true is false, so this is false || false, which is false.

11. The answer is D.

- The outer loop is going to execute 10 times: j = 0 through j = 9.
- The inner loop is trickier. The number of times it executes depends on j.
  - The first time through, j = 0, so k = 10, 9, 8 . . . 1 (10 times).
  - The second time through, j = 1, so k = 10, 9, 8 . . . 2 (9 times).
  - The third time through, j = 2, so k = 10, 9, 8 . . . 3 (8 times).
  - There's a clear pattern here, so we don't need to write them all out. We do have to be careful that we stop at the right time though.
  - The last time through the loop, j = 9, so k = 10 (1 time).
  - Adding it all together: 10 + 9 + 8 + 7 + 6 + 5 + 4 + 3 + 2 + 1 = 55

12. The answer is A.

- b1 and b2 are both true. The OR (||) in the while condition requires that at least one of them be true, so we enter the loop for the first time.
  - 7 > 4 so we execute b2 = ! b2. That's a tricky little statement that is used to flip the value of a boolean. Since b2 is true, !b2 is false, and we assign that value back into b2, which becomes false.
  - We skip the else clause and subtract one from x.
  - At the bottom of the loop: b1 = true, b2 = false, x = 6.
- The while condition now evaluates to (true || false), which is still true, so we execute the loop.
  - 6 > 4 so we flip b2 again. b2 is now true.
  - Skip the else clause, subtract one from x.
  - At the bottom of the loop: b1 = true, b2 = true, x = 5.
- The while condition now evaluates to (true || true), which is true, so we execute the loop.
  - 5 > 4 so we flip b2 again. b2 is now false.
  - Skip the else clause, subtract one from x.
  - At the bottom of the loop: b1 = true, b2 = false, x = 4.
- The while condition now evaluates to (true || false), which is still true, so we execute the loop.
  - This time the if condition is false, so we execute the else clause, which flips the condition of b1 instead of b2. b1 is now false.
  - Subtract one from x.
  - At the bottom of the loop: b1 = false, b2 = false, x = 3.
- Now when we go to evaluate the while condition, we have (false || false). Do not execute the loop. Print the value of x, which is 3.

13. The answer is C.

- Option I
  - As we enter the loop for the first time, i = 1 and that's what gets printed. Then we multiply by 2, i = 2.
  - We print 2, multiply by 2, i = 4.
  - We print 4, multiply by 2, i = 8.
  - We print 8, multiply by 2, i = 16, which fails the loop condition so we exit the loop.
  - Final output: 1 2 4 8
- Option II
  - We can see by examining the for loop that we will execute the loop 4 times: i = 4, 3, 2, 1.
  - Each time through the loop we will print (4 – i) * 2 + 1.
  - (4 – 4) * 2 + 1 = 1
  - (4 – 3) * 2 + 1 = 3
  - (4 – 2) * 2 + 1 = 5
  - (4 – 1) * 2 + 1 = 7
  - Final output: 1 3 5 7
- Option III
  - We can see by examining the for loop that we will execute the loop 4 times: i = 0, 1, 2, 3.
  - Each time through the loop we will print (int)Math.pow(2, i). (Math.pow returns a double, so we cast to an int before printing.)
  - $2^0 = 1$
  - $2^1 = 2$
  - $2^2 = 4$
  - $2^3 = 8$
  - Final output: 1 2 4 8
- Options I and III have the same output.

14. The answer is A.

- The first time through the loop: count = 6.0 and num = 0.
  - Looking at the if condition: 6.0 != 0, but 0 / 6 is 0 → false, so we skip the if clause and go back to the start of the loop.
- Back to the top of the loop: count is still 6.0, but num = 1.
  - Looking at the if condition: 6.0 ! = 0 && 1 / 6.0 > 0 → true, so we execute the if clause and count = 6.0 – 1 = 5.0.
- Back to the top of the loop: count = 5.0 and num = 2.
  - Looking at the if condition: 5.0 ! = 0 && 1 / 5.0 > 0 → true so we execute the if clause and count = 5.0 – 2 = 3.0.
- Back to the top of the loop: count = 3.0 and num = 3.
  - Looking at the if condition: 3.0 ! = 0 && 3 / 3.0 > 0 → true so we execute the if clause and count = 3.0 – 3 = 0.0.
- Back to the top of the loop: count = 0.0 and num = 4.
  - Looking at the if condition: Here's where it gets interesting. If we execute 4 / 0.0, we are going to crash our program with an ArithmeticExceptionError, but that's not going to happen. The first part of the condition, count != 0 is false, and Java knows that false AND anything is false, so it doesn't even bother to execute the second part of the condition. That is called short-circuiting, and it is often used for exactly this purpose. The condition is false, the if clause doesn't get executed.
- Back to the top of the loop: count = 0.0 and num = 5 so our loop is complete and we print the value of count, which is 0.0.

15. The answer is A.

- The only thing we can do is to trace the code carefully line by line until we have the answer.
- n = 0. Print 0 % 4 which is 0
    - 0 % 5 = 0 so we execute the else, now n = 3.
- Print 3 % 4, which is 3
    - 3 % 5 = 3 so we execute the else, now n = 6.
- Print 6 % 4, which is 2
    - 6 % 5 = 1 so we execute the else, now n = 9.
- Print 9 % 4, which is 1
    - 9 % 5 = 1 so we execute the else, now n = 12.
- Print 12 % 4, which is 0
    - 12 % 5 = 2 so finally we execute the if, now n = 16.
- Print 16 % 4, which is 0
    - 16 % 5 = 1 so we execute the else, now n = 19.
- Print 19 % 4, which is 3
    - 19 % 5 is 4 so we execute the else, now n = 22 which completes the loop.
- We have printed: 0 3 2 1 0 0 3

16. The answer is C.

- To convert from decimal to hexadecimal (base 16), we need to remember the place values for base 16. A base 16 number has the 1's place on the right (like every base), preceded by the 16's place (and then, if needed, the 162 or 256 place, and so on).

16	1
?	?

- Our number is less than 256, so we start with the 16's place. How many times does 16 go into 93? 16 * 5 = 80. So we put a 5 in the 16's place.

16	1
5	?

- 93 – 80 = 13, so we have 13 left. We need to put 13 in the 1's place. That's D in hex. So our answer is 5D.

16	1
5	D = 13

- Double-check by converting back: (16 * 5) + (13 * 1) = 93

17. The answer is B.

- Remember that the place values for a two-digit number written in base 16 are: the 16's place followed by the 1's place. Also remember that A in base 16 has the decimal value 10, and F has the decimal value 15.

16	1
A = 10	F = 15

- (16 * 10) + (15 *1) = 175

18. The answer is E.

- We can do this by converting the base 10 number to bases 2, 8, and 16, or we can convert each of the given numbers back to base 10. Do whichever way is more comfortable for you. Most people prefer going *to* base 10, so let's do it that way.

I. Line up the digits with the base 2 place values:

64	32	16	8	4	2	1
1	1	0	1	1	0	0

Add up the place values that have a 1 under them:
64 + 32 + 8 + 4 = 108
This one works!

II. Line up the digits with the base 8 place values:

64	8	1
1	5	4

Since they aren't all 1's this time, we have to multiply:
(1 * 64) + (5 * 8) + (4 * 1) = 64 + 40 + 4 = 108
This one works!

III. Line up the digits with the base 16 place values:

16	1
6	C

First we need to figure out what that C means. A = 10, B = 11, so C = 12.
Multiply (6 * 16) + (12 * 1) = 108
This one works!

- All three results are equivalent to 108 base 10.

Classes and Objects (Basic Version)

IN THIS CONCEPT

Summary: This is a big one. If you don't understand this concept, you will fail the AP Computer Science A Exam. This chapter introduces you to the most fundamental object-oriented features of Java. It explains the relationship between an object and a class. It also describes the simplest form of a class and the minimal requirements needed to make one. More advanced aspects of classes and objects will be discussed in Concept 7.

Key Ideas

- Java is an object-oriented programming language whose foundation is built on classes and objects.
- A class describes the characteristics of any object that is created from it.
- An object is a virtual entity that is created using a class as a blueprint.
- Objects are created to store and manipulate information in your program.
- An instance variable is used to store an attribute of an object.
- A method is an action that an object can perform.
- The keyword new is used to create an object.
- A reference variable stores the address of the object, not the object itself.

Overview of the Relationship Between Classes and Objects

The intent of an **object-oriented programming language** is to provide a framework that allows a programmer to manipulate information after storing it in an object. The following paragraphs describe the relationship between many different vocabulary terms that are necessary to understand Java as an object-oriented programming language. The terms are so related that I wanted to describe them all together rather than in separate pieces.

> **Warning: Proceed with Caution**
>
> The next set of paragraphs includes many new terms. You may want to reread this section many times. If you expect to earn a 5 on the exam, you must master everything that follows. Challenge yourself to learn it so well that you can explain this entire chapter to someone else.

Java classes represent things that are nouns (people, places, or things). A **class** is the blueprint for constructing all **objects** that come from it. The **attributes** of the objects from the class are represented using **instance variables**. The values of all of the instance variables determine the **state** of the object. The state of an object changes whenever any of the values of the instance variables change. **Methods** are the virtual actions that the objects can perform. They may either **return** an answer or provide a service.

Objects are created using the class that holds the blueprint of the object. The class contains a piece of code called a **constructor**, which tells the computer how to build (construct) an object from that class. The keyword **new** is used whenever the programmer wants to create an object from a class. When new is followed by the name of a constructor of a class, a new object is created. The action of constructing an object is also referred to as **instantiating** an object and the object is referred to as an **instance** of the class. In addition to creating the new object, a **reference** to the object is created. The **object reference variable** holds the memory address of the newly created object and is used to call methods that are contained in the object's class.

This is a really important concept in Java, so here are a few examples of classes. Your challenge is to think of other examples as you wander through your day.

Class Name	Instance Variables (Attributes)	Methods (Actions)
FacebookAccount class	profileName, profilePicture, friendList, wallPosts	postMessageOnWall, addAFriend, sendMessage, postPicture
Song class	title, artist, genre, length	play, pause, stop
CellPhone class	contactList, batteryLevel, imageFiles, musicFiles	call, text, takePicture, hangUp, redial

The `class` Declaration

When you design your own class, you must create a file that contains the **class declaration**. The name of the class is identified in this declaration. By Java naming convention, a class name always starts with an uppercase letter and is written in camel case.

Example

Create a class called the Circle class:

```
public class Circle                    // Class declaration
{

}
```

Instance Variables

The virtual attributes that describe an object of a class are called its **instance variables.** A class can have as many instance variables of any data type as it wants as long as they describe some characteristic or feature of an object of the class. Unlike local primitive variables, default values are assigned to primitive instance variables when they are created (int variables are 0, doubles are 0.0, and booleans are false). The phrase **has-a** refers to the instance variables of a class as in: an objectName has-a instanceVariableName.

Example

Every Circle object has-a radius so create an instance variable called radius:

```
public class Circle                    // Class declaration
{
   private double radius;              // Instance variable
}
```

`private` Versus `public` Visibility

The words **private** and **public** are called **visibility modifiers** (or **access level modifiers**). Using the word private in an instance variable declaration ensures that other classes do not have access to the data stored in the variable without asking the object to provide it. Using the word public in the class declaration ensures that any programmer can make objects from that class.

`private` Versus `public`

On the AP Computer Science A Exam, always give instance variables private access visibility. This ensures that the data in these variables is hidden.

Constructors

Constructors are the builders of the virtual objects from a class. They are used in combination with the keyword new to create an object from the class. Constructors have the same name as the class and are typically listed near the top of a class.

No-Argument Constructor

The constructor is the code that is called when you create a new object. When you define a new class, it comes with **no-argument constructor**. Depending on the IDE that you use, you may or may not see it. It is generally recommended that you write one anyway, especially when you are first learning classes.

> **The No-Argument Constructor**
>
> The no-argument constructor gets its name because no information is passed to this constructor when creating a new object. It is the constructor that allows you to make a generic object from a class.

Parameterized Constructors and Parameters

If you will know some information about an object prior to creating it, then you may want to write a **parameterized constructor** in your class. You use the parameterized constructor to give initial values to the instance variables when creating a brand-new object. The parameters that are defined in the parameterized constructor are placed in a **parameter list** and are on the same line as the declaration of the constructor.

The process of sending the initial values to the constructor is called **passing a parameter** to the constructor. The actual value that is passed to the constructor is called an **argument** (or **actual parameter**) and the variable that receives the value inside the constructor is called the **formal parameter**. You include a line of code in the parameterized constructor that assigns the value of the formal parameter to the instance variable.

If you choose to write a parameterized constructor for a class, then the default, no-argument constructor vanishes. This is why it is generally recommended that you simply write your own.

Example

Write the no-argument constructor and one parameterized constructor for the Circle Class. The parameterized constructor must have a parameter variable for the radius of a circle.

```
public class Circle                         // Class declaration
{
    private double radius;                  // Instance variable

    public Circle()                         // No-argument constructor
    {
        // The radius gets a default value of 0.0
    }

    public Circle(double rad)               // Parameterized constructor
    {
        radius = rad;                       // The radius is set to rad
    }
}
```

Parameters in the Parameter List

A parameter is a variable that is located in a parameter list in the constructor declaration. A pair of parentheses after the name of the constructor encloses the entire parameter list.

For example, the parameterized constructor for the Circle class has one parameter variable in its parameter list. The name of the parameter is rad (short for radius) and its data type is a double. The no-argument constructor for the Circle class does not have any parameters in its parameter list.

```
public Circle()                  // no parameters in the parameter list
{
}
public Circle(double rad)        // one parameter in the parameter list
{
   radius = rad;
}
```

Methods

The real power of an object-oriented programming language takes place when you start to manipulate objects. A **method** defines an action that allows you to do these manipulations. A method has two options. It may simply perform a service of some kind, or it may compute a value and give the answer back. The action of giving a value back is called **returning** a value.

Methods that simply perform some action are called **void** methods. Methods that return a value are called **return** (or **non-void**) methods. Return methods must define what data type they will be returning and include a return statement in the method. A return method can return any kind of data type.

Using the word **public** in the **method declaration** ensures that the method is available for other objects from other classes to use.

The `return` Statement on the AP Computer Science A Exam

Methods that are supposed to return a value must have a reachable return statement. Methods that are void do not have a return statement. This is a big deal on the free-response questions of the AP Computer Science A Exam as you will lose points if you include a return statement in a void method or a constructor.

Accessor and Mutator Methods

When first learning how to write methods for a class, beginning programmers learn two types of methods:

1. Methods that allow you to access the values stored in the instance variables
2. Methods that allow you to modify the values of the instance variables

Methods that return the value of an instance variable are called **accessor** (or **getter**) methods. Methods that change the value of an instance variable are called **mutator** (or **modifier** or **setter**) methods. Methods that don't perform either one of these duties are simply called methods.

Is My Carbonated Soft Drink a Soda, a Pop, or a Coke?

These all refer to the same thing, it just depends on where you live. In a similar way, the methods that retrieve the instance variables of a class can be referred to as accessor or getter methods. The methods that change the value of an instance variable can be called modifier, mutator, or setter methods.

Declaring a Method

Methods are declared using a line of code called a **method declaration** statement. The declaration includes the access modifier, the return type, the name of the method, and the parameter list. The **method signature** only includes the name of the method and the parameter list.

General Form of a Method Declaration

```
accessModifier returnType methodName(parameterList)
```

One-Track Mind

Methods should have only one goal. The name of the method should clearly describe the purpose of the method.

Example

Write an accessor method for the radius, a mutator method for the radius, and a method that calculates and returns the area of the circle.

```
public class Circle                          // Class declaration
{
   private double radius;                    // Instance variable

   public Circle()                           // No-argument constructor
   {
      // The radius gets a default value of 0.0
   }

   public Circle(double rad)                 // Parameterized constructor
   {
      radius = rad;                          // The radius is set to rad
   }

   public double getRadius()                 // Accessor method
   {
      return radius;
   }

   public void setRadius(double rad)         // Mutator method
   {
      radius = rad;                          // The radius is set to rad
   }

   public double getArea()                   // Method that calculates area
   {
      return 3.14 * radius * radius;         // A simple way to compute area
   }
}
// End of Circle class
```

Java Ignores Spaces

Java ignores all spacing that isn't relevant. The compiler doesn't care how pretty the code looks in the IDE; it just has to be syntactically correct. On the AP exam, you will sometimes see code that has been streamlined to save space on the page.

For example, the getRadius method for the Circle class could appear like this on the exam:

```
public double getRadius()
{   return radius;   }
```

Putting It All Together: The Circle and CircleRunner Classes

I'm now going to combine the Circle class that we just created with another class to demonstrate code that follows an example of two classes interacting in an object oriented setting. The Circle class defined how to make a Circle object. The CircleRunner class is where the programmer creates virtual Circle objects and then manipulates them.

Summary of the Circle Class

Every circle has-a radius, so the Circle class defines an instance variable called radius. The class also has two constructors as ways to build a Circle object. The no-argument constructor is used whenever you don't know the radius at the time that you create the Circle object. The radius gets the default value for a double, which is 0.0. The parameterized constructor is used if you know the radius of the circle at the time you construct the Circle object. You pass the radius when you call the parameterized constructor and the constructor assigns the value of its parameter to the instance variable.

The Circle class also has three methods:

1. The first method, getRadius, returns the radius from that object. Since the method is *getting* the radius for the programmer, it is called an accessor (or getter) method.
2. The second method, setRadius, is a void method that sets the value of the radius for the object. It is referred to as a modifier (or mutator or setter) method. This method has a parameter, which is a variable that receives a value that is passed to it when the method is called (or invoked) and assigns this value to the instance variable.
3. Finally, the third method, getArea, calculates and returns the area of the circle using the object's radius.

Summary of the CircleRunner Class

The Circle class can't do anything all by itself. To be useful, Circle objects are created in a class that is *outside* of the Circle class. Each object that is created from a class is called an **instance of the class** and the act of constructing an object is called **instantiating** an object. A class that contains a **main method** is sometimes called a **runner** class. Other times, they are called **client** programs. The main method is the point of entry for a program. That means that when the programmer runs the program, Java finds the main method first and executes it. You will not have to write main methods on the AP exam.

Notice that the Circle class does not contain a **public static void main(String[] args)**. The purpose of the Circle class is to define how to build and manipulate a Circle object; however, Circle objects are created and manipulated *in a different class* that contains a main method.

How to Make a New Object from a Class

When you want to make an object from a class, you have to call one of the class's constructors. However, you may have more than one constructor, including the no-argument constructor or one or more parameterized constructors. The computer decides which constructor to call by looking at the parameter list in the declaration of each constructor. Whichever constructor matches the call to the constructor is used to create the object.

Examples

```
// This code uses the no-argument constructor (no parameters)
Circle circle1 = new Circle();

// This code uses the parameterized constructor  (1 parameter)
Circle circle2 = new Circle(4);

// This code causes a compile-time error because there isn't
// a Circle constructor that contains a String in its parameter list
Circle circle3 = new Circle("Fred");
```

Example

Write a runner class that instantiates three Circle objects. Make the first Circle object have a radius of 10, while the second and third Circles will get the default value. Manipulate the radii of the Circles and print their areas.

```
// This class creates and manipulates three Circle objects
public class CircleRunner
{
   public static void main(String[] args)
   {
      Circle myCircle = new Circle(10);          // myCircle's radius = 10.0
      Circle hisCircle = new Circle();           // hisCircle's radius = 0.0
      Circle herCircle = new Circle();           // herCircle's radius = 0.0

      System.out.println(myCircle.getArea());    // prints 314.0
      System.out.println(hisCircle.getArea()); // prints 0.0

      hisCircle.setRadius(5);                   //hisCircle's radius = 5.0
      herCircle.setRadius(2 * hisCircle.getRadius());
                                                 //herCircle's radius = 10.0

     System.out.println(hisCircle.getArea()); // prints 78.5
     System.out.println(herCircle.getArea()); // prints 314.0
   }
}
```

```
OUTPUT
314.0
0.0
78.5
314.0
```

Every Object from the Same Class Is a Different Object

Every object that is constructed from a class has its own instance variables. Two *different* objects can have the same **state**, which means that their instance variables have the same values.

In this example, the CircleRunner class creates three different Circle objects called myCircle, hisCircle, and herCircle. Notice that the myCircle object has a radius of 10. This is because when the Circle object created it, the computer looked for the parameterized constructor and when it found it, assigned the 10 to the parameter rad. For the hisCircle and herCircle objects, the no-argument constructor was called and the instance variable, radius, was given the default value of 0.0.

Remember that myCircle, hisCircle, and herCircle all refer to different objects that were all created from the same Circle class and each object has its own radius. When each object calls the getArea method, it uses *its own* radius to compute the area.

The Dot Operator

The dot operator is used to call a method of an object. The name of the object's reference variable is followed by a decimal point and then the name of the method.

```
objectReference.methodName();
```

Understanding the Keyword new When Constructing an Object

Beginning Java programmers often have a hard time understanding why the name of the class is used twice when creating an object, such as the Circle objects in the CircleRunner. Let's analyze the following line of code to find out what's really happening.

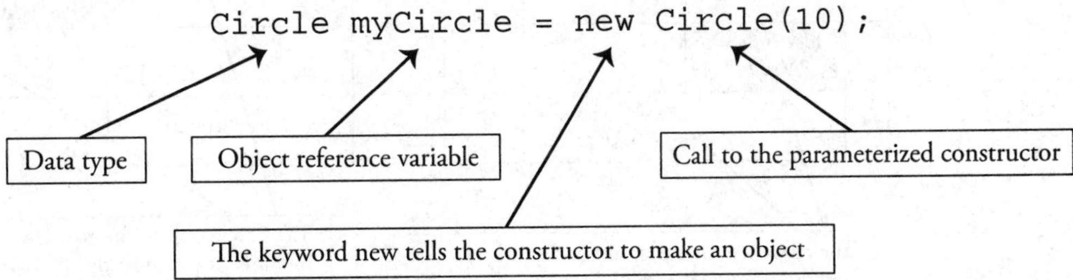

The first use of the word Circle is the data type of the variable. It's similar to how we defined an int or a double variable. The variable myCircle is called a **reference variable** of type Circle. It's similar to a primitive variable in that it holds a value, but it's *way* different in the fact that it *does not* hold the Circle object itself. The value that myCircle holds is the memory address of the soon-to-be-created Circle object.

The second use of the word Circle(10) is a call to the parameterized constructor of the Circle class. The keyword **new** tells the computer, "Hey we are about ready to make a new object." The computer looks for a constructor that has a single parameter that matches the data type of the 10. Once it finds this constructor, it gives the 10 to the parameter, rad. The code in the constructor assigns the instance variable, radius, the value of the parameter rad. Then, voilá! The radius of the Circle object is 10.

Finally, the computer looks for a place to store the Circle object in memory (RAM) and when it finds a location, it records the memory address. The construction of the Circle is now complete, and the address of the newly formed Circle object is assigned to the reference variable, myCircle.

Now, anytime you want to manipulate the object that is referenced by myCircle, such as give it a new radius, you are actually telling myCircle to go find the object that it is referencing (also called, *pointing to*), and then set the radius for that object.

The Reference Variable

The reference variable does not hold the object itself. It holds the address of where the object is stored in RAM (random access memory).

The Reference Variable Versus the Actual Object

The following diagram is meant to help you visualize the relationship between object reference variables and the objects themselves in the computer's memory. Note: The memory addresses are simulated and only meant to serve as an example of addresses in the computer's memory.

Question 1: What happens after these two lines of code are executed?

```
Circle hisCircle = new Circle(20);
Circle herCircle = new Circle(15);
```

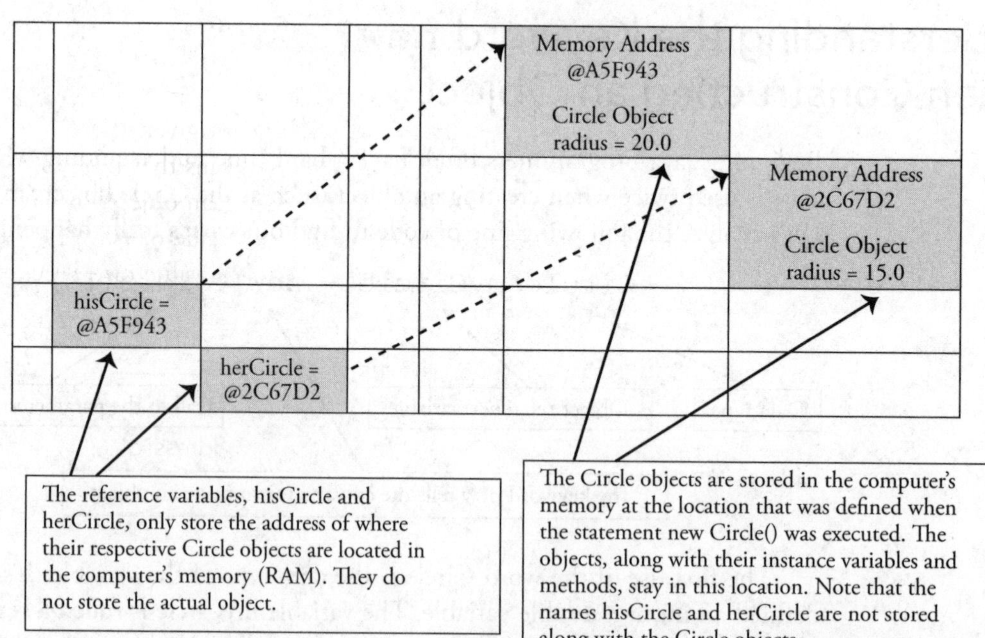
Memory Address
@A5F943
Circle Object
radius = 20.0
Memory Address
@2C67D2
Circle Object
radius = 15.0
hisCircle =
@A5F943
herCircle =
@2C67D2
The reference variables, hisCircle and herCircle, only store the address of where their respective Circle objects are located in the computer's memory (RAM). They do not store the actual object.
The Circle objects are stored in the computer's memory at the location that was defined when the statement new Circle() was executed. The objects, along with their instance variables and methods, stay in this location. Note that the names hisCircle and herCircle are not stored along with the Circle objects.

Question 2: What do you suppose happens after this line of code is executed?

```
herCircle = hisCircle;
```

If you are following the diagram correctly, you will see that herCircle "takes on the value" of the address of the hisCircle reference variable. Now, herCircle contains the same address that hisCircle contains. This means that both reference variables contain the same address of the same object and are pointing to the same object. The word **aliasing** is used to describe the situation when two different object reference variables contain the same address of an object.

Look at the object that used to be referenced by herCircle. Nobody is pointing to it. It has been detached from the rest of the world and will be **garbage collected** by Java. This means that Java will delete it when it is good and ready.

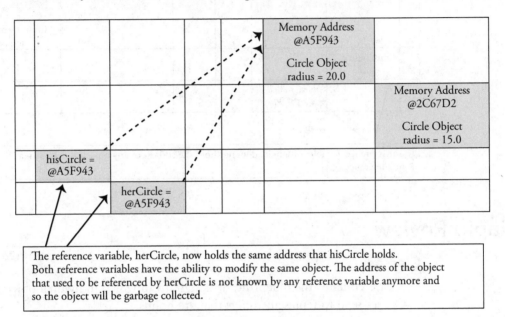

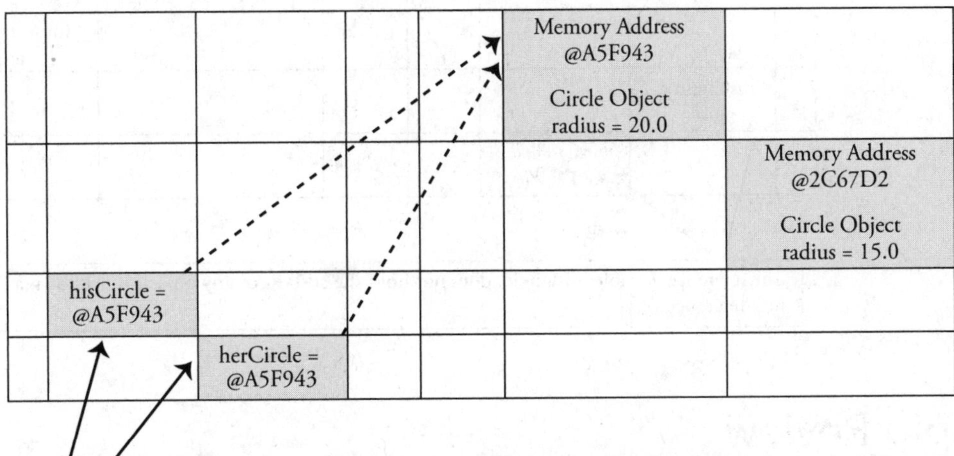

The `null` Reference

When an object reference variable is created but not given the address of an object, it is considered to have a **null reference**. This simply means that it is supposed to know where an object lives, but it does not. Whenever you create an object reference variable and don't assign it the address of an actual object, it has the value of **null**, which means that it has no value. You can also assign a reference variable the value of null to erase its memory of its former object.

The `null` Reference

A reference variable that does not contain the address of any object is called a null reference.

Question 3: What do you suppose happens when this statement is executed?

```
hisCircle = null;
```

The reference variable hisCircle is now null; it has no value. This means that it does not contain the address of any object. The reference variable herCircle still holds the address of a Circle object and so it can access methods of that object.

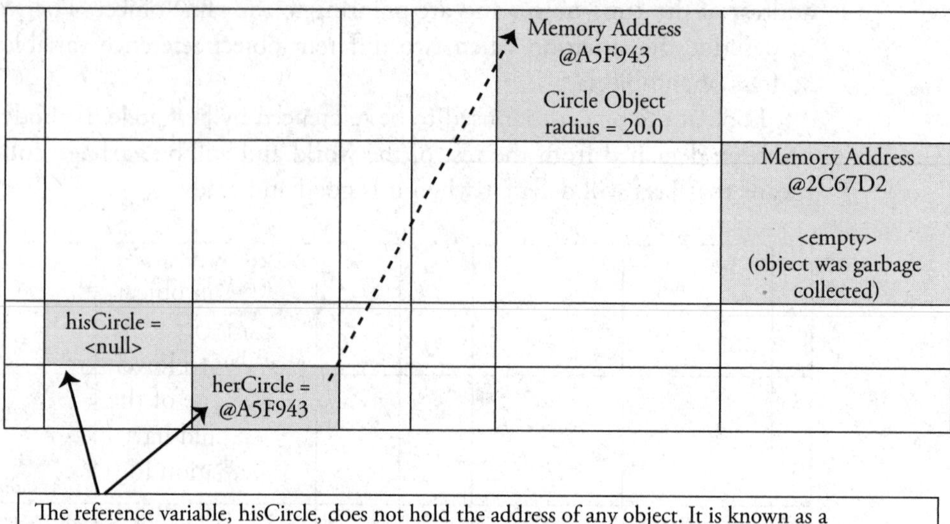

› Rapid Review

Classes

- A class contains the blueprint, or design, from which individual objects are created.
- Classes tend to represent things that are nouns.
- A class declaration (or class definition) is the line of code that declares the class.
- By naming convention, class names begin with an uppercase letter and use camel case.
- The two access modifiers that are tested on the AP Computer Science A Exam are public and private.
- All classes should be declared public.
- A public class means that the class is visible to all classes everywhere.

Instance Variables

- Instance variables (or fields) are properties that all objects from the same class possess. They are the attributes that help distinguish one object from another in the same class.
- All instance variables should be declared private.
- A class can have as many instance variables as is appropriate to describe all the attributes of an object from the class.
- By naming convention, all instance variables begin with a lowercase letter and use camel case.
- All primitive instance variables receive default values. The default value of an int is 0, the default value of a double is 0.0, and the default value for a boolean is false.
- The current values of all of the instance variables of an object determine the state of the object.
- The state of the object is changed whenever any of the instance variables are modified.

Constructors

- Every class, by default, comes with a no-argument (or empty) constructor.
- Parameterized constructors should be created if there are values that should be assigned to the instance variables at the moment that an object is created.
- A class can have as many parameterized constructors as is relevant.
- The main reason to have a parameterized constructor is to assign values to the instance variables.
- Other code may be written in the constructor.
- A parameter is a variable that receives values and is declared in a parameter list of a method or constructor.

Methods

- The methods of a class describe the actions that an object can perform.
- Methods tend to represent things that are verbs.
- By naming convention, method names begin with a lowercase letter and use camel case.
- The name of a method should describe the purpose of the method.
- Methods should not be multi-purpose. They should have one goal.
- A method declaration describes pertinent information for the method such as the access modifier, the return type, the name of the method, and the parameter list.
- Methods that don't return a value are called void methods.
- Methods that return a value of some data type are called return methods.
- Methods can return any data type.
- Return methods must include a reachable return statement.
- Methods that return the value of an instance variable are called accessor (or getter) methods.
- Methods that modify the value of an instance variable are called mutator (or modifier or setter) methods.
- The data type of all parameters must be defined in the parameter list.

Objects

- An object from a class is a virtual realization of the class.
- Objects store their own data.
- An instance of a class is the same thing as an object of a class.
- The word "instantiate" is used to describe the action of creating an object. Instantiating is the act of constructing a new object.
- The keyword *new* is used whenever you want to create an object.
- An object reference variable stores the memory address of where the actual object is stored. It can only reference one object.
- The reference variable does not store the object itself.
- Two reference variables are equal only if they refer to the exact same object.
- The word null means no value.
- A null reference describes a reference variable that does not contain an address of any object.
- You can create as many objects as you want from one class. Each is a different object that is stored in a different location in the computer's memory.

› Review Questions

Basic Level

1. The relationship between a class and an object can be described as:

(A) The terms class and object mean the same thing.
(B) A class is a program, while an object is data.
(C) A class is the blueprint for an object and objects are instantiated from it.
(D) An object is the blueprint for a class and classes are instantiated from it.
(E) A class can be written by anyone, but Java provides all objects.

2. Which of the following is a valid constructor declaration for the `Cube` class?

I. `public static Cube()`
II. `public void Cube()`
III. `public Cube()`
IV. `public Cube(int side)`

(A) I only
(B) II only
(C) III only
(D) I and II only
(E) III and IV only

Questions 3–6 refer to the following class.

```
public class Student
{
    private String name;
    private double gpa;

    public Student(String newName, double newGPA)
    {
        name = newName;
        gpa = newGpa;
    }

    public String getName()
    {
        return name;
    }

    /*  Additional methods not shown  */
}
```

3. Which of the following could be used to instantiate a new `Student` object called `sam`?

(A) `Student sam = new Student();`
(B) `sam = new Student();`
(C) `Student sam = new Student("Sam Smith", 3.5);`
(D) `new Student sam = ("Sam Smith", 3.5);`
(E) `new Student(sam);`

4. Which of the following could be used in the `StudentTester` class to print the name associated with the `Student` object instantiated in problem 3?

(A) `System.out.println(getName());`
(B) `System.out.println(sam.getName());`
(C) `System.out.println(getName(sam));`
(D) `System.out.println(Student.name);`
(E) `System.out.println(getName(Student));`

5. Which of the following is the correct way to write a method that changes the value of a `Student` object's `gpa` variable?

(A)
```
public double setGpa(double newGpa)
{
    return gpa;
}
```
(B)
```
public double setGpa()
{
    return gpa;
}
```
(C)
```
public void setGpa ()
{
    gpa++;
}
```
(D)
```
public setGpa(double newGpa)
{
    gpa = newGpa;
}
```
(E)
```
public void setGpa(double newGpa)
{
    gpa = newGpa;
}
```

Advanced Level

6. Consider the following code segment.

```
Student s1 = new Student("Emma Garcia", 3.9);
Student s2 = new Student("Alice Garrett", 3.2);
s2 = s1;
s2.setGpa(4.0);
System.out.println(s1.getGpa());
```

What is printed as a result of executing the code segment?

(A) `3.2`
(B) `3.9`
(C) `4.0`
(D) Nothing will be printed. There will be a compile-time error.
(E) Nothing will be printed. There will be a run-time error.

7. Free-Response Practice: Dream Vacation Class

Everyone fantasizes about taking a dream vacation.

Write a full `DreamVacation` class that contains the following:

a. An instance variable for the name of the destination
b. An instance variable for the cost of the vacation (dollars and cents)
c. A no-argument constructor
d. A parameterized constructor that takes in both the name of the vacation and the cost
e. Accessor (getter) methods for both instance variables
f. Modifier (setter) methods for both instance variables

› Answers and Explanations

Bullets mark each step in the process of arriving at the correct solution.

1. The answer is C.
   - A class is a blueprint for constructing objects of the type described by the class definition.

2. The answer is E.
   - The general form of a constructor declaration is:

   ```
   public NameOfClass(optional parameter list)
   ```

   - Notice that there is no return type, not even void.
   - In this example, the name of the class is Cube. III is a correct declaration for a no-argument constructor and IV is a correct declaration for a constructor with one parameter.

3. The answer is C.
   - The general form of an instantiation statement is:

   ```
   Type variableName = new Type(parameter list);
   ```

   - In this example, the type is Student and the variable name is sam.
   - We have to look at the constructor code to see that the parameter list consists of a String followed by a double.
   - The only answer that has the correct form with the parameters in the correct order is:

   ```
   Student sam = new Student("Sam Smith", 3.5);
   ```

4. The answer is B.

- We do not have direct access to the instance variables of an object, because they are private. We have to ask the object for the data we need with an accessor method. In this case, we need to call the getName method.
- We need to ask the specific object we are interested in for its name. The general syntax for doing that is:

```
variableName.methodName();
```

- Placing the call in the print statement gives us:

```
System.out.println(sam.getName());
```

5. The answer is E.

- A modifier or setter method has to take a new value for the variable from the caller and assign that value to the variable. The new value is given to the method through the use of a parameter.
- A modifier method does not return anything. Why would it? The caller has given the value to the method. The method doesn't have to give it back.
- So we need a void method that has one parameter. It has to change the value of gpa to the value of the parameter. Option E shows the only method that does that.

6. The answer is C.

- The instantiation statements give us 2 objects: s1 and s2.
- s1 references (points to) a Student object with the state information: name = "Emma Garcia", gpa = 3.9
- s2 references (points to) a Student object with the state information: name = "Alice Garrett", gpa = 3.2
- When we execute the statement s2 = s1, s2 changes its reference. It now references the exact same object that s1 references, with the state information: name = "Emma Garcia", gpa = 3.9
- s2.setGpa(4.0); changes the gpa in the object that both s1 and s2 reference, so when s1.getGpa(); is executed, that's the value it retrieves and that's the value that is printed.

7. On the AP exam, you are often asked to write an entire class, like you were in this question. Check your answer against our version below. The comments are to help you read our solution. Your solution does not need comments.

> **Make Variable Names Meaningful**
> It's OK if you use different variable names from what we use in the solutions to the chapter questions. Just be sure that the names are meaningful.

```
public class DreamVacation
{
    // Here are the private instance variables (parts a and b).
    private String destination;
    private double cost;

    // Here is the no-argument constructor (part c).
    public DreamVacation()
    {
    }

    // Here is the parameterized constructor (part d).
    public DreamVacation(String myDestination, double myCost)
    {
        destination = myDestination;
        cost = myCost;
    }

    // Here are the two accessor methods (part e).
    // Their names contain the word "get" since it is standard practice.
    public String getDestination()
    {
        return destination;
    }

    public double getCost()
    {
        return cost;
    }

    // Here are the two modifier methods (part f).
    // Their names contain the word "set" since it is standard practice.
    public void setDestination(String myDestination)
    {
        destination = myDestination;
    }

    public void setCost(double myCost)
    {
        cost = myCost;
    }
}
```

The `String` Class

IN THIS CONCEPT

Summary: When people think of the word *data*, they usually think "numbers." However, non-numerical data can be stored in a computer too. In this concept, you will learn how letters, words, and even sentences are stored in variables. String objects can store alphanumerical data, and the methods from the String class can be used to manipulate this data.

Key Ideas

- Strings are used to store letters, words, and other special characters.
- To concatenate strings means to join them together.
- The String class has many methods that are useful for manipulating String objects.

The `String` Variable

How does Twitter know when a tweet has more than 280 characters? How does the computer know if your username and password are correct? How are Facebook posts stored? What we need is a way to store letters, characters, words, or even sentences if we want to make programs that are useful to people. The **String** data type helps solve these problems.

When we wanted to store numbers so that we can retrieve them later, we created either an int or double variable. Now that we want to store letters or words, we need to create **String variables**. A String can be a single character or group of characters like a word or a sentence. **String literals** are characters surrounded by double quotation marks.

General Way to Make a `String` Variable and Assign It a Value

```
String nameOfString = "characters that make up a String literal";
```

Examples

Make some String variables and assign them values:

```
String myName = "Captain America";       // myName is "Captain America"
String jennysNumber = "867-5309";        // numbers are treated as characters
```

The `String` Object

String variables are actually objects. The String class is unique because String objects can be created by simply declaring a String variable like we did above or by using the constructor from the String class. The reference variable knows the address of the String object. The object is assigned a value of whatever characters make up the String literal.

General Way to Make a `String` Object Using the Constructor from the `String` Class

```
String nameOfString = new String("characters that make a string");
```

Example 1

Make a String object and assign it a value:

```
String myName = new String("Captain America"); // myName is Captain America
```

Example 2

Create a String object, but don't give it an initial value (use the empty constructor from the String class). The String object contains an **empty string**, which means that the value exists (it is not null); however, there are no characters in the String literal and it has a length of 0 characters.

```
String yourName = new String();    // yourName contains an empty string
```

Example 3

Create a String reference variable but don't create a String object for it. The String reference variable is assigned a value of **null** because there isn't an address of a String object for it to hold.

```
String theirName;    // theirName is null
```

null String Versus Empty String

A **null** string is a string that has no value and no length. It does not hold an address of a String object.

An **empty** string is a string that has a value of "" and its length is zero. There is no space inside of the double quotation marks, because the string is empty.

It may seem like they are the same thing, but they are not. The difference is subtle. The empty string has a value and the null string does not. Furthermore, you can perform a method on an empty string, but *you cannot perform a method on a null string.*

Any attempt to execute a method on a null string will result in a run-time error called a **NullPointerException**.

A Visual Representation of a String Object

This is a visual representation of what a string looks like in memory. The numbers below the characters are called the **indices** (plural of **index**) of the string. Each index represents the location of where each of the characters lives within the string. Notice that the first index is zero. An example of how we read this is: the character at index 4 of myFavoriteFoodGroup is the character "c". Also, *spaces* count as characters. You can see that the character at index 3 is the space character.

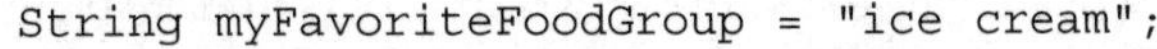

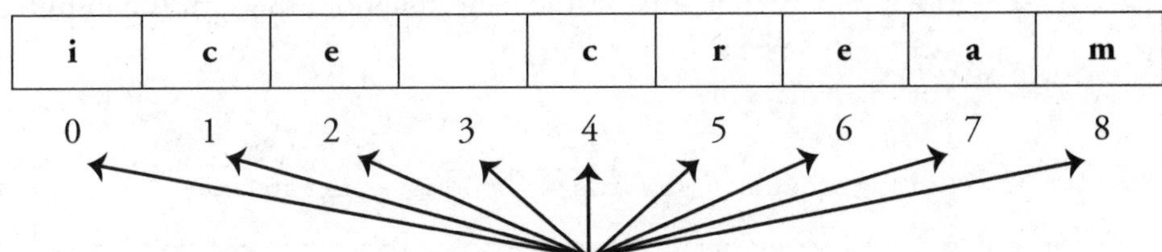

These numbers represent the index of each character in the String object.

String Concatenation

Now that you know how to create a string and give it a value, let's make lots of them and string them together (Ha!). The process of joining strings together is called **concatenation.** The + and += signs are used to concatenate strings.

Example 1

Concatenate two strings using +

```
String firstName = "Harry";
String lastName = "Potter";
String entireName = firstName + lastName;   // entireName is "HarryPotter"
```

Example 2

Concatenate two strings using +=

```
String firstName = "Harry";
String lastName = "Potter";
lastName += firstName;                      // lastName is "PotterHarry"
```

Example 3

Concatenate a number and a string using +

```
String firstName = "Harry";
int swords = 8;
String result = firstName + swords;                    // result is "Harry8"
```

Example 4

Incorrectly concatenate a string and the sum of two numbers using +. Notice that the sum is not computed.

```
String firstName = "Harry";
int swords = 8;
int brooms = 4;
String result = firstName + swords + brooms;          // result is "Harry84"
```

Example 5

Correctly concatenate a string and the sum of two numbers using +

```
String firstName = "Harry";
int swords = 8;
int brooms = 4;
String result = firstName + (swords + brooms);        // result is "Harry12"
```

Example 6

Concatenate the sum of two numbers and a string in a different order. Notice that when the numbers come first in the concatenation, the sum is computed.

```
String firstName = "Harry";
int swords = 8;
int brooms = 4;
String result = swords + brooms + firstName;          // result is "12Harry"
```

Concatenation

Strings can be joined together to form a new string using concatenation. The + sign or the += operation may be used to join strings.

You can even join an int or a double along with a string. These numbers will be rendered useless for any mathematical purpose because they are turned into strings.

The Correct Way to Compare Two `String` Objects

Everyone reading this book has had to log into a computer at some time. How does the computer know if the user typed the username and password correctly? The answer is that the software uses an operation to compare the input with the information in a database.

In Java, we can determine if two strings are equal to each other. This means that the two strings have exactly the same characters in the same order.

The Correct Way to Compare Two `String` Objects

Use the ***equals*** method or the **compareTo** method to determine if two String variables contain the exact same character sequence.

Important String Methods

The equals Method

The **equals** method returns a boolean answer of true if the two strings are identical and it returns an answer of false if the two strings differ in any way. The equals method is **case-sensitive**, so, for example, the uppercase "A" is different than the lowercase "a". The ! operator can be used in combination with the equals method to compare if two strings are not equal.

Example 1

Determine if two strings are equal:

```
String username1 = "Cat Lady";
String username2 = "cat lady";                       // notice the lowercase letters
System.out.println(username1.equals("Cat Lady")); // prints true
System.out.println(username2.equals("Cat Lady")); // prints false
```

Example 2

Determine if two strings are not equal:

```
String pet1 = "Cat";
String pet2 = "Dog";
System.out.println(!pet1.equals(pet2));       // prints true
```

Never Use == to Compare Two String Objects

We used the double equals (==) to compare if two numbers were the same. However, == does not correctly tell you if two strings are the same. What's worse is that comparing two strings using == does not throw an error, it simply compares whether or not the two String references point to the same object. Even worse is that sometimes it appears to work correctly. Just don't do it!

```
String username = "CatLady";
String first = "Cat";
String last = "Lady";
if (username == first + last)
{
   // Comparison will be false.
   // NEVER USE == TO COMPARE TWO STRINGS!
}
```

The compareTo Method

The second way to compare two strings is to use the **compareTo** method. Instead of returning an answer that is a boolean like the equals method did, the compareTo method returns an integer. The result of the compareTo method is a positive number, a negative number, or zero. This integer describes how the two strings are ordered **lexicographically**; a case-sensitive ordering of words similar to a dictionary but using the **UNICODE** codes of each character.

The `compareTo` Method

The compareTo method returns an integer that describes how the two strings compare.

```
int result = firstString.compareTo(secondString);
```

- The answer is zero if the two strings are exactly the same.
- The answer is negative if firstString comes before secondString (lexicographically).
- The answer is positive if firstString comes after secondString (lexicographically).

Example 1

The answer is zero if the two strings are exactly the same:

```
String string1 = "Hello";
String string2 = "Hello";
int result = string1.compareTo(string2);       // result is 0
```

Example 2

The answer is negative when the first string comes before the second string (lexicographically):

```
String string1 = "Hello";
String string2 = "Kitty";
int result = string1.compareTo(string2);       // result is negative
```

Example 3

The answer is positive when the first string comes after the second string (lexicographically):

```
String string1 = "Hello";
String string2 = "Kitty";
int result = string2.compareTo(string1);       // result is positive
```

Example 4

Uppercase letters come before their lowercase counterparts. "Hello" comes before "hello" when comparing them using lexicographical order:

```
String string1 = "Hello";
String string2 = "hello";
int result = string1.compareTo(string2);       // result is negative
```

Fun Fact: *Unicode is a worldwide computing standard that assigns a unique number to nearly every possible character (including symbols) for every language. There are more than 120,000 characters listed in the latest version of Unicode. For example, the uppercase letter "A" is 41 (hexadecimal), the lowercase letter "a" is 61 (hexadecimal), and "ñ" is F1 (hexadecimal).*

The `length` Method

The **length** method returns the number of characters in the specified string (including spaces).

Example 1

Find the length of a string:

```
String tweet = "I'm going to rock this AP CS exam.";
int result = tweet.length();                  // result is 34
```

Example 2

Find the length of a string that is empty, but not null (it has "" as its value):

```
String pokemonCard = new String();
int result = pokemonCard.length();              // result is 0
```

Example 3

Attempt to find the length of a string that is null:

```
String pokemonCard;                             // no String created
int result = pokemonCard.length();              // error: NullPointerException
```

The `NullPointerException`

Attempting to use a method on a string that has not been initialized will result in a run-time error called a nullPointerException. The compiler is basically telling you that there isn't an object to work with, so how can it possibly perform any actions with it?

The `indexOf` Method

To find if a string contains a specific character or group of characters, use the **indexOf** method. It searches the string and if it locates what you are looking for, it returns the index of its location. Remember that the first index of a string is zero. If the indexOf method doesn't find the string that you are looking for, it returns -1.

Example 1

Find the index of a specific character in a string:

```
String sport = "Soccer";
int result = sport.indexOf("e");          // result is 4
```

Example 2

Find the index of a character in a string when it appears more than once:

```
String sport = "Soccer";
int result = sport.indexOf("c");          // result is 2 (finds only
                                          // the first "c")
```

Example 3

Attempt to find the index when the character is not in the string:

```
String sport = "Team";
int result = sport.indexOf("i");          // result is -1
                                          // there is no "i" in Team (ha)
```

Example 4

Find the index of a string inside a string:

```
String sport = "Soccer";
String findMe = "ce";
int result = sport.indexOf(findMe);       // result is 3
```

The substring Method

The **substring** method is used to extract a specific character or group of characters from a string. All you have to do is tell the substring method where to start extracting and where to stop.

The substring method is **overloaded**. This means that there is more than one version of the method.

Overloaded Method	Description
`substring(int index)`	Starts extracting at index and stops extracting when it reaches the last character in the string
`substring(int firstIndex, int secondIndex)`	Starts extracting at firstIndex and stops extracting at secondIndex − 1

Example 1

Extract every character starting at a specified index from a string:

```
String motto = "May the force be with you.";
String result = motto.substring(14);           // result is "be with you."
```

Example 2

Extract one character from a string:

```
String motto = "May the force be with you.";
String result = motto.substring(2,3);          // result is "y"
```

Example 3

Extract two characters from a string:

```
String motto = "May the force be with you.";
String result = motto.substring(5,7);          // result is "he"
```

Example 4

Extract the word "force" from a string:

```
String motto = "May the force be with you.";
String result = motto.substring(8,13);         // result is "force"
```

Example 5

Use numbers that don't make any sense:

```
String motto = "May the force be with you.";
String result = motto.substring(72);  //error: StringIndexOutOfBoundsException
```

The StringIndexOutOfBoundsException

If you use an index that doesn't make any sense, you will get a run-time error called a **StringIndexOutOfBoundsException**. Using an index that is greater than or equal to the length of the string or an index that is negative will produce this error.

Example 6

Extract the word "the" from the following string by finding and using the first and second spaces in the string:

```
String motto = "May the force be with you.";
int firstSpace = motto.indexOf(" ");
int secondSpace = motto.indexOf(" ",firstSpace + 1);
String word = motto.substring(firstSpace + 1, secondSpace); //word is "the"
```

Helpful Hint for the `substring` Method

To find out how many characters to extract, subtract the first number from the second number.

```
String myString = "I solemnly swear that I am up to no good.";
String result = myString.substring(5,13);        // result is "emnly sw"
```

Since 13 − 5 = 8, the result contains a string that is a total of eight characters.

Note: When the start index and end index are the same number, the substring method does not return a character of the string; it returns the empty string "".

A `String` Is Immutable

String variables are actually references to String objects, but they don't operate the same way as normal object reference variables.

Every time you make a change to a String variable, Java actually tosses out the old object and creates a new one. You are never actually making changes to the original String object and *resaving* the same object with the changes. Java makes the changes to the object and then creates a brand-new String object and that is what you end up referencing.

Example

Demonstrate how a String variable is immutable through concatenation. Even though it seems like you are modifying the String variable, you are not. A completely brand-new String object is created and the variable is now referencing it:

```
String str = "CS";      // str references a String object "CS"
str = str + " Rules"; // str now references a new String object "CS Rules"
```

Escape Sequences

How would you put a double quotation mark inside of a string? An error occurs if you try because the double quotes are used to define the start and end of a String literal. The solution is to use an **escape sequence.**

Escape sequences are special codes that are sent to the computer that say, "Hey, watch out, something different is coming." They use a backslash followed by a special character. Escape sequences are commonly used when concatenating strings to create a single string. Here are the most common escape sequences.

Purpose	Sequence	Example	`System.out.print(result)`
include a double quote	`\"`	`String result = "\"OMG\"";`	`"OMG"`
tab	`\t`	`String result = "a\tb\tc";`	`a b c`

new line	`\n`	`String result = "a\nb\nc";`	`a`
			`b`
			`c`

backslash	`\\`	`String result = "I am a backslash: \\";`	`I am a backslash: \`

› Rapid Review

Strings

- The String class is used to store letters, numbers, or other symbols.
- A String literal is a sequence of characters contained within double quotation marks.
- You can create a String variable by declaring it and directly assigning it a value.
- You can create a String variable using the keyword new along with the String constructor.
- Strings are objects from the String class and can utilize methods from the String class.
- Strings can be concatenated by using the + or += operators.
- Only compare two String variables using the equals method or the compareTo method.
- Strings should never be compared using the == comparator.
- Every character in a string is assigned an index. The first index is zero. The last index is one less than the length of the string.
- Escape sequences are special codes that are embedded in String literals to perform such tasks as issuing a new line or a tab, or printing a double quote or a backslash.
- A null string does not contain any value. It does not reference any object.
- If a string is null, then no methods can be performed on it.
- A NullPointerException is thrown if there is any attempt to perform a method on a null string.
- An empty string is not the same as a null string.
- An empty string has a value of "" and a length of zero.

String Methods

- The equals(String other) method returns true if the two strings that are being compared contain exactly the same characters.
- The compareTo(String other) method compares two strings lexicographically.
- Lexicographical order is a way of ordering words based on their Unicode values.
- The length() method returns the number of characters in the string.
- The indexOf(String str) method finds and returns the index value of the first occurrence of str within the String object that called the method. The value of -1 is returned if str is not found.
- The substring(int index) method returns a new String object that begins with the character at index and extends to the end of the string.
- The substring(int beginIndex, int endIndex) method returns a new String object that begins at the beginIndex and extends to the character at endIndex − 1.
- Strings are immutable. A new String object is created each time changes are made to the value of a string.
- A StringIndexOutOfBoundsException error is thrown for any attempt to access an index that is not in the bounds of the String literal.

› Review Questions

Basic Level

1. Consider the following code segment.

```
String newString = "Welcome";
String anotherString = "Home";
newString = anotherString + "Hello";
System.out.println(newString);
```

What is printed as a result of executing the code segment?

(A) `HomeHello`
(B) `HelloHello`
(C) `WelcomeHello`
(D) `Welcome Hello`
(E) `Home Hello`

2. What is printed as a result of executing the following statement?

```
System.out.println(7 + 8 + (7 + 8) + "Hello" + 7 + 8 + (7 + 8));
```

(A) `7878Hello7878`
(B) `7815Hello7815`
(C) `30Hello30`
(D) `30Hello7815`
(E) `7815Hello30`

3. Consider the following method.

```
public String mixItUp(String entry1, String entry2)
{
    entry2 = entry1;
    entry1 = entry1 + entry2;
    String entry3 = entry1 + entry2;
    return entry3;
}
```

What is returned by the call `mixItUp("AP", "CS")`?

(A) `"APAP"`
(B) `"APCSAP"`
(C) `"APAPAP"`
(D) `"APCSCS"`
(E) `"APAPAPAP"`

4. Consider the following code segment.

```
String str1 = "presto chango";
String str2 = "abracadabra";

int i = str1.indexOf("a");
System.out.println(str2.substring(i, i + 1) + i);
```

What is printed as a result of executing the code segment?

(A) `p0`
(B) `r9`
(C) `r10`
(D) `ra9`
(E) `ra10`

5. Consider the following code segment.

```
String s1 = "avocado";
String s2 = "banana";
System.out.println(s1.compareTo(s2) + " " + s2.compareTo(s1));
```

Which of these could be the result of executing the code segment?

(A) `0 0`
(B) `-1 -1`
(C) `-1 1`
(D) `1 -1`
(E) `1 1`

6. Which of the following returns the last character of the string `str`?

(A) `str.substring(0);`
(B) `str.substring(0, str.length());`
(C) `str.substring(length(str));`
(D) `str.substring(str.length() - 1);`
(E) `str.substring(str.length());`

Advanced Level

7. Consider the following code segment.

```
String ex1 = "example";
String ex2 = "elpmaxe";
ex1 = ex1.substring(1, ex1.indexOf("p"));
ex2 = ex2 + ex2.substring(3, ex1.length()) + ex2.substring(3, ex2.length());
String ex3 = ex1.substring(1) + ex2.substring(ex2.indexOf("x"), ex2.length() - 2);
System.out.println(ex3);
```

What is printed as a result of executing the code segment?

(A) `amxem`
(B) `amxema`
(C) `amxepma`
(D) `ampxepm`
(E) The program will terminate with a `StringIndexOutOfBoundsException`

8. Consider the following method.

```
public boolean whatDoesItDo(String st)
{
    for (int i = 0; i < st.length() / 2; i++)
    {
        String sub1 = st.substring(i, i + 1);
        String sub2 = st.substring(st.length() - i - 1, st.length() - i);
        if (!sub1.equals(sub2))
        {
            return false;
        }
    }
    return true;
}
```

What is the purpose of the method `whatDoesItDo`?

(A) The method returns `true` if `st` has an even length, `false` if it has an odd length.
(B) The method returns `true` if `st` contains any character more than once.
(C) The method returns `true` if `st` is a palindrome (spelled the same backward and forward), `false` if it is not a palindrome.
(D) The method returns `true` if `st` begins and ends with the same letter.
(E) The method returns `true` if the second half of `st` contains the same sequence of letters as the first half, `false` if it does not.

9. Consider the following method.

```
public int firstLetterIndex(String s2)
{
    String s1 = "abcdefghijklmnopqrstuvwxyz";
    return s1.indexOf(s2.substring(s2.length() - 2));
}
```

What is returned as a result of the call `firstLetterIndex("on your side")`?

(A) `-1`
(B) `0`
(C) `3`
(D) `4`
(E) `5`

10. Which print statement will produce the following output?

```
/\what    time
"is it?"//\\
```

(A) `System.out.print("/\\what\ttime\n\"is it?\"//\\\\");`
(B) `System.out.print("//\\what\ttime\n\"is it?\"///\\\\");`
(C) `System.out.print("/\\what\time\n\"is it?\"//\\\\");`
(D) `System.out.print("/\\what\ttime\n"is it?"//\\\\");`
(E) `System.out.print("//\\what\time + \"is it?\"///\\\\");`

〉 Answers and Explanations

Bullets mark each step in the process of arriving at the correct solution.

1. The answer is A.

- When two strings are concatenated (added together), the second one is placed right after the first one (do not add a space).

```
newString = anotherString + "Hello"
          = "Home" + "Hello" = "HomeHello"
```

2. The answer is D.

- This question requires that you understand the two uses of the + symbol.
- First, execute what is in the parentheses. Now we have:

```
System.out.println(7 + 8 + 15 + "Hello" + 7 + 8 + 15);
```

- Now do the addition left to right. That's easy until we hit the string:

```
System.out.println(30 + "Hello" + 7 + 8 + 15);
```

- When you add a number and a string in any order, Java turns the number into a string and then concatenates the string:

```
System.out.println("30Hello" + 7 + 8 + 15);
```

- Now every operation we do will have a string and a number, so they all become string concatenations, not number additions.

```
System.out.println("30Hello7815");
```

3. The answer is C.

- To answer this problem, you need to understand that String objects are immutable (the only other immutable objects in our Java subset are Integer and Double, which we will learn about in Concept 4).
- When the method begins, entry1 = "AP" and entry2 = "CS".
- Then the assignment statement reassigns entry2 so that both variables now refer to the String object containing "AP".
- When we execute the next statement, entry1 = entry1 + entry2, we would expect that to change for both variables, since they both reference the same object, but String objects don't work that way. Whenever a string is modified, a new String object is created to hold the new value. So now, entry1 references "APAP" but entry2 still references "AP".
- We put those together, and “APAPAP” is returned.

4. The answer is B.

- The indexOf method returns the index location of the parameter.
- The letter "a" has index 9 in "presto chango" (remember to start counting at 0).
- The substring will start at index 9 and stop *before* index 10, meaning it will return only the letter at index 9 in "abracadabra".
- Once again, start counting at 0. Index 9 of "abracadabra" is "r".
- Concatenate that with i which equals 9. Print "r9".

5. The answer is C.

   - CompareTo takes the calling string and compares it to the parameter string. If the calling string comes *before* the parameter string, it returns a negative number. If it comes *after* the parameter string, it returns a positive number. If they are equal, it returns 0.
   - The first compareTo is "avocado".compareTo("banana") so it returns -1.
   - The second compareTo switches the order to "banana".compareTo("avocado") so it returns 1.
   - Note that although we often only care about whether the answer is positive or negative, the actual number returned has meaning. Because "a" and "b" are one letter apart, 1 (or -1) is returned. If the example had been "avocado" and "cucumber", 2 (or -2) would have been returned.

6. The answer is D.

   - Let's take the word "cat" as an example. It has a length of 3, but the indices of its characters are 0, 1, 2, so the last character is at length − 1.
   - We need to start our substring at str.length() − 1. We could say:

   ```
   str.substring(str.length() - 1, str.length());
   ```

   but since we want the string all the way to the end we can use the single parameter version of substring.

7. The answer is B.

   - `ex1.substring(1, ex1.indexOf("p"))` starts at the character at index 1 (remember we start counting at 0) and ends before the index of the letter "p". ex1 now equals "xam".
   - Let's take this one substring at a time. At the beginning of the statement, ex2 = "elpmaxe".
     - `ex2.substring(3, ex1.length())` The length of ex1 is 3, so this substring starts at the character at index 3 and ends *before* the character at index 3. It stops before it has a chance to begin and returns the empty string. Adding that to the current value of ex2 does not change ex2 at all.
     - `ex2.substring(3, ex2.length())` starts at the character at index 3 and goes to the end of the string, so that's "maxe".
     - Add that to the current value of ex2 and we have "elpmaxemaxe".
   - We are going to take two substrings and assign them to ex3.
     - The first one is isn't too bad. ex1 = "xam", so ex1.substring(1) = "am". Remember that if substring only has 1 parameter, we start there and go to the end of the string.
     - The second substring has two parameters. Let's start by figuring out what the parameters to the substring are equal to.
     - Remember ex2 = "elpmaxemaxe" so indexOf("x") is 5.
     - The second parameter is ex2.length() − 2. The length of ex2 = 11, subtract 2 and the second parameter = 9.
     - Now we can take the substring from 5 to 9, or from the first "x" to just before the second "x", which gives us "xema".
     - Add the two substrings together: "amxema"

8. The answer is C.

- This method takes a String parameter and loops through the first half of the letters in the string.
- The first substring statement assigns the letter at index i to sub1. You may have to run through an example by hand to discover what the second substring does.
- If our string is "quest":
  - the first time through the loop, i = 0, so sub1 = "q" and sub2 = st.substring(5 – 0 – 1, 5 – 0) = "t",
  - the second time, i = 1, so sub1 = "u" and sub2 = st.substring(5 – 1 – 1, 5 – 1) = "s".
- This loop is comparing the first letter to the last, the second to the second to last, and so on, which will tell us if the string is a palindrome.

9. The answer is C.

- Let's figure this out step by step. The length of "on your side" is 12 (don't forget to count spaces), so `s1.indexOf(s2.substring(s2.length() - 2))` simplifies to `s1.indexOf(s2.substring(10))`
- That gives us the last two letters of s2, so our statement becomes `s1.indexOf("de")`
- Can we find "de" in s1? Yes. It starts at index 3 and that is what is returned.

10. The answer is A.

- Let's take this piece by piece. A forward slash "/" does not require an escape sequence, but a backslash needs to be doubled since "\" is a special character. Our print sequence needs to start: `/\\`
- The next interesting thing that happens is the tab before "time". The escape sequence for tab is `"\t"`. Our print sequence is now: `/\\what\ttime` (Don't be confused by the double t. One belongs to "\t" and the other belongs to "time".)
- Now we need to go to the next line. The newline escape sequence is `"\n"`
- We also need escape sequences for the quote symbols, so now our print sequence is: `/\\what\ttime\n\"is it\"`
- Add the last few slashes, remembering to double the "\" and we've got our final result. `/\\what\ttime\n\"is it?\"//\\\\`

The `Math`, `Integer`, and `Double` Classes

IN THIS CONCEPT

Summary: The Java API (application programming interface) contains a library of thousands of useful classes that you can use to write all kinds of programs. Because the Java API is huge, the AP Computer Science A Exam only tests your knowledge of a subset of these classes. This concept provides practice working with three of those classes: the Math, Integer, and Double classes.

Key Ideas

- The Java API contains a library of thousands of classes.
- The Math class contains methods for performing mathematical calculations.
- The Integer class is for creating and manipulating objects that represent integers.
- The Double class is for creating and manipulating objects that represent decimal numbers.

The Java API and the AP Computer Science A Exam Subset

Want to make a trivia game that poses random questions? Want to flip a coin 1 million times to test the laws of probability? Want to write a program that uses the quadratic formula to do your math homework? If so, you will want to learn about the Java API.

The Java **API**, or **application programming interface**, describes a set of classes and files for building software in Java. It shows how all the Java features work and interact. Since the Java API is gigantic, only a subset of the Java API is tested on the AP Computer Science A Exam. All of the classes described in this subset are required. The Appendix contains the entire Java subset that is on the exam.

The `Math` Class

The **Math class** was created to help programmers solve problems that require mathematical computations. Most of these methods can be found on a calculator that you would use in math class. In fact, if you think about how you use your calculator, it will make it easier to understand how the Math class methods work. The Math class also includes a random number generator (RNG) to help programmers solve problems that require random events.

Compared to the other classes in the AP Computer Science subset, the Math class is special in that it contains only **static** methods. This means that you don't need to create an object from the class in order to use the methods in the class. All you have to do is use the class name followed by the dot operator and then the method name.

The `Math` Class Doesn't Need an Object

The methods and fields of the Math class are accessed by using the class name followed by the method name.

```
Math.methodName();
```

Important `Math` Class Methods

The Math class contains dozens of awesome methods for doing mathematical work. The AP exam only requires you to know a small handful of them.

Mathematical Operation	Method Name in Java
Absolute Value	`abs`
Square Root	`sqrt`
Base Raised to a Power	`pow`
Random Number Generator	`random`

Example 1

Demonstrate how to use the absolute value method on an integer:

```
int result = Math.abs(-8);              // result is 8
```

Example 2

Demonstrate how to use the absolute value method on a double:

```
double result = Math.abs(-8.4);         // result is 8.4
```

Example 3

Demonstrate how to use the square root method. The sqrt method will compute the square root of any number that is greater than or equal to 0. The result is always a double, even if the input is a perfect square number.

```
double result = Math.sqrt(49);              // result is 7.0
```

Example 4

Demonstrate how to use the power method. The pow method always returns a double.

```
double result = Math.pow(5,3); // result is 125.0 (5 raised to the 3rd power)
```

Example 5

Demonstrate how to use the random method.

```
double result = Math.random(); // result is a double in the interval [0.0, 1)
```

Example 6

Demonstrate how to access the built-in value of pi from the Math class.

```
double result = Math.PI;         // PI is not a method, it is a public field
```

Random Number Generator in the `Math` Class

The random number generator from the Math class returns a double value in the range from 0.0 to 1.0, including 0.0, but not including 1.0. Borrowing some notation from math, we say that the method returns a value in the interval [0.0, 1.0).

The word *inclusive* means to include the endpoints of an interval. For example, the interval [3,10] means all of the numbers from 3 to 10, including 3 and 10. Written mathematically, it's the same as $3 \le x \le 10$. Finally, you could say you are describing all the numbers between 3 and 10 (inclusive).

Using `Math.random()` to Generate a Random Integer

The random method is great for generating a random double, but what if you want to generate a random integer? The secret lies in multiplying the random number by the total number of choices and then casting the result to an integer.

General Form for Generating a Random Integer Between 0 and Some Number

```
int result = (int)(Math.random() * (total # of choices));
```

General Form for Generating a Random Integer Between Two Positive Numbers

```
int high = /*  some number greater than zero  */
int low = /*  some number greater than zero and less than high  */

int result = (int)(Math.random() * (high - low + 1)) + low;
```

Example 1

Generate a random integer between 0 and 12 (inclusive):

```
int result = (int)(Math.random() * 13);           // 13 possible answers
```

Example 2

Generate a random integer between 4 and 20 (inclusive):

```
int result = (int)(Math.random() * 17) + 4;       // 17 possible answers
```

Example 3

Pick a random character from a string. Use the length of the string in the calculation.

```
String quote = "And may the odds be ever in your favor.";
int index = (int)(Math.random() * quote.length());
String result = quote.substring(index, index + 1); //random character
```

Example 4

Simulate flipping a coin a million times:

```
int heads = 0, tails = 0;
for (int i = 0; i < 1000000; i++)
{
   int flip = (int)(Math.random() * 2) + 1;     // flip is either 1 or 2
   if (flip == 1)
      heads++;                                 // flip is a 1, so increment heads
   else
      tails++;                                 // flip is a 2, so increment tails
}
System.out.println("Total number of heads:  " + heads);
System.out.println("Total number of tails:  " + tails);
```

The Correct Way to Compare If Two `double` Values Are Equal

It is never a good idea to compare if two double values are equal using the `==` sign. Calculations with a double can contain rounding errors. You may have seen this when printing the result of some kind of division and you get something like this as a result: `7.00000000002`.

The correct way to compare if two doubles are equal is to use a *tolerance*. If the difference between the two numbers is less than or equal to the tolerance, then we say that the two numbers are *close enough* to be considered equal. Since we might not know which of the two numbers is bigger, subtract the two numbers and find the absolute value of the result. This way, the difference between the two numbers is always positive.

General Form for Comparing If Two `double` Values Are Equal

```
double a =  /*  Some double  */
double b =  /*  Some double  */
double tolerance = 0.00001;  // The 0.00001 varies with the application

boolean closeEnough = Math.abs(a - b) <= tolerance;
```

Example

Determine if two doubles are "equal" using a very small tolerance:

```
double num1 = 3.99999;
double num2 = 4.0;
double tolerance = 0.0001;
boolean result = Math.abs(num1 - num2) <= tolerance;    // result is true
```

Avoid Using the == Sign When Comparing double Values

Rounding errors cause problems when trying to determine if two double values are equal. Never use == to compare if two doubles are equal. Instead, determine if the difference between the two doubles is close enough for the two doubles to be considered equal.

The Integer Class

The **Integer** class has the power to turn a primitive int into an object. This class is called a **wrapper class** because the class *wraps* the primitive int into an object. A major purpose for the Integer class will be revealed in the Data Structures concept. Java can convert an Integer object back into an int using the intValue method.

Example 1

Create an Integer object from an int.

```
int num = 5;
Integer myInteger = new Integer(num); // myInteger is an object, value 5
```

Example 2

Obtain the value of a Integer object using the intValue( ) method.

```
int num = 8;
Integer myInteger = new Integer(8);    // myInteger is an object, value 8
int result = myInteger.intValue();     // result is a primitive, value 8
```

Example 3

Attempt to create an Integer object using a double value.

```
double num = 5.67;
Integer myInteger = new Integer(num);    // compile-time error
```

Example 4

Attempt to create an Integer object without using a value.

```
Integer myInteger = new Integer();       // compile-time error
```

Fields of the Integer Class

MAX_VALUE and **MIN_VALUE** are public **fields** of the Integer class. A field is a property of a class (like an instance variable). These two are called **constant** fields and their values cannot be changed during run-time. By naming convention, constants are typed using only uppercase and underscores. Also, they don't have a pair of parentheses, because they are not methods.

Why is there a maximum or minimum integer? Well, Java sets aside 32 bits for every int. So, the largest integer that can be stored in 32 bits would look like this: 01111111 11111111 11111111 11111111. Notice that the *leading* bit is 0. That's because the first bit is used to assign the sign (positive or negative value) of the integer. If you do the conversion to decimal, this binary number is equal to 2,147,483,647; the largest integer that can be stored in an Integer object. It's also equal to one less than 2 to the 31st.

If you add one to this maximum number, the result causes an **arithmetic overflow** and the new number is pretty much meaningless. Worse yet, the overflow *does not* cause a run-time error, so your program doesn't crash. Moral of the story: be careful when you are using extremely large positive or small negative integers.

Example 1
Obtain the value of the largest Integer:

```
int result = Integer.MAX_VALUE;       // result is 2147483647 (2^31 - 1)
```

Example 2
Obtain the value of the smallest Integer:

```
int result = Integer.MIN_VALUE;      // result is -2147483648 (-2^31)
```

MAX_VALUE and MIN_VALUE Are Constants

A **constant** in java is a value that is defined by a programmer and cannot be changed during the running of the program. The two values, MAX_VALUE and MIN_VALUE, are constants of the Integer class that were defined by the creators of Java. By naming convention, constants are always typed in uppercase letters and underscores.

Fun Fact: *Java has different data types for storing numbers. The* ***long*** *data type uses 64 bits and the largest number it can store is 9,223,372,036,854,775,807. This is not on the AP exam.*

The Double Class

The **Double** class is a wrapper class for primitive doubles. The Double class is used to turn a double into an object.

Example 1
Create a Double object from a double.

```
double num = 5.67;
Double myDouble = new Double(num);         // myDouble is an object, value 5.67
```

Example 2
Obtain the value of a Double object by using the doubleValue() method.

```
double num = 6.125;
Double myDouble = new Double(6.125);
double result = myDouble.doubleValue();   // result is an object, value 6.125
```

Example 3
Attempt to create a Double object without using a value.

```
Double myDouble = new Double();            // compile-time error
```

Summary of the Integer and Double Classes

Feature	Description
intValue	Integer method that returns the value of the Integer object
Integer.MAX_VALUE	Public field (constant) of the Integer class that is the largest integer that can be stored in Java (2,147,483,647)
Integer.MIN_VALUE	Public field (constant) of the Integer class that is the smallest integer that can be stored in Java (-2,147,483,648)
doubleValue	Double method that returns the value of the Double object

❯ Rapid Review

Math Class

- The Math class has static methods, which means you don't create a Math object in order to use the methods or fields from the class.
- The dot operator is used to call the methods from the Math class: Math.methodName().
- The Math.abs(int x) method returns an int that is the absolute value of x.
- The Math.abs(double x) method returns a double that is the absolute value of x.
- The Math.sqrt(double x) method returns a double that is the square root of x.
- The Math.pow(double base, double exponent) method returns a double that is the value of base raised to the power of exponent.
- The Math.random() method returns a double that is randomly chosen from the interval [0.0, 1.0).
- By casting the result of a Math.random() statement, you can generate a random integer.
- Avoid comparing double values using the == sign. Instead, use a tolerance.

Integer and Double Classes

- The Integer and Double classes are called wrapper classes, since they *wrap* a primitive int or double value into an object.
- An Integer object contains a single field whose type is int.
- The Integer(int value) constructor constructs an Integer object using the int value.
- The intValue() method returns an int that is the value of the Integer object.
- The Integer.MAX_VALUE constant is the largest possible int in Java.
- The Integer.MIN_VALUE constant is the smallest possible int in Java.
- The Double(double value) constructor constructs a Double object using the double value.
- The doubleValue() method returns a double value that is the value of the Double object.

› Review Questions

Basic Level

1. Which statement assigns a random integer from 13 to 27 inclusive to the variable `rand`?

(A) `int rand = (int)(Math.random() * 13) + 27;`
(B) `int rand = (int)(Math.random() * 27) + 13;`
(C) `int rand = (int)(Math.random() * 15) + 13;`
(D) `int rand = (int)(Math.random() * 14) + 27;`
(E) `int rand = (int)(Math.random() * 14) + 13;`

2. After executing the code segment, what are the possible values of the variable `var`?

```
int var = (int) Math.random() * 1 + 10;
```

(A) All integers from 1 to 10
(B) All real numbers from 1 to 10 (not including 10)
(C) 10 and 11
(D) 10
(E) No value. Compile-time error.

3. Which of the following statements generates a random integer between -23 and 8 (inclusive)?

(A) `int rand = (int)(Math.random() * 32 - 23);`
(B) `int rand = (int)(Math.random() * 32 + 8);`
(C) `int rand = (int)(Math.random() * 31 - 8);`
(D) `int rand = (int)(Math.random() * 31 - 23);`
(E) `int rand = (int)(Math.random() * 15 + 23);`

4. Consider the following code segment.

```
int r = 100;
int answer = (int)(Math.PI * Math.pow(r, 2));
System.out.println(answer);
```

What is printed as a result of executing the code segment?

(A) `30000`
(B) `31415`
(C) `31416`
(D) `31415.9265359`
(E) Nothing will print. The type conversion will cause an error.

5. Consider the following code segment.

```
int a = /* value supplied within program */
int b = /* value supplied within program */
if (/* missing condition */)
{
    int result = a + b;
}
```

Which of the following should replace `/* missing condition */` to ensure that `a + b` will not be too large to be held correctly in result?

(A) `a + b  <  Integer.MAX_VALUE`
(B) `a + b <= Integer.MAX_VALUE`
(C) `Integer.MIN_VALUE + a <= b`
(D) `Integer.MAX_VALUE - a <= b`
(E) `Integer.MAX_VALUE - a >= b`

Advanced Level

6. Consider the following code segment.

```
double tabulate = 100.0;
int repeat = 1;
while (tabulate > 20)
{
    tabulate = tabulate - Math.pow(repeat, 2);
    repeat++;
}
System.out.println(tabulate);
```

What is printed as a result of executing the code segment?

(A) `20.0`
(B) `6.0`
(C) `19.0`
(D) `9.0`
(E) `-40.0`

7. Consider the following code segment.

```
int rand = (int)(Math.random() * 5 + 3);
rand = (int)(Math.pow(rand, 2));
```

After executing the code segment, what are the possible values for the variable `rand`?

(A) 3, 4, 5, 6, 7
(B) 9, 16, 25, 36, 49
(C) 5, 6, 7
(D) 25, 36, 49
(E) 9

8. Consider the following code segment.

```
int val1 = 10;
int val2 = 7;
val1 = (val1 + val2) % (val1 - val2);
val2 = 3 + val1 * val2;
int val3 = (int)(Math.random() * val2 + val1);
```

After executing the code segment, what is the possible range of values for `val3`?

(A) All integers from 2 to 17
(B) All integers from 2 to 75
(C) All integers from 75 to 76
(D) All integers from 2 to 74
(E) All integers from 2 to 18

9. Assume `double dnum` has been correctly declared and initialized with a valid value. What is the correct way to find its absolute value, rounded to the nearest integer?

(A) `(int) Math.abs(dnum - 0.5);`
(B) `(int) Math.abs(dnum + 0.5);`
(C) `(int) (Math.abs(dnum) + 0.5);`
(D) `Math.abs((int)dnum - 0.5);`
(E) `Math.abs((int)dnum) - 0.5;`

10. Which of these expressions correctly computes the positive root of a quadratic equation assuming coefficients `a, b, c` have been correctly declared and initialized with valid values? Remember, the equation is:

$$\frac{-b + \sqrt{b^2 - 4ac}}{2a}$$

(A) `(-b + Math.sqrt (Math.pow(b, 2)) - 4ac) / 2a`
(B) `(-b + Math.sqrt (Math.pow(b, 2) - 4ac)) / 2a`
(C) `(-b + Math.sqrt (Math.pow(b, 2) - 4 * a * c)) / 2 * a`
(D) `(-b + Math.sqrt (Math.pow(b, 2)) - 4 * a * c) / (2 * a)`
(E) `(-b + Math.sqrt (Math.pow(b, 2) - 4 * a * c)) / (2 * a)`

❯ Answers and Explanations

Bullets mark each step in the process of arriving at the correct solution.

1. The answer is C.

- The basic form for generating a random int between high and low is:

```
int result = (int)(Math.random() * (high - low + 1)) + low;
```

- Our "high" is 27, and our "low" is 13, so the answer is C.
- Be sure to place the parentheses correctly. They can go after the multiplication or after the addition, but not after the Math.random().

2. The answer is D.

- This problem doesn't have the necessary parentheses discussed in the explanation for problem 1.
- As a result, the (int) cast will cast only the value of Math.random(). Since Math.random() returns a value between 0 and 1 (including 0, not including 1), truncating will *always* give a result of 0.
- Multiplying by 1 still gives us 0, and adding 10 gives us 10.

3. The answer is A.

- This problem is similar to problem 1.
- Fill in (high − low + 1) = (8 − (-23)) + 1 = 32 and low = -23 and the answer becomes `(int)(Math.random() * 32 - 23).`
- Notice that this time the parentheses were placed around the whole expression, not just the multiplication.

4. The answer is B.

-
```
(int)(Math.PI * Math.pow(r, 2))
    = (int)(3.1415926... * 10000)
    = (int)(31415.926...)
    = 31415
```

5. The answer is E.

- Integer.MAX_VALUE is a constant that represents the greatest value that can be stored in an int. In this problem, we need to make sure that a + b is not greater than that value.
- Solution b seems like the logical way to do this, but if a + b is too big to fit in an int, then we can't successfully add them together to test if they are too big. a + b will overflow.
- We need to do some basic algebra and subtract a from both sides giving us:

```
b < Integer.MAX_VALUE - a
```

- The left and right sides of the expression have been switched, so we need to flip the inequality. Flipping the < gives us >=.

6. The answer is D.

- When we enter the loop, tabulate = 100 and repeat = 1
  - tabulate = 100 – Math.pow(1, 2) = 99; increment repeat to 2
- 99 > 20 so we enter the loop again
  - tabulate = 99 – Math.pow(2, 2) = 95; increment repeat to 3
- 95 > 20
  - tabulate = 95 – Math.pow(3, 2) = 86; increment repeat to 4
- 86 > 20
  - tabulate = 86 – Math.pow(4, 2) = 70; increment repeat to 5
- 70 > 20
  - tabulate = 70 – Math.pow(5, 2) = 45; increment repeat to 6
- 45 > 20
  - tabulate = 45 – Math.pow(6, 2) = 9; increment repeat to 7
- 9 < 20 so we exit the loop and print 9

7. The answer is B.

- This problem is like problem 1 only backward. If (high – low + 1) = 5 and low = 3, then high = 7. The first line gives us integers from 3 to 7 inclusive.
- The second line squares all those numbers, giving us: 9, 16, 25, 36, 49.

8. The answer is E.

- Taking it line by line, val1 = 10, val2 = 7 so:

```
val1 = (val1 + val2) % (val1 - val2) = 17 % 3 = 2
val2 = 3 + val1 * val2 = 3 + 2 * 7 = 17
```

- The next statement becomes:

```
val3 = (int) (Math.random() * 17 + 2)
```

- We know low = 2 and (high – low + 1) = 17, so high = 18.
- val3 will contain integers from 2 to 18 inclusive.

9. The answer is C.

- The easiest way to solve this is to plug in examples. We need to make sure we test both positive and negative numbers, so let's use 2.9 and -2.9. The answer we are looking for in both cases is 3.
- solution a:

```
(int) Math.abs(2.9 - 0.5) = (int) Math.abs(2.4) = (int) 2.4 = 2
```

  NO (but interestingly, this works for -2.9)
- solution b:

```
(int) Math.abs(2.9 + 0.5) = (int) Math.abs(3.4) = (int) 3.4 = 3
```

  YES, but does it work in the negative case?

```
(int) Math.abs(-2.9 + 0.5)= (int) Math.abs(-2.4) = (int) 2.4 = 2
```

  NO, keep looking
- solution c:

```
(int)(Math.abs(2.9) + 0.5) = (int)(2.9 + 0.5) = (int) 3.4 = 3
```

  YES, but does it work in the negative case?

```
int)(Math.abs(-2.9) + 0.5) = (int)(2.9 + 0.5) = (int) 3.4 = 3
```

  YES, Found it!

10. The answer is E.

- Options A and B are incorrect. They use the math notation 4ac and 2a for multiplication instead of 4 * a * c and 2 * a.
- Solution C is incorrect. This option divides by 2 and then multiplies that entire answer by a, which is not what we want. 2 * a needs to be in parentheses.
- Look carefully at D and E. What's the difference? The only difference is the placement of the close parenthesis, either after the (b, 2) or after the 4 * a * c. Which should it be? The square root should include both the Math(b, 2) and the 4 * a * c, so the answer is E.

Data Structures

IN THIS CONCEPT

Summary: Sometimes a simple variable is not enough. To solve certain problems, you need a list of variables or objects. The array, the two-dimensional array, and the ArrayList are three complex data structures that are tested on the AP Computer Science A Exam. You must learn the advantages and disadvantages of each of these data structures, know how to work with them, and be able to choose the best one for the job at hand.

Key Ideas

- An array is a data structure that can store a list of variables of the same data type.
- An array is not resizable once it is created.
- A two-dimensional (2-D) array is a data structure that is an array of arrays.
- The 2-D array simulates a rectangular grid with coordinates for each location based on the row and the column.
- An ArrayList is a data structure that can store a list of objects from the same class.
- An ArrayList resizes itself as objects are added to and removed from the list.
- To traverse a data structure means to visit each element in the structure.
- The enhanced for loop (for-each loop) is a special looping structure that can be used by either arrays or ArrayLists.

What Is a Data Structure?

Have you ever dreamed about having the high score in the "Top Scores" of a game? How does Facebook keep track of your friends? How does Vine know what video to loop next? All of these require the use of a complex data structure.

Simple Data Structure Versus Complex Data Structure

As programmers, we use **data structures** to store information. A primitive variable is an example of a **simple data structure** as it can store a single value to be retrieved later. A variable that can store a list of other variables is called a **complex data structure**.

Simple Data Structure
(can hold one thing)

`String myGame`
`"Mario Kart"`

Complex Data Structure
(can hold a list of things)

`String[] myGames`
`"Mario Kart"`
`"Minecraft"`
`"Pokemon"`
`"Tetris"`

Thinking of ways to represent physical objects by using virtual objects can be challenging. Really awesome programmers have the ability to visualize complex real-world situations and design data structures to represent these physical objects in a virtual way. In the drawing below, the programmer is trying to figure out a way to represent the choices from the lunch menu in a program by creating a complex data structure.

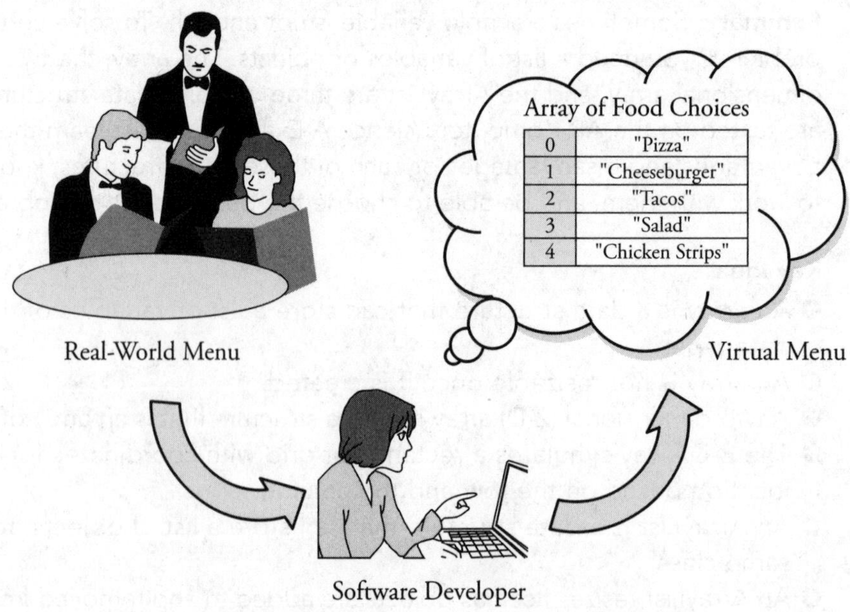

Three Important Complex Data Structures

There are three complex data structures that are tested on the AP Computer Science A Exam:

- The array (or one-dimensional array)
- The 2-D array
- The ArrayList

Each of these will be explained in detail later in this concept, but for now, I want to give you an overview of which data structure would be the best choice for a given situation. You will want to become a master at deciding the best data structure for a particular situation. This normally comes with experience, but here are a few examples as to which one is the best under what circumstances.

Situation	Recommendation	Justification
A program helps an elevator know what floor it is on.	Array	The number of floors is fixed. The elevator can go to any floor by knowing what number it is.
Facebook keeps track of how many friends you have.	ArrayList	The number of friends you have on Facebook may increase or decrease. You are allowed to add or remove anyone at any time regardless of where they are in the list.
Your program is going to simulate chess, Candy Crush, or 2048.	2-D Array	Each of these games can be simulated on either a square or rectangular grid in which the row and column are used to find out what is in each cell.
A cell phone keeps track of text messages.	ArrayList	The number of text messages on a cell phone can increase or decrease. You can even delete all of the messages.
A program keeps track of what classes you have each period of the school day.	Array	The number of class periods in the school day is fixed. Each period is assigned a value (1st hour is math, 2nd hour is science, etc.).

The Array

Definition of an Array

The non-programming definition of an array is "an impressive display of a particular thing." We use the word in common language when we say things like, "Wow, look at the wide *array* of phone cases at the mall!" It means that there is a group or list of things that are of the same kind.

An **array (one-dimensional array)** in Java is a **complex data structure** because it can store a **list** of primitives or objects. It also stores them in such a way that each of the items in the list can be referenced by an **index**. This means that the array assigns a number to each of the slots that holds each of the values. By doing this, the programmer can easily find the value that is stored in a slot, change it, print it, and so on. The values in the array are called the **elements** of the array. An array can hold any data type, primitive or object.

Declaring an Array and Initializing It Using a Predefined List of Data

There are two ways to create an array on the AP Computer Science A Exam. The first technique is used when you already know the values that you are going to store in the list and you have no reason to add any more items. **A pair of brackets**, [ ], tells the compiler to create an array, not just a regular variable.

> **General Form for Creating an Array Using a Predefined List of Data**
>
> `dataType[] nameOfArray = {dataValue1, dataValue2, dataValue3, . . . };`
>
> Notice the pair of brackets [ ] after dataType. The brackets signify this variable is a reference to an array object.

Example

Declare an array of String variables that represent food choices at a restaurant:

```
String[] foodChoices = {"Pizza", "Cheeseburger", "Tacos", "Salad",
     "Chicken Strips"};
```

The graphic at the right is a visual representation of the foodChoices array. The strings representing the actual food choices are the **elements** in the array (Pizza, Cheeseburger, etc.) and each slot is assigned a number, called the **index** (0, 1, 2, etc.). The index is a number that makes it easy for the programmer to know what element is placed in what slot. The first index is always zero. The last index is always one less than the length of the array.

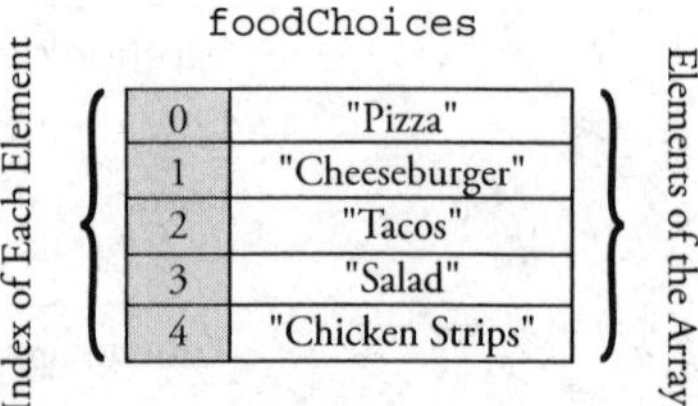

How to "Speak" Array

When talking about elements of an array, we say things like, "Pizza is the first element in the foodChoices array and its index is zero" or "The element at index 2 of the foodChoices array is Tacos."

The `length` Field

After an array is created and initialized, you can ask it how many slots it has by using the **length** field. Think of the length as an instance variable for an array. Notice that the length does not have a pair of parentheses after it, because it is not a method.

Example

Find the length of the foodChoices array:

```
int result = foodChoices.length;             // result is 5 (not 4!)
```

> **The `length` of an Array**
>
> The **length** field of an array returns the total number of slots that are set aside for that array. It does not return the total number of valid elements in the array. Also, since the first index of an array is always zero, the length of the array is always one more than the last index.

Declaring an Array Using the Keyword new

The second technique for creating an array is used when you know the length of the list, but you may not know all of the values that will go in the list. Again, a pair of **brackets**, [ ], is used to tell the compiler that you want to create an array, not just a regular variable.

> **General Form for Creating an Array Using the Keyword new**
>
> ```
> dataType[] nameOfArray = new dataType[numberOfSlots];
> ```
>
> The number of slots is fixed and all of the elements in the array get the default value for the data type. You can also declare just an array reference variable and then use the keyword new to create the array later.
>
> ```
> line 1:  dataType[] nameOfArray;                      // no array created
> line 2:  nameOfArray = new dataType[numberOfSlots];   // array is created
> ```

> **No Resizing an Array**
>
> An array cannot be resized after it is created.

Example

Create an array that will represent all of your semester grade point averages for all four years of high school. Let's assume there are two semesters per year. This makes a total of eight semesters. So, let's declare an array to hold eight double values and call the array myGPAs.

```
double[] myGPAs = new double[8];
```

myGPAs	
0	0.0
1	0.0
2	0.0
3	0.0
4	0.0
5	0.0
6	0.0
7	0.0

Ouch, all zeros. That hurts. No worries, we can change that. You see, when an array is created in this manner, all of the elements become whatever the **default value** is for that data type. The default value for a double is 0.0.

Now that the array is created, let's put some valid numbers into it. In order to replace a value in an array, you need to know the correct index.

Let's let index 0 represent the first semester of your freshman year and let's suppose you earned a GPA of 3.62.

```
myGPAs[0] = 3.62;  // puts 3.62 in slot 0
```

And, let's suppose you earned a 3.75 for the second semester of your sophomore year.

```
myGPAs[3] = 3.75;  // puts 3.75 in slot 3
```

myGPAs		
0	3.62	freshman
1	0.0	
2	0.0	sophomore
3	3.75	
4	0.0	junior
5	0.0	
6	0.0	senior
7	0.0	

> **Smart, but Not so Smart**
>
> The computer has no idea that these numbers represent the GPAs for a student in school. As far as it's concerned, it is just storing a bunch of numbers in an array.

Example

Pick a random color from an array of color names:

```
String[] colors = {"Red","Orange","Yellow","Green","Blue","Indigo","Violet"};
String randomColor = colors[(int)(Math.random() * colors.length)];
System.out.println("My favorite color is " + randomColor);
```

```
OUTPUT
Blue        (Note: The color is chosen randomly and can be different for each run.)
```

The `ArrayIndexOutOfBoundsException`

If you ever accidently use an index that is too big (greater than or equal to the length of the array) or negative, you will get a run-time error called an **ArrayIndexOutOfBoundsException**.

Example

You declare an array of Circle objects and attempt to print the radius of the last circle. But you get an ArrayIndexOutOfBoundsException because your index is too large. Note: The length of the array is 6. The index of the last Circle object is 5. Then, write it the correct way:

```
Circle[] arr = new Circle[6];
double radius = arr[arr.length].getRadius();          // Run-time error
double result = arr[arr.length - 1].getRadius();      // Correct way
```

Common Error

A common error is to accidentally set the index of an array to be the length of the array. The first index of the array is zero, so the last index of the array is **length – 1**.

```
nameOfArray[nameOfArray.length - 1]    // Last element in an array
nameOfArray[nameOfArray.length]        // ArrayIndexOutOfBoundsException
```

Traversing an Array

To **traverse** an array means to move through each slot in the array, one at a time. The most common way to perform a **traversal** is to start with the first index and end with the last index. But that's not the only way. You could start with the last index and move toward the first. You also could do all of the even indices first and then all of the odds. You get what I'm saying? Traversing an array just means that you visit every element at every index in an array.

Print the Contents of a One-Dimensional Array by Traversing It

Example

Print out the names of all your friends on Facebook using a for loop. By traversing the array from the first element to the last element, you can print each friend's name.

```
String[] faceBookFriends = {"Sierra", "Ciara", "Siarra", "Ciera"};
for (int i = 0; i < faceBookFriends.length; i++)
{
   System.out.println(faceBookFriends[i]);
}
```

```
OUTPUT
Sierra
Ciara
Siarra
Ciera
```

Off-by-One Error

This is a logic error that every programmer has committed at some point. It means that in a loop of some kind, your index is off by one number. You either went one index too high or one index too low.

The Enhanced `for` Loop (the `for-each` Loop)

The **enhanced for loop** (aka the **for-each loop**) can also be used to traverse an array. It does not work the same way as the standard for loop. The for-each loop always starts at the beginning of the array and ends with the last element in the array. As the name indicates, the for-each loop iterates through each of the elements in the array, and *for each* element, the temporary variable "takes on" its value. The temporary variable is only a copy of the actual value and any changes to the temporary variable are not reflected in the actual value.

General Form `for` the Enhanced for Loop (the `for-each` Loop)

```
dataType[] arrayName =  /* array filled in some way */

for (dataType temporaryVariable : arrayName)
{
   // instructions that use temporaryVariable
}
```

There is no loop control variable for the enhanced for loop (it is a **black box**). You only have access to the elements in the array, not the index. Also, the data type of the temporary variable **must** be the same as the data type of the array.

Traversing a One-Dimensional Array Using the Enhanced `for` Loop

Example

Print out the names of all your friends on Facebook using a for-each loop. Note: friend is a temporary variable that is of the same data type as the data stored in the array.

```
String[] faceBookFriends = {"Jordan", "Jordin", "Jordyn"};
for (String friend : faceBookFriends)
{
   System.out.println(friend);
}
```

```
OUTPUT
Jordan
Jordin
Jordyn
```

Mistake When Printing the Contents of an Array

Beginning programmers make the mistake of trying to print the contents of an array by printing the array reference variable. The result that is printed is *garbage*. This is because the reference variable only contains the address of where the array is; it does not store the elements of the array.

```
String[] faceBookFriends = {"Jordan", "Jordin", "Jordyn"};
System.out.println(faceBookFriends);   // Don't do this, it doesn't work
```

OUTPUT

```
@1765f32          // This code does not print the contents of the array
```

Note: The best way to print the contents of an array is to traverse the array using either a for loop or a for-each loop.

The 2-D Array

Definition of a 2-D Array

A **two-dimensional (2-D) array** is a complex data structure that can be visualized as a rectangular **grid** made up of **rows** and **columns**. Technically, it is **an array of arrays**. It can store any kind of data in its slots; however, each piece of data has to be of the same data type.

Two-dimensional arrays are actually fun to work with and are very practical when creating grid-style games. Many board games are based on a grid: checkers, chess, Scrabble, etc. Many apps are based on grids too: CandyCrush, 2048, Ruzzle. I'm sure you can think of others.

Declaring a 2-D Array and Initializing It Using a Predefined List of Data

When you know the values that you want to store in the 2-D array, you can declare the array and store the values in it immediately. This operation is just like what we did for the one-dimensional array. **Two pairs of brackets**, [ ][ ], are used to tell the computer that it is not a regular variable.

General Form for Creating a 2-D Array Using a Predefined List of Data

```
dataType[][] nameOf2DArray = { {value1, value2, value3},
                               {value4, value5, value6},
                               {. . . , . . . , . . . } };
```

value1	value2	value3
value4	value5	value6
. . .	. . .	. . .

Note: Pairs of curly braces are used to wrap the first row, then the second row, and so on. If you look closely at this visual, you will see how a 2-D array is really an array of arrays.

Example

Declare a 2-D array that represents a CandyCrush board with four rows and three columns. Notice that the four rows are numbered 0 through 3 and the three columns are numbered 0 through 2.

```
String[][] candyBoard = {{"Jelly Bean", "Lozenge", "Lemon Drop"},
                         {"Gum Square", "Lollipop Head", "Jujube Cluster"},
                         {"Lozenge", "Lollipop Head", "Lemon Drop"},
                         {"Jelly Bean", "Lollipop Head", "Lozenge"}};
```

This is the visual representation of what the candyBoard array looks like inside the computer.

	0	1	2
0	"Jelly Bean"	"Lozenge"	"Lemon Drop"
1	"Gum Square"	"Lollipop Head"	"Jujube Cluster"
2	"Lozenge"	"Lollipop Head"	"Lemon Drop"
3	"Jelly Bean"	"Lollipop Head"	"Lozenge"

Rows and Columns of the 2-D Array

Every cell in a 2-D array is assigned a pair of coordinates that are based on the row number and the column number. The first pair of brackets represents the row number and the second pair of brackets represents the column number. The rows and columns both begin at zero and end with one less than the number of rows or columns.

Example

Change the "Lozenge" in row 2 and column 0 to be a "Lemon Drop".

```
candyBoard[2][0] = "Lemon Drop";
```

	0	1	2
0	"Jelly Bean"	"Lozenge"	"Lemon Drop"
1	"Gum Square"	"Lollipop Head"	"Jujube Cluster"
2	"Lemon Drop"	"Lollipop Head"	"Lemon Drop"
3	"Jelly Bean"	"Lollipop Head"	"Lozenge"

Declaring a 2-D Array Using the Keyword new

A 2-D array object can be created using the keyword **new**. Every cell in the 2-D array is filled with the default value for its data type.

General Form for Creating a 2-D Array Using the Keyword new

```
dataType[][] nameOfArray = new dataType[numberOfRows][numberOfColumns];
```

Example

Declare a 2-D array that represents a Sudoku Board that has nine rows and nine columns. Put the number 6 in row 2, column 7:

```
int[][] mySudokoBoard = new int[9][9];   // the order is always [rows][columns]
mySudokoBoard[2][7] =  6;                // puts a 6 in row 2, column 7
```

9 Columns

9 Rows

	0	1	2	3	4	5	6	7	8
0	0	0	0	0	0	0	0	0	0
1	0	0	0	0	0	0	0	0	0
2	0	0	0	0	0	0	0	6	0
3	0	0	0	0	0	0	0	0	0
4	0	0	0	0	0	0	0	0	0
5	0	0	0	0	0	0	0	0	0
6	0	0	0	0	0	0	0	0	0
7	0	0	0	0	0	0	0	0	0
8	0	0	0	0	0	0	0	0	0

How to Refer to the Rows and Columns in a 2-D Array

The position, `myBoard[0][5]`, is read as *row 0, column 5.*

The position, `myBoard[3][0]`, is read as *row 3, column 0.*

Using the `length` Field to Find the Number of Rows and Columns

The number of rows in a 2-D array is found by accessing the length field. The number of columns in a 2-D array is found by accessing the length field on the name of the array *along with one of the rows* (it doesn't matter which row).

Example 1

Retrieve the number of rows from a 2-D array:

```
double[][] myBoard = new double[8][3];
int result = myBoard.length;          // result is 8 (number of rows)
```

Example 2

Retrieve the number of columns from a 2-D array:

```
double[][] myBoard = new double[8][3];
int result1 = myBoard[0].length;      // result1 is 3 (number of columns)
int result2 = myBoard[5].length;      // result2 is also 3
```

Accessing a 1-D Array from Within the 2-D Array

Consider this 2-D array declaration:

```
double[][] myBoard = new double[8][3];
```

`myBoard[0]` is a 1-D array that consists of the row with index 0.
`myBoard[5]` is a 1-D array that consists of the row with index 5.

Traversing a 2-D Array in Row-Major Order

Traversing a 2-D array means to visit every cell in the grid. This can be done in many different ways, but for the AP Exam, you need to know two specific ways.

Row-Major order is the process of traversing a 2-D array in the manner that English-speaking people read a book. Start at the top left cell and move toward the right along the first row until you reach the end of the row. Then start at the next row at the left-most cell, and then move along to the right until you reach the end of that row. Repeat this process until you have visited every cell in the grid and finish with the bottom-right cell.

Example

Traverse an array in row-major order using a nested for loop. Start with row 0 and end with the last row. For each row, start with column 0 and end with the last column.

```
String[][] myGrid =  /* 2-D array provided by the user */
for (int row = 0; row < myGrid.length; row++)                    // every row
{
   for (int column = 0; column < myGrid[0].length; column++)   // every column
   {
     System.out.println(myGrid[row][column]);                   // every cell
   }
}
```

Traversing a 2-D Array in Column-Major Order

Column-Major order is the process of traversing a 2-D array by starting at the top left cell and moving downward until you reach the bottom of the first column. Then start at the top of the next column and work your way down until you reach the bottom of that column. Repeat this until you have visited every cell in the grid and finish with the bottom-right cell.

Example

Traverse an array in column-major order using a nested for loop. Start with column 0 and end with the last column. For each column, start with row 0 and end with the last row.

```
String[][] myGrid =   /* 2-D array provided by the user */
for (int column = 0; column < myGrid[0].length; column++)   // every column
{
   for (int row = 0; row < myGrid.length; row++)            // every row
   {
      System.out.println(myGrid[row][column]);              // every cell
   }
}
```

> **`ArrayIndexOutOfBoundsException`**
>
> As with a 1-D array, if you use an index that is not within the 2-D array, you will get an ArrayIndexOutOfBoundsException.

Fun Fact: *Java provides support for multi-dimensional arrays, such as 3-D arrays; however, the AP Computer Science A Exam does not require you to know about them.*

The `ArrayList`

Definition of an `ArrayList`

An ArrayList is a **complex data structure** that allows you to add or remove objects from a list and it changes size automatically.

Declaring an `ArrayList` Object

An ArrayList is an object of the ArrayList class. Therefore, to create an ArrayList, you need to use the keyword new along with the constructor from the ArrayList class. You also need to know the data type of the objects that will be stored in the list. The ArrayList uses a pair of **angle brackets**, < and >, to enclose the class name of the objects it will store.

> **Two General Forms for Declaring an `ArrayList`**
>
> ```
> ArrayList<ClassName> nameOfArrayList   = new ArrayList<ClassName>();
> ```
> or
> ```
> List<ClassName> nameOfArrayList   = new ArrayList<ClassName>();
> ```
>
> Note: Questions on the AP Exam will use either **ArrayList** or **List** for the data types of the reference variables for an ArrayList object. **List** will be explained in Concept 9.

The graphic that follows is a visual representation of what the ArrayList looks like in memory.

Example

Declare an ArrayList of Circle objects. Please note that I am referring back to the Circle Class from Concept 2 and that the memory address is simulated.

```
ArrayList<Circle> myCircles = new ArrayList<Circle>();
```

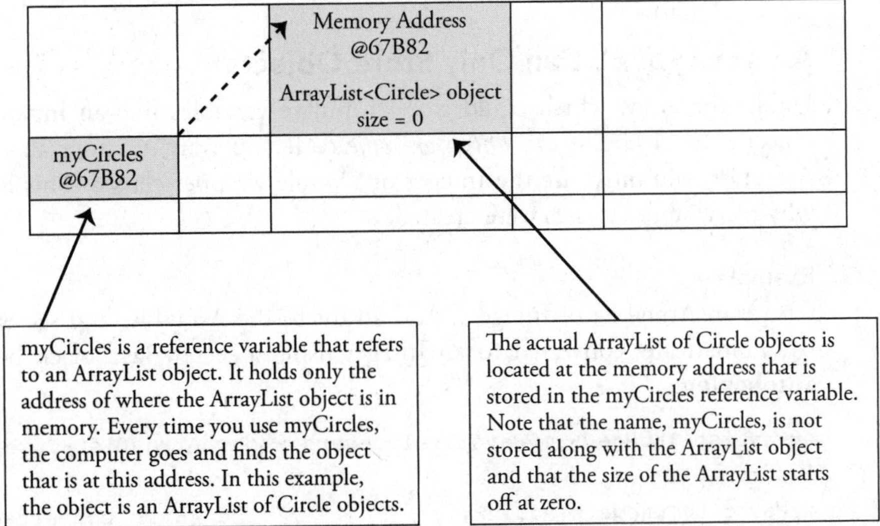
Memory Address
@67B82
ArrayList<Circle> object
size = 0
myCircles
@67B82
myCircles is a reference variable that refers to an ArrayList object. It holds only the address of where the ArrayList object is in memory. Every time you use myCircles, the computer goes and finds the object that is at this address. In this example, the object is an ArrayList of Circle objects.
The actual ArrayList of Circle objects is located at the memory address that is stored in the myCircles reference variable. Note that the name, myCircles, is not stored along with the ArrayList object and that the size of the ArrayList starts off at zero.

An `ArrayList` Always Starts Out Empty

When you create an ArrayList object, it is empty, meaning that there are no items in the list. It's like when your mom starts to make a "To Do" list and she writes the words "To Do" on the top of a piece of paper. The list is created but there is nothing in the list.

An `ArrayList` Is Resizable

When your mom writes, "Go grocery shopping" or "Buy awesome video game for favorite child" on her To Do list, the size of the list grows. As she completes a task on the list, she crosses it out and the size of the list shrinks. This is exactly how an ArrayList is resized.

> **Automatic Resizing**
>
> An awesome feature of the ArrayList is its ability to resize itself as elements are added to or removed from the list. The size() method (explained later) immediately recalculates how many elements are in the list.

An ArrayList Requires an import Statement

The ArrayList is not part of the built-in Java language package, so you have to let the compiler know that you plan on creating an ArrayList by putting an import statement prior to the class declaration.

```
import java.util.ArrayList;
```

You will not need to write any import statements on the AP exam.

An ArrayList Can Only Store Objects

Unlike the array, which could store primitive variables like an int or double, as well as objects, the *ArrayList can only store objects.* If you want to store an int or double in an ArrayList, you must use the Integer or Double wrapper classes. This is one of the reasons why the wrapper classes were created.

Example

Create an ArrayList of Integers. Add an int to the ArrayList and secretly watch as the int is automatically converted to an Integer using a secret, Java black box technique called **autoboxing.**

```
ArrayList<Integer> myFavoriteIntegers = new ArrayList<Integer>();
int num = 45;
myFavoriteIntegers.add(num);          // The 45 is converted to an Integer
```

Important ArrayList Methods

The ArrayList class comes with a large array of methods (see my pun). These methods make it easy to work with the objects inside the ArrayList. The AP Computer Science A Exam does not require you to know all of the methods from the ArrayList class; however, it does require you to know a subset of them.

The add Method

There are two **add** methods for the ArrayList. The **add(E object)** method **appends** the object to the end of the list. This means that it adds the object to the end of the list. It also returns the value true. The size of the ArrayList is automatically updated to reflect the addition of the new element.

Example

Create an ArrayList of Circle objects. Add three Circle objects to the ArrayList where the first has a radius of 8, the second doesn't provide a radius so the radius gets the default value of zero, and finally, the last circle has a radius of 6.5. Note: This example will be used in the explanations for the other ArrayList methods.

```
ArrayList<Circle> myCircles = new ArrayList<Circle>();

line 1: myCircles.add(new Circle(8));
line 2: myCircles.add(new Circle());
line 3: myCircles.add(new Circle(6.5));
```

After line 1 is executed: There is one Circle object in the ArrayList.

index	myCircles
0	Circle with radius 8.

⟵ A Circle object is added to the ArrayList.

After line 2 is executed: There are two Circle objects in the ArrayList.

index	myCircles
0	Circle with radius 8.
1	Circle with radius 0.

After line 3 is executed: There are three Circle objects in the ArrayList.

index	myCircles
0	Circle with radius 8.
1	Circle with radius 0.
2	Circle with radius 6.5.

> **What's with `E`?**
>
> The data type **E** is known as a **generic type**. It simply means that you can put any kind of data type here. I like to say, "The method takes **E**very kind of data type."

Another `add` Method

The **add(int index, E object)** method inserts the object into the position index in the ArrayList, shifting the object that was previously at position index and each of the objects after it over one index. The index for each of the objects affected by the **add** is incremented. The method does not return a value.

> **An `ArrayList` Is Like the Lunch Line at School**
>
> An ArrayList can be visualized like the lunch line at school. Imagine there are 10 students in line and they are numbered 0 through 9. Person 0 is the first person in line.
>
> Suppose the unthinkable happens and a student walks up and cuts the line. They have just inserted themselves into the list. If the *cutter* is now the third person in line, then they have just performed an **add(2, "cutter")**. This action impacts everyone who is in line *after* the cutter. The person who used to be in position 2, is now in position 3. The person who used to be at position 3 is now in position 4, and so on. The index for each person behind the cutter was *incremented by one*.

```
line 4:  myCircles.add(1, new Circle(4));
```

After line 4 is executed: The Circle object with a radius of 4 was inserted into the position with an index of 1 (the second position). All of the Circle objects after it had to move over one slot.

index	myCircles
0	Circle with radius 8
1	Circle with radius 4
2	Circle with radius 0
3	Circle with radius 6.5

⟵ The add(index, object) method inserts an object into the list at specific index. All objects after it move over to accommodate it.

The `size` Method

The **size()** method returns the number of items in the ArrayList. Notice that this is different from the **length** field of the array, which tells you how many slots were set aside and not the actual number of valid items stored in the array.

```
line 5: int howManyCircles = myCircles.size();     // howManyCircles is 4
```

IndexOutOfBoundsException

As in the 1-D array and the 2-D array, you will get an error if you try to access objects outside of the range of the list. The error when doing this with an ArrayList is called the IndexOutOfBoundsException.

```
myCircles.add(88, new Circle());          //  IndexOutOfBoundsException
```

The index, 88, is not in the range of 0 ≤ index ≤ myCircle.size().

The `remove` Method

The **remove(int index)** method deletes the object from the list that is at index. Each of the objects after this shifts down one index. The method also returns a reference to the object that was removed from the list.

```
line 6: Circle someCircle = myCircles.remove(2); // someCircle's radius is 0
```

After line 6 is executed: the Circle object that used to be in the slot at index 2 is removed (the circle with a radius of 0). The Circle objects that were positioned after it all move down one slot and someCircle now points to the Circle object that was removed.

index	myCircles
0	Circle with radius 8.
1	Circle with radius 4.
2	Circle with radius 6.5.

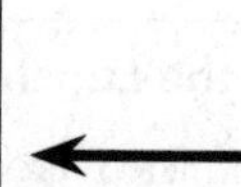

The remove(2) method deleted the object that was at index 2. It returns a reference to the object that used to be at index 2.

The `get` Method

The **get(int index)** method returns the object that is located in the ArrayList at position index. It doesn't remove the object; it just returns a copy of the object reference. This way, an alias is created (more than one object reference pointing to the same object). The value of index must be greater than or equal to zero and less than the size of the ArrayList.

Example

Get the Circle object at index 2 and assign it to a different Circle reference variable:

```
line 7: Circle someCircle = myCircles.get(2);    // someCircle's radius is 6.5
```

The `set` Method

The **set(int index, E object)** method replaces the object in the ArrayList at position index with object. It returns a reference to the object that was previously at index.

Example

Replace the Circle object at index 0 with a new Circle object with a radius of 20:

```
line 8: Circle c = myCircles.set(0, new Circle(20)); // c's radius is 8
```

index	myCircles
0	Circle with radius 20.
1	Circle with radius 4.
2	Circle with radius 6.5.

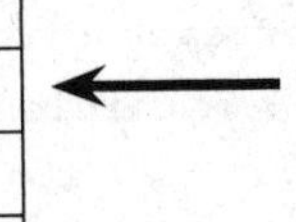

The set(0, Circle(20)) call replaces the object at index 0. It returns a reference to the object that was previously at index 0.

`length` Versus `size()` Versus `length()`

To find the number of slots in an array, use the length field.
To find the number of objects in an ArrayList, use the size() method.
To find the number of characters in a string, use the length() method.

Traversing an `ArrayList` Using a `for` Loop

The following code uses a for loop to print the area of each of the Circle objects in the ArrayList. The get method is used to retrieve each Circle object and then the getArea method is used on each of these Circle objects. Note that the **dot operator** is used *twice* in the same instruction. First, to get the current circle in the ArrayList, and second to get the area for that circle.

```
// print the area of each circle in the ArrayList using a standard for loop
for (int i = 0; i < myCircles.size(); i++)
{
   System.out.println(myCircles.get(i).getArea());    // print all the areas
}
```

```
OUTPUT
1256.636
50.265
132.732
```

Traversing an `ArrayList` Using the Enhanced `for` Loop

The enhanced for loop (for-each loop) can be used with an ArrayList. The following code prints the area of each of the Circle objects in the ArrayList. Notice that the temporary variable, circle, is the same data type as the objects in the ArrayList. In contrast to the general for loop shown above, the enhanced for loop does not use a loop control variable. Therefore, there is *no need for the get(i)* method since the variable circle is a copy of each of the Circle objects, one at a time and the getArea method can be used directly with circle.

```
// print the area of each circle in the ArrayList using a for-each loop
for (Circle circle : myCircles)
{
   System.out.println(circle.getArea());   // no index used in for-each loop
}
```

```
OUTPUT
1256.636
50.265
132.732
```

Printing the Contents of an `ArrayList`

Unlike an array, the contents of an ArrayList can be displayed to the console by printing the reference variable. The contents are enclosed in brackets and separated by commas.

```
ArrayList<Double> myDoubles = new ArrayList<Double>();
myDoubles.add(23.5);
myDoubles.add(50.1);
myDoubles.add(7.5);
System.out.println(myDoubles);    // print the contents of the ArrayList
```

```
OUTPUT
[23.5, 50.1, 7.5]
```

Note: If you want to format the output when printing the contents of an ArrayList, you should traverse the ArrayList.

Do Not Use `remove()` in the Enhanced `for` Loop

Never use the enhanced for loop (for-each) loop to remove an item from an ArrayList. You need the index to perform this process and the for-each loop doesn't use one. However, if you have to perform a remove, use an index variable with a while loop. Increment the index to get to the next element in the ArrayList. But, if you perform a remove, then don't increment the index.

```
int index = 0;
while (index < myCircles.size())
{
   if (myCircles.get(index) <= 0)
      myCircles.remove(index);      // remove any circle whose radius <= 0
   else
      index++;          // only increment index when not removing a circle
}
```

❯ Rapid Review

The Array

- An array is a complex data structure that can store a list of data of the same data type.
- An array is also referred to as a one-dimensional (or 1-D) array.
- An array can be created by assigning it a predetermined list.
- An array can be created by using the keyword new.
- An array can store either primitive data types or object reference types.
- Even though arrays are objects, they do not have methods.
- An array has one public field, called length.
- The length of the array refers to how many elements can be stored in the array and is decided when the array is created.
- The length of an array cannot be resized once it is declared. That is, an array cannot be made shorter or longer once it is created.
- The length field returns the total number of locations that exist in the array, and not the number of meaningful elements in the array.
- An element of an array refers to the item that is stored in the array.
- The index of an array refers to the position of an element in an array.
- The first index of an array is 0. The last index is its length − 1.
- Unused indices in an array automatically receive the default value for that data type.
- If an array contains objects, their default values are null. You need to instantiate an object for each element of the array before they can be used.
- Using an index that is not in the range of the array will throw an ArrayIndexOutOfBoundsException.

Traversing an Array

- Traversing an array refers to the process of visiting every element in the array.
- There are many options for traversing an array; however, the most common way is to start at the beginning and work toward the end.
- Based on the situation, there are times where you may want to traverse an array starting at the end and work toward the beginning.
- Sometimes you may want to traverse only a section of the array.
- When using a for loop to traverse an array, be sure to use length − 1 rather than length when accessing the last element in the array.
- The enhanced for loop iterates through each element in the array without using an index.
- The enhanced for loop is also known as the for-each loop.
- The enhanced for loop always starts with the first element and ends with the last.
- Use an enhanced for loop when you don't need to know the index for each element and you don't need to change the element in any way.

The 2-D Array

- A 2-D array is a complex data structure that can store a grid of data of the same type.
- The 2-D array is actually an array of arrays.
- The rows and columns are numbered starting with zero.
- The top row is row zero. The left-most column is column zero.
- The access to each cell in the 2-D array is always [*row*][*column*].
- The indices of the top-left cell are [0][0].
- The index of the bottom-right cell is [*numberOfRows* − 1][*numberOfColumns* − 1].
- Two-dimensional arrays can store primitive data or object data.
- To store a value in a 2-D array: arrayName[*rowNumber*][*columnNumber*] = value;
- To retrieve the number of rows in a 2-D array, use arrayName.length.
- To retrieve the number of columns in a rectangular 2-D array, use arrayName[0].length. Note: Any valid row number works instead of 0.
- Using an index that is not in the range of the 2-D array will throw an ArrayIndexOutOfBoundsException.

Traversing a 2-D Array

- To traverse a 2-D array means to visit each element in every row and column.
- The most common way to traverse a 2-D array is called Row-Major order.
- Row-Major order starts in the upper left, location [0][0], and travels to the right until the end of the row is reached. Then the next move is to the next row [1][0]. The process is repeated until the right-most column in the bottom row is reached.
- The second most common way to traverse a 2-D array is called Column-Major order.
- Column-Major order also starts in the upper left location [0][0]; however, it travels down until the end of the first column is reached. Then, the next move is to the top of the next column [0][1]. The process is repeated until the bottom row in the right-most column is reached.
- All of the 2-D arrays on the AP Computer Science A Exam will be rectangular.

The ArrayList

- The ArrayList is a complex data structure that can store a list of objects.
- The two general forms for declaring an ArrayList are:

```
ArrayList<ClassName> arrayListName = new ArrayList<ClassName>();
List<ClassName> arrayListName = new ArrayList<ClassName>();
```

- An ArrayList can only hold a list of objects; it cannot hold a list of primitive data.
- Use the Integer or Double classes to make an ArrayList of int or double values.
- The initial size of an ArrayList is 0.
- The add(E object) method appends the object to the end of the ArrayList. It also returns true.
- To append means to add on to the end of a list.
- The add(int index, E object) method inserts object at position index (note: index must be in the interval: [0,size]). As a result of this insertion, the elements at position index and higher move 1 index further from the 0 index.
- The get(int index) method returns a reference to the object that is at index.
- The set(int index, E object) method replaces the element at position index with object and returns the element that was formerly at index.
- The remove(int index) method removes the element at position index, and subsequently subtracts one from the indices of the elements at positions index + 1 and greater. It also returns the element that was removed.
- The size() method returns an int that represents the number of elements that are currently in the ArrayList.
- The size of the ArrayList grows and shrinks by either adding or removing elements.
- The size() method adjusts accordingly as elements are added or removed.
- Using an index that is not in the range of the ArrayList will throw an IndexOutOfBoundsException.
- The ArrayList requires an import statement since it is not in the standard library, but this will never be required on the AP exam.

Traversing an ArrayList

- To traverse an ArrayList means to visit each of the elements in the list.
- There are many ways to traverse an ArrayList, but the most common way is to start at the beginning and work toward the end.
- If a for loop is used to traverse an ArrayList, then the get() method will be required to gain access to each of the elements in the ArrayList.
- If an enhanced for loop is used to traverse an ArrayList, then the get() method is not required to gain access to each of the elements in the ArrayList, since each object is automatically retrieved by the loop, one at a time.
- If you need to remove elements from an ArrayList, you may consider starting at the end of the list and working toward the beginning.

› Review Questions

Basic Level

1. Consider the following code segment.

```
int number = 13;
int[] values = {0, 1, 2, 3, 4, 5};
for (int val : values)
{
    number += val;
}
System.out.println(number);
```

What is printed as a result of executing the code segment?

(A) 0
(B) 13
(C) 26
(D) 28
(E) 56

2. Consider the following code segment.

```
int total = 3;
int[] values = {8, 6, 4, -2};
total += values[total];
total += values[total];
System.out.println(total);
```

What is printed as a result of executing the code segment?

(A) 0
(B) 1
(C) 3
(D) 7
(E) 16

3. Assume that cities is an `ArrayList<String>` that has been correctly constructed and populated with the following items.

```
["Oakland", "Chicago", "Milwaukee", "Seattle", "Denver", "Boston"]
```

Consider the following code segment.

```
cities.remove(2);
cities.add("Detroit");
cities.remove(4);
cities.add("Cleveland");
```

What items does `cities` contain after executing the code segment?

(A) ["Cleveland", "Detroit", "Oakland", "Chicago", "Seattle", "Denver", "Boston"]
(B) ["Oakland", "Milwaukee", "Seattle", "Boston", "Detroit", "Cleveland"]
(C) ["Oakland", "Chicago", "Seattle", "Boston", "Detroit", "Cleveland"]
(D) ["Oakland", "Milwaukee", "Denver", "Boston", "Detroit", "Cleveland"]
(E) ["Oakland", "Chicago", "Seattle", "Denver", "Detroit", "Cleveland"]

4. Consider the following code segment.

```
int[] values = {7, 9, 4, 1, 6, 3, 8, 5 };
for (int i = 3; i < values.length; i += 2)
{
    System.out.print(values[i - 2]);
}
```

What is printed as a result of executing the code segment?

(A) `91`
(B) `746`
(C) `913`
(D) `79416385`
(E) `416385`

5. Consider the following code segment.

```
List<String> subjects = new ArrayList<String>();
subjects.add("French");
subjects.add("History");
subjects.set(1, "English");
subjects.add("Art");
subjects.remove(1);
subjects.set(2, "Math");
subjects.add("Biology");
System.out.println(subjects);
```

What is printed as a result of executing the code segment?

(A) `[French, English, Math, Biology]`
(B) `[French, Art, Biology]`
(C) `[French, English, Art, Math, Biology]`
(D) `[French, Math, Biology]`
(E) `IndexOutOfBoundsException`

6. Free-Response Practice: Modifying a 2-D array

Write a method that takes a 2-D array as a parameter and returns a new 2-D array.

The even rows of the new array should be exactly the same as the array passed in.

The odd rows of the new array should be replaced by the contents of the row above.

For example, if the input array is:

1	5	4	8	7	3
2	4	3	5	7	6
2	4	3	3	6	0
9	8	9	9	4	1

Then the returned array should be:

1	5	4	8	7	3
1	5	4	8	7	3
2	4	3	3	6	0
2	4	3	3	6	0

Here is the declaration for your method.

```
public int[][] modify(int[][] arr)
```

Advanced Level

7. Consider the following code segment.

```
int total = 3;
ArrayList<Integer> integerList = new ArrayList<Integer>();
for (int k = 7; k < 11; k++)
{
    integerList.add(k + 3);
}
for (Integer i : integerList)
{
    total += i;
    if (total % 2 == 1)
    {
        total -= 1;
    }
}
System.out.println(total);
```

What is printed as a result of executing the code segment?

(A) `34`
(B) `37`
(C) `46`
(D) `47`
(E) `49`

8. Consider the following code segment.

```
int[] nums = {0, 0, 1, 1, 2, 2, 3, 3};
for (int i = 3; i < nums.length - 1; i++)
{
    nums[i + 1] = i;
}
```

What values are stored in array `nums` after executing the code segment?

(A) `ArrayIndexOutOfBoundsException`
(B) `{2, 2, 3, 3, 4, 4, 5, 5}`
(C) `{0, 0, 1, 3, 5, 7, 9, 11}`
(D) `{0, 0, 1, 1, 3, 4, 5, 6}`
(E) `{0, 0, 1, 2, 2, 3, 4, 5}`

9. Assume that `msg` is an `ArrayList<String>` that has been correctly constructed and populated with the following items.

```
["can", "i", "delete", "words", "starting", "with", "letters",
     "between", "n", "and", "z"]
```

Which code segment removes all `String` objects starting with a letter from the second half of the alphabet (n-z) from the `ArrayList`?

Precondition: all String objects will be lowercase

Postcondition: msg will contain only String objects from the first half of the alphabet (a–m)

I.
```
for (int i = 0; i < msg.size(); i++)
{
    if (msg.get(i).compareTo("n") >= 0)
    {
        msg.remove(i);
    }
}
```

II.
```
for (int i = msg.size() - 1; i >= 0; i--)
{
    if (msg.get(i).compareTo("n") >= 0)
    {
        msg.remove(i);
    }
}
```

III.
```
int i = 0;
while (i < msg.size())
{
    if (msg.get(i).compareTo("n") >= 0)
    {
        msg.remove(i);
    }
    else
    {
        i++;
    }
}
```

(A) I only
(B) I and II only
(C) II and III only
(D) I and III only
(E) I, II, and III

10. Free-Response Practice: Filling a 2-D array

Write a method that will fill in a 2-D boolean array with a checkerboard pattern of alternating `true` and `false`. The upper left corner (0, 0) should always be `true`.

Given a grid size of 3 × 4, the method should return:

true	false	true	false
false	true	false	true
true	false	true	false

The method will take the number of rows and columns as parameters and return the completed array. Here is the method declaration and the array declaration:

```
public static boolean[][] makeGrid(int rows, int cols)
{
    boolean[][] grid;

    /* to be implemented */
}
```

› Answers and Explanations

Bullets mark each step in the process of arriving at the correct solution.

1. The answer is D.

- This is a for-each loop. Read it like this: "For each int (which I will call val) in values. . . ."
- The loop will go through each element of the array values and add it to number. Notice that number starts at 13.

```
13 + 0 + 1 + 2 + 3 + 4 + 5 = 28
```

2. The answer is D.

- values[total] = values[3] = -2. Remember to start counting at 0.
- 3 + -2 = 1, now total = 1.
- values [total] = values[1] = 6.
- 1 + 6 = 7 and that is what is printed.

3. The answer is E.

- Let's picture the contents of our ArrayList in a table:

0	1	2	3	4	5
Oakland	Chicago	Milwaukee	Seattle	Denver	Boston

- After the remove at index 2, we have this (notice how the indices have changed):

0	1	2	3	4
Oakland	Chicago	Seattle	Denver	Boston

- After adding Detroit, we have:

0	1	2	3	4	5
Oakland	Chicago	Seattle	Denver	Boston	Detroit

- After the remove at index 4, we have:

0	1	2	3	4
Oakland	Chicago	Seattle	Denver	Detroit

- And then we add Cleveland:

0	1	2	3	4	5
Oakland	Chicago	Seattle	Denver	Detroit	Cleveland

4. The answer is C.

- Let's lay out the array along with its indices.

index	0	1	2	3	4	5	6	7
contents	7	9	4	1	6	3	8	5

- The first time through the loop, i = 3.
  - values[3–2] = values [1] = 9 Print it.
- Next time through the loop, i = 5.
  - values [5–2] = values [3] = 1 Print it.
- Next time through the loop, i = 7.
  - Remember that even though the last index is 7, the length of the array is 8.
  - values [7–2] = values[5] = 3 Print it.
- We exit the loop having printed "913".

5. The answer is E.
   - Let's make tables to represent what happens as each statement is executed.

```
subjects.add("French");
```

0
French

```
subjects.add("History");
```

0	1
French	History

```
subjects.set(1, "English");
```

0	1
French	English

```
subjects.add("Art");
```

0	1	2
French	English	Art

```
subjects.remove(1);
```

0	1
French	Art

```
subjects.set(2, "Math");
```

Oops! Can't do it. Trying will generate an `IndexOutOfBoundsException`.

6. We can find the even rows by using the condition (row % 2 == 0). In other words, if I can divide the row number by 2 and not have a remainder, it's an even row. Even rows just get copied as is. Odd rows get copied from the row above [row – 1].

   Here is the completed method.

```
public int[][] modify(int[][] arr) {

    int[][] newArr = new int[arr.length][arr[0].length];

    for (int row = 0; row < arr.length; row++)
    {
        for (int col = 0; col < arr[row].length; col++)
        {
            if (row % 2 == 0)
            {
                newArr[row][col] = arr[row][col];           // even row
            }
            else
            {
                newArr[row][col] = arr[row - 1][col];      // odd row
            }
        }
    }
    return newArr;
}
```

7. The answer is C.

- After the initial for loop, the contents of the ArrayList are: [10, 11, 12, 13].
- Let's consider what the for-each loop is doing. For each Integer (called i) in integerList
  - add it to total (notice that total starts at 3).
  - (total % 2 == 1) is a way of determining if a number is odd. If total is odd, subtract 1 from it.
- Executing the for-each loop gives us:
  - total = 3 + 10 = 13 . . . odd → 12
  - total = 12 + 11 = 23 . . . odd → 22
  - total = 22 + 12 = 34 . . . even!
  - total = 34 + 13 = 47 . . . odd → 46 and our final answer.

8. The answer is D.

- On entering the loop, nums = {0, 0, 1, 1, 2, 2, 3, 3} and i = 3. Set nums[4] to 3.
- Now nums = {0, 0, 1, 1, 3, 2, 3, 3} and i = 4. Set nums[5] to 4.
- Now nums = {0, 0, 1, 1, 3, 4, 3, 3} and i = 5. Set nums[6] to 5.
- Now nums = {0, 0, 1, 1, 3, 4, 5, 3} and i = 6. Set nums[7] to 6.
- Now nums = {0, 0, 1, 1, 3, 4, 5, 6} and i = 7. nums.length − 1 = 7, so exit the loop.

9. The answer is C.

- Using a loop to remove items from an ArrayList can be very tricky because when you remove an item, all the items after the removed item change indices. That's why option I does not work. When item 3 "words" is removed, "starting" shifts over and becomes item 3, but the loop marches on to item 4 (now "with") and so "starting" is never considered and never removed.
- Working backward through a loop works very well when removing items from an ArrayList because only items after the removed item will change indices, and you will have already looked at those items. Option II works and is the simplest solution.
- Option III also works because it only increments the loop counter when an item is not removed. If an item is removed, the loop counter remains the same, so the item that gets shifted down ("starting" in the example above) is not skipped. Option III uses a while loop because it changes the loop index i, and it is bad programming style to mess with the loop index inside a for loop.

10. There are many ways to solve this problem. This implementation uses the fact that grid[row][col] is true when row + col is even.

```
public static boolean[][] makeGrid(int rows, int cols)
{
    boolean[][] grid;
    grid = new boolean[rows][cols];          //initializes all elements to false
    for (int r = 0; r < rows; r++)
    {
        for (int c = 0; c < cols; c++)
        {
            if ((r + c) % 2 == 0)            // r + c is even
            {
                grid[r][c] = true;
            }
        }
    }
    return grid;
}
```

Here is a tester class so you can see if the way you wrote the method was also correct. Copy in your code. Try it with all shapes and sizes of grid to test it thoroughly.

```
public class GridBuilder
{
    public static void main(String[] args)
    {
        boolean[][] grid = makeGrid(3, 4); //change (3,4) to test
        for (int r = 0; r < grid.length; r++)
        {
            for (int c = 0; c < grid[0].length; c++)
            {
                System.out.print(grid[r][c] + "\t");
            }
            System.out.println();
        }
    }

    public static boolean[][] makeGrid(int rows, int cols)
    {
        boolean[][] grid;
        //put your code in here
    }
}
```

Algorithms (Basic Version)

IN THIS CONCEPT

Summary: Software developers use algorithms extensively to write the code that will automate a process in a computer program. The algorithms described in this concept are some of the fundamental algorithms that you should know for the AP Computer Science A Exam.

Key Ideas

- An algorithm is a set of steps that when followed complete a task.
- Pseudocode is a hybrid between an algorithm and actual Java code.
- Efficient code performs a task in the fastest way.
- The swap algorithm exchanges the values of two variables.
- The copy algorithm makes a duplicate of a data structure.
- The sequential search algorithm finds a value in a list.
- The accumulate algorithm traverses a list, adding each value to a total.
- The find-highest algorithm traverses and compares each value in the list to determine which one is the largest.

How We Use Algorithms

In our daily lives, we develop and use algorithms all the time without realizing that we are doing it. For example, suppose your grandma calls you because she wants to watch *Game of Thrones* using the HBO GO app on her smart TV, but she can't figure out how to do it. You would respond to her by saying, "Hi, Grandma, well, the first thing you have to do is turn your TV on," then yada, yada, yada until finally you say, "Press play."

What you have just done is developed a process for how to solve a problem. Now, if she insists that you write the steps down on a notecard so she can follow it anytime she wants to, then, you will be writing the algorithm for "How to watch *Game of Thrones* on grandma's smart TV."

> Give a man a fish and he will eat for a day; teach a man to fish and he will eat for a lifetime.
> —Chinese proverb

Why Algorithms Are Important

The word **algorithm** can be simply defined as "a process to be followed to solve a problem." Computer programmers use algorithms to teach the computer how to automate a process.

The number of algorithms to learn is infinite, so rather than attempting to learn all possible algorithms, my goal is to teach you *how to write an algorithm*. This will be handy on the AP Computer Science A Exam when you have to either analyze or generate code to accomplish something you've never done before. The two algorithm concepts in this book teach you how to think like a software developer.

> Give developers a line of code and they will smile for a day; teach developers to write their own algorithms and they will smile for a lifetime.
> —Coding proverb

Algorithm Versus Pseudocode Versus Real Java Code

Algorithms are great as directions for how to do something; however, they can't be typed into a computer as a real Java program. The integrated development environment (IDE) needs to have the correct syntax so that the compiler can interpret your code properly. Can you imagine typing in the directions for Grandma as lines of code?

So, to solve this problem, programmers translate algorithms into **pseudocode**, which is a code-like language. This pseudocode is then translated into real code that the computer will understand.

Algorithm → **Pseudocode** → **Java code**

This process is really powerful because as programming languages change, the algorithms can stay the same. In the examples that follow, I've written the code in Java, but it could've easily been written in C++, Python, etc.

> **Turning Your Ideas into Code**
>
> Pseudocode is an inner step when turning your ideas into code. It is code-like but does not use the syntax of any programming language.

The Swap Algorithm

Swapping is the concept of switching the values that are stored in two different variables. The ability to swap values is essential for sorting algorithms (explained in Concept 12). To complete a swap, you need to use a temporary variable to store one of the values.

> **Problem:** Swap the values inside two integer variables. For example, if a = 6 and b = 8, then make a = 8 and b = 6.

Algorithm:
Step 1: Let a = 6 and b = 8
Step 2: Create a temporary variable called c
Step 3: Assign c the value of a
Step 4: Assign a the value of b
Step 5: Assign b the value of c
Step 6: Now, the value of a is 8 and b is now 6

Pseudocode:
set a = 6
set b = 8
c = a
a = b
b = c

Java Code:

```
int a = 6;
int b = 8;
int c;
c = a;
a = b;
b = c;
```

The Copy Algorithm for the Array and `ArrayList`

Making a copy of an array (or ArrayList) means to create a brand-new object that is a duplicate of the original. This is different from creating a new array variable (or ArrayList variable) that references the original array (or ArrayList). That was called making an alias. It is easy for new programmers to make the mistake of thinking that these two are the same.

> **Problem:** Suppose you have an array of integers. Make a new array object that is a copy of this array. The two arrays should contain the same exact values but not be the same object. Then, modify your code to work with an ArrayList of numbers.

Algorithm:
Step 1: Create an array that is the same length as the original
Step 2: Look at each of the values in the original array one at a time
Step 3: Assign the value in the copy to be the corresponding value in the original
Step 4: Continue until you reach the last index of the original array

Pseudocode:

Create an array called duplicate that has the same length as the original array
for (iterate through each element in the original array)
{
    set duplicate[index] = original[index]
}

Java Code 1: Java code using an array (`for` loop)

```
int[] original = {23, 51, 14, 50};               // These numbers are for example
int[] duplicate = new int[original.length];   // create duplicate array

for (int i = 0; i < original.length; i++)     // iterate through the original
{
   duplicate[i] = original[i];                  // copy the values one at a time
}
```

Java Code 2: Java code using an array (`for-each` loop)

```
int[] original = {23, 51, 14, 50};
int[] duplicate = new int[original.length];
int index = 0;                          // create an index for duplicate array

for (int temp : original)               // temp is a copy of the actual object
{
   duplicate[index] = temp;             // for-each loop does not use an index
   index++;                             // update index for the duplicate array
}
```

Java Code 3: Java code using an `ArrayList` (`for` loop)

```
ArrayList<Integer> original = new ArrayList<Integer>();
original.add(23);
original.add(51);
original.add(14);
original.add(50);
ArrayList<Integer> duplicate = new ArrayList<Integer>();

for (int i = 0; i < original.size(); i++)
{
   duplicate.add(orginal.get(i)); // get the original and add it to duplicate
}
```

Java Code 4: Java code using an `ArrayList` (`for-each` loop)

```
ArrayList<Integer> original = new ArrayList<Integer>();
original.add(23);
original.add(51);
original.add(14);
original.add(50);
ArrayList<Integer> duplicate = new ArrayList<Integer>();

for (Integer temp : original)
{
   duplicate.add(temp);         // add each value in original to duplicate
}
```

Copying an Array Reference Versus Copying an Array

The following code demonstrates a common mistake when attempting to copy an array.

```
int[] original = {23, 51, 14, 50};
int[] copy = original;
```

This code makes the reference variable copy refer to the same object that original refers to. This means that if you change a value in copy, it also changes the value in original. This code *does not create a new object* that is a duplicate of the original. This code makes both reference variables refer to the same object (the object that original refers to).

The Sequential Search Algorithm

When you search iTunes for a song, how does it actually find what you are looking for? It may use a sequential search to find the name of a song. A **sequential search** is the process of looking at each of the elements in a list, one at a time, until you find what you are looking for.

General Problem: Search for "Sweet Caroline" in a list of song titles.

Refined Problem: Write a method that allows you to search an array of song titles for a specific song title.

Final Problem: Write a method called searchForTitle that has two parameters: a string array and a string that represents the search target. The method should return the index of the target string if it is found and -1 if it does not find the target String. Note: I provide two solutions to solve this problem and compare their efficiency afterward.

The Search Target

When you search for a specific item in a list, the item you are looking for is often referred to as the search target.

Solution 1: Look at every item in the list.

Algorithm 1:

Step 1: Create a variable called foundIndex and set it equal to -1
Step 2: Look at each of the titles in the array one at a time
Step 3: Compare each title to the target title
Step 4: If the title is equal to the target title, then assign foundIndex to value of the current index
Step 5: Continue looking until you reach the end of the list
Step 6: Return the value of foundIndex

Pseudocode 1:

```
set foundIndex = -1
for (iterate through every element in the list)
{
  if (title = target title)
  {
    set foundIndex = current index
  }
}
return the value of foundIndex
```

Java Code 1:

```
public int searchForTitle(String[] titles, String target)
{
    int foundIndex = -1;
    for (int i = 0; i < titles.length; i++)
    {
        if (titles[i].equals(target))
        {
            foundIndex = i;                 // remember this index
        }
    }
    return foundIndex;                      // returns the foundIndex
}
```

Solution 2: Stop looking if you find the search target.

Algorithm 2:

Step 1: Look at each of the titles in the array one at a time
Step 2: Compare the target title to each of the titles
Step 3: If the title is equal to the target title, then stop looking and return the current index
Step 4: Continue looking until you reach the end of the list
Step 5: Return -1

Pseudocode 2:

```
index = 0
while (not at the end of the list)
{
  if (title = target title)
  {
    return index
  }
  index++
}
return -1
```

Java Code 2:

```
public int searchForTitle(String[] titles, String target)
{
    int index = 0;
    while (index < titles.length)
    {
        if (titles[index].equals(target))
        {
            return index;                       // target was found
        }
        index ++;                               // target was not found yet
    }
    return -1;    // after searching entire list, the target was not found
}
```

Efficiency

Analyze the two solutions that searched for the title of a song.

The second algorithm is *more efficient* than the first because the method stops looking for the title as soon as it finds it. The first algorithm is *less efficient* than the second because it continues to compare each of the titles in the list against the search target even though it may have already found the search target.

> **Efficiency**
>
> An efficient algorithm does its job in the fastest way it can. Efficient programs are optimized to reduce CPU (central processing unit) time and memory usage.

The Accumulate Algorithm

Suppose you have a list of all the baseball players on a team and you want to find the total number of hits that the team had. As humans, we can do this quite easily. Just add them up. But how do you teach a computer to add up all of the hits?

> **General Problem:** Add up all of the hits for a baseball team.
>
> **Refined Problem:** Write a method that finds the sum of all of the hits in an array of hits.
>
> **Final Problem:** Write a method called findSum that has one parameter: an int array. The method should return the sum of all of the elements in the array. Then, modify your code to work with a 2-D array of int values and also an ArrayList of integers.

Algorithm:

Step 1: Create a variable called sum and set it equal to zero
Step 2: Look at each of the elements in the list
Step 3: Add the element to the sum
Step 4: Continue until you reach the end of the list
Step 5: Return the sum

Pseudocode:

```
set sum = zero
for (iterate through all the elements in the list)
{
    sum = sum + element
}
return sum
```

Code 1: Java code using an array (`for-each` loop)

```
public int findSum(int[] arr)
{
    int sum = 0;                                   // initializing the sum to 0
    for (int value : arr)                          // for every int in arr
    {
       sum += value;                               // Add the element to the sum
    }
    return sum;                                    // return the sum
}
```

Code 2: Java code using an array (`for` loop)

```
public int findSum(int[] arr)
{
    int sum = 0;
    for (int i = 0; i < arr.length; i++)
    {
       sum += arr[i];
    }
    return sum;
}
```

Code 3: Java code using a 2-D array (Row-Major order)

```
public int findSum(int[][] arr)
{
    int sum = 0;                                        // initialize sum to 0
    for (int row = 0; row < arr.length; row++)          // do every row
    {
        for (int col = 0; col < arr[row].length; col++)
        {                                          // do every column in the row
            sum += arr[row][col];                  // add value in cell to sum
        }
    }
    return sum;                                    // return sum
}
```

Code 4: Java code using an ArrayList (`for-each` loop)

```
public int findSum(ArrayList<Integer> arr)
{
    int sum = 0;
    for (int value : arr)
    {
        sum += value;
    }
    return sum;
}
```

Code 5: Java code using an ArrayList (`for` loop)

```
public int findSum(List<Integer> arr)
{
    int sum = 0;
    for (int i = 0; i < arr.size(); i++)
    {
        sum += arr.get(i);
    }
    return sum;
}
```

The Find-Highest Algorithm

What is the highest score on your favorite video game? As more people play a game, how does the computer figure out what is the highest score in the list?

General Problem: Find the highest score out of all the scores for a video game.

Refined Problem: Write a method that finds the highest score in a list of scores.

Final Problem: Write a method called findHigh that has one parameter: an int array. The method should return the largest value in the array. Then, modify your code to work with a 2-D array of int values and also an ArrayList of Integers.

Solution 1: Let the highest be the first element in the list.

Algorithm:
Step 1: Create a variable called high and set it to the first element in the list
Step 2: Look at each of the scores in the list
Step 3: If the score is greater than the high score, then make it be the new high score
Step 4: Continue until you reach the end of the list
Step 5: Return the high score

Pseudocode:

```
set high = first score
for (iterate through all the scores in the list)
{
  if (score > high)
  {
    high = score
  }
}
return high
```

Code 1: Java code using an array (`for` loop)

```
public int findHigh(int[] arr)
{
    int high = arr[0];                          // set high = first element
    for (int i = 1; i < arr.length; i++)        // start with the 2nd element
    {
        if (arr[i] > high)                      // if element is greater than high
            high = arr[i];                      // set high equal to that element
    }
    return high;                                // return the highest in the array
}
```

Code 2: Java code using an array (`for-each` loop)

```
public int findHigh(int[] arr)
{
    int high = arr[0];
    for (int value : arr)
    {
        if (value > high)
        {
            high = value;
        }
    }
    return high;
}
```

Code 3: Java code using a 2-D array (Row-Major order)

```
public int findHigh(int[][] arr)
{
    int high = arr[0][0];
    for (int row = 0; row < arr.length; row++)
    {
        for (int col = 0; col < arr[row].length; col++)
        {
            if (arr[row][col] > high)
            {
                high = arr[row][col];
            }
        }
    }
    return high;
}
```

Code 4: Java code using an `ArrayList` (`for-each` loop)

```
public int findHigh(ArrayList<Integer> arr)
{
    int high = arr.get(0);
    for (int value : arr)
    {
        if (value > high)
        {
            high = value;
        }
    }
    return high;
}
```

Code 5: Java code using an `ArrayList` (`for` loop)

```
public int findHigh(ArrayList<Integer> arr)
{
    int high = arr.get(0);
    for (int i = 0; i < arr.size(); i++)
    {
        if (arr.get(i) > high)
        {
            high = arr.get(i);
        }
    }
    return high;
}
```

Solution 2: Let the highest be an extremely low value.

There is an alternative solution to the high/low problem that you should know. To execute this alternative, we modify the first step in the previous algorithm.

Step 1: Create a variable called high and set it to *a really, really small number*.
This algorithm works as long as the really, really small number is guaranteed to be smaller than at least one of the numbers in the list. In Java, we can use the public fields from the Integer class to solve this problem.

```
int high = Integer.MIN_VALUE;
```

Likewise, to solve the *find-lowest* problem, you could set the low to be a really, really big number.

```
int low = Integer.MAX_VALUE;
```

Searching for the Smallest Item in a List

If we changed the previous example to find the *lowest* score in the list, then the comparator in the if-statement would be changed to less than, <, instead of greater than, >.

› Rapid Review

Algorithms and Pseudocode

- The purpose of this concept is to teach you how to translate an idea into code.
- An algorithm is a sequence of steps that, when followed, produces a specific result.
- Algorithms are an extremely important component of software development since you are teaching the computer how to do something all by itself.
- There are an infinite number of algorithms.
- Pseudocode is a hybrid between an algorithm and real code.
- Pseudocode does not use correct syntax and is not typed into an IDE.
- Pseudocode helps you translate an algorithm into real code since it is code-like and more refined than an algorithm.
- Programmers translate algorithms into pseudocode and then translate pseudocode into real code.

Swap Algorithm

- The swap algorithm allows you to switch the values in two different variables.
- To swap the values in two different variables, you must create a temporary variable to store one of the original variables.
- To swap num1 and num2, set temp = num1, let num1 = num2, and then let num2 = temp.

Copy Algorithm

- The copy algorithm is used to create a duplicate version of a data structure.
- It looks at each element in the original list and assigns it to the duplicate list.
- The duplicate is not an alias of the original; it is a new object that contains all of the same values as the original.

Sequential Search Algorithm

- The sequential search algorithm searches a list to find a search target.
- It looks at each element in the list one at a time comparing the element to the search target.

Accumulate Algorithm

- The accumulate algorithm finds the sum of all the items in a list.
- It looks at each element in the list one at a time, adding the value to a total.

Find-Highest Algorithm

- The find-highest algorithm finds the largest value in a list.
- It looks at each element in the list one at a time, comparing the value to the current high value.
- Common ways to implement this algorithm include initializing the highest value with the first value in the list or to an extremely small negative number.
- The find-highest algorithm can be modified to find the lowest item in a list.

❯ Review Questions

1. Replace only highs and lows.

Write a method that replaces the highest values and the lowest values in a list. The method takes one parameter: an int array. The range of the numbers is 0–1000 (inclusive). If a value in the array is greater than 750, replace the value with 1000. If a value in the array is less than 250, replace it with 0. The method should return an int array of the same length as the parameter array and leave the original array untouched.

The array passed to the method:

634	521	786	899	509	235	750	806	142	645

The array returned by the method:

634	521	1000	1000	509	0	750	1000	0	645

```
public int[] replaceHighAndLow(int[] arr)
{
    // Write the implementation
}
```

2. Determine if the numbers in an `ArrayList` are always increasing.

Write a method that determines if the values in an ArrayList are always increasing. The method takes one parameter: an ArrayList of Integer. The method returns true if the elements in the array are continually increasing and false if the elements are not continually increasing.

```
ArrayList<Integer> listOfIntegers = /* contains values from table below  */
boolean result = isIncreasing(listOfIntegers);    // result is true
```

The ArrayList passed to the method:

34	35	36	37	38	40	42	43	51	52

```
public boolean isIncreasing(ArrayList<Integer> arr)
{
    // Write the implementation
}
```

3. Find all the songs that contain the word "Love" in the title.

One of the most popular themes in music lyrics is love. Suppose a compilation of thousands of songs is stored in a grid. Count the number of songs that contain the word "Love" in the title.

Write a method that counts the number of song titles that contains a certain string in the title. The method has two parameters: a 2-D array of String objects and a target string. The method should return an integer that represents the number of strings in the 2-D array that contains the target string.

```
String[][] songs = /*  2-d array fill with the values in the table below  */
int result = findCount(songs, "Love");                     // result is 5
```

The 2-D array passed to the method:

"We Are the Champions"	"You Shook Me All Night Long"	"We Found Love"
"Bleeding Love"	"Stairway to Heaven"	"Won't Get Fooled Again"
"I'd Do Anything for Love"	"Stupid Crazy Love"	"Love in This Club"
"Since U Been Gone"	"One More Time"	"Walk This Way"

```
public int findCount(String[][] arr, String target)
{
    // Write the implementation
}
```

4. Determine the relative strength index of a stock.

The relative strength index (RSI) of a stock determines if the stock is overpriced or underpriced. The range of the RSI is from 0 to 100 (inclusive). If the value of the RSI is greater than or equal to 70, then the stock is determined to be overpriced and should be sold.

Write a method that takes an array of RSI values for a specific stock and determines when the stock is overpriced. The method has one parameter: a double array. The method should return a new array of the same length as the parameter array. If the value in the RSI array is greater than 70, set the value in the overpriced array to 1. If the RSI value is less than or equal to 70, set the value in the overpriced array to 0.

The RSI array that is sent to the method:

55.6	63.2	68.1	70.1	72.4	73.9	71.5	68.3	67.1	66.2

The overpriced array that is returned by the method:

0	0	0	1	1	1	1	0	0	0

```
public int[] overpriced(double[] rsiValues)
{
    // Write the implementation
}
```

5. Fill an array with even numbers.

Write a method that fills an array with random even numbers. The method takes two parameters: the number of elements in the array and the range of the random numbers (0–number inclusive).

```
int[] result = onlyEvens(10, 100);  //result contains ten even #'s [0,100]
```

result contains the following (results will change each time program is run):

56	96	88	30	0	92	12	98	4	28

```
public static int[] onlyEvens(int arraySize, int range)
{
    // Write the implementation
}
```

6. Determine if the rate is increasing.

If a stock price is increasing, that means the value of it is going up. If the *rate* that the stock price is increasing is increasing, it means that the price is rising at a faster rate than it was the day before. Write a method that determines if the rate at which a stock price increases is increasing. The method has one parameter: an ArrayList of Double values that represent the stock prices. Return true if the rate that the stock is increasing is increasing, and false if the rate is not increasing.

```
ArrayList<Double> prices = /* contains the values from the table below  */
boolean result = rateIsIncreasing(prices);    // result is true
```

stockPrices contains the following:

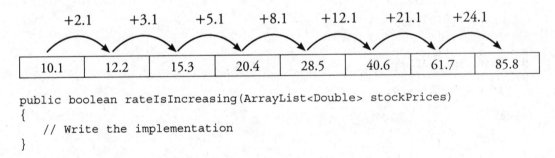

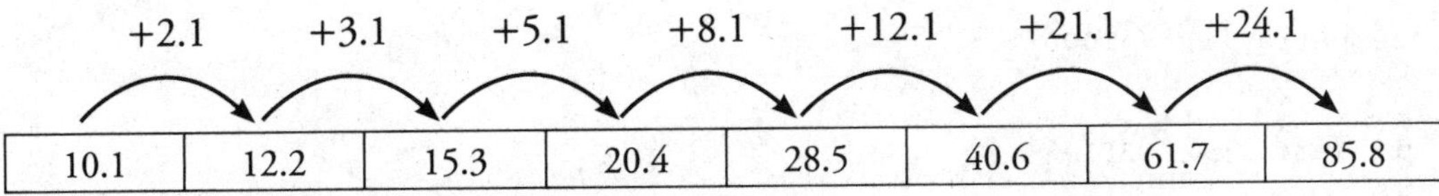

10.1	12.2	15.3	20.4	28.5	40.6	61.7	85.8

```
public boolean rateIsIncreasing(ArrayList<Double> stockPrices)
{
    // Write the implementation
}
```

› Answers and Explanations

1. Replace only highs and lows.

Algorithm:
Step 1: Create a new array called temp that is the same length as the original
Step 2: Look at each of the scores in the list
Step 3: If the score is greater than 750, then store 1000 in the temp array
Step 4: If the score is less than 250, then store 0 in the temp array
Step 5: If the score is between 250 and 750, then store the score in the temp array
Step 6: Continue until you reach the end of the list
Step 7: Return the temp array

Pseudocode:

```
create a temporary array
for (iterate through all the scores in the original array)
{
if (score > 750)
{
  tempArray[index] = 1000
}
else
{
  if (score < 250)
  {
    tempArray[index] = 0
  }
  else
  {
    tempArray[index] = score
  }
}
return tempArray
```

Java code:

```java
public int[] replaceHighAndLow(int[] arr)
{
    int[] tempArray = new int[arr.length];

    for (int i = 0; i < arr.length; i++)
    {
        if (arr[i] > 750)
        {
            tempArray[i] = 1000;
        }
        else
            if (arr[i] < 250)
            {
                tempArray[i] = 0;
            }
            else
            {
                tempArray[i] = arr[i];
            }
    }
    return tempArray;
}
```

2. Determine if the numbers in an `ArrayList` are always increasing.

Algorithm:

Step 1: Look at each of the scores in the list starting at the second element
Step 2: If the first score is less than or equal to the second score, then return false
Step 3: Advance to the next element and continue until you reach the end of the list
Step 4: Return true

Pseudocode:

```
for (iterate through all the scores in the original list starting at the second element)
{
  if (list[index – 1] <= list[index])
  {
    return false
  }
}
return true
```

Java code:

```
private static boolean isIncreasing(ArrayList<Integer> arr)
{
    for (int i = 1; i < arr.size(); i++)
    {
        if (arr.get(i) <= arr.get(i - 1))      // if current <= previous
        {
            return false;                      // list is not increasing
        }
    }
    return true;
}
```

3. Find all the songs that contain the word "Love" in the title.

Algorithm:
Step 1: Initialize a count variable to zero
Step 2: Look at each of the scores in the 2-D array (using Row-Major order)
Step 3: If the song title contains the target String, then increment the count
Step 4: Continue until you reach the end of the list
Step 5: Return count

Pseudocode:

```
for (iterate through all the rows in the 2-D array)
{
    for (iterate through all the columns in the 2-D array)
    {
      if (list[row][column] contains target String)
      {
        count = count + 1
      }
    }
}
return count
```

Java code:

```
private static int findCount(String[][] arr, String target)
{
    int count = 0;
    for (int r = 0; r < arr.length; r++)
    {
        for (int c = 0; c < arr[r].length; c++)
        {
            if (arr[r][c].indexOf(target) != -1)      // target is in title
            {
                count++;
            }
        }
    }
    return count;
}
```

4. Determine the relative strength index of a stock.

Algorithm:
Step 1: Create an array called temp that is the same length as the parameter array
Step 2: Look at each of the scores in the array
Step 3: If the score > 70, then set the corresponding element of the temp array to 1; otherwise set it to 0
Step 4: Continue until you reach the end of the list
Step 5: Return temp

Pseudocode:

```
create an integer array called temp that is the same length as the parameter array
for (iterate through all the elements in the parameter array)
{
   if (parameter array[index] > 70)
   {
      temp[index] = 1
   }
   else
   {
      temp[index] = 0
   }
}
return temp
```

Java code:

```
public int[] overpriced(double[] rsiValues)
{
    int[] temp = new int[rsivalues.length];

    for (int i = 0; i < rsivalues.length; i++)
    {
        if (rsivalues[i] >= 70)
        {
            temp[i] = 1;
        }
        else
        {
            temp[i] = 0;
        }
    }
    return temp;
}
```

5. Fill an array with randomly chosen even numbers. There are several ways to do this. Here is one.

Algorithm:
Step 1: Create an array called temp that is the same length as the parameter array
Step 2: Loop as many times as the length of the parameter array
Step 3: Generate a random number in the appropriate range
Step 4: While the random number is not even, generate a random number in the appropriate range
Step 5: Put the random number in the list
Step 6: Continue until you complete the correct number of iterations
Step 7: Return temp

Pseudocode:
```
create an integer array called temp that is the same length as the parameter array
for (iterate as many times as the length of the parameter array)
{
    create a random number in the interval from 0 to parameter range
    while (the newly created random number is not even)
    {
        generate a new random value and assign it to the random number
    }
    temp[index] = random number
}
return temp
```

Java code:
```
public int[] onlyEvens(int arraySize, int range)
{
    int[] evens = new int[arraySize];

    for (int i = 0; i < arraySize; i++)
    {
        int number = (int)(Math.random() * range);
        while (number % 2 != 0)                      // as long as the number is odd
        {
            number = (int)(Math.random() * range);   // generate a new number
        }
        evens[i] = number;                       // add the even number to the array of evens
    }
    return evens;
}
```

6. Determine if the rate is increasing.

Algorithm:

Step 1: Iterate through the parameter array starting at the 3rd element
Step 2: If (the third element minus the second element) is less than or equal to the (second element minus the first element), then return false
Step 3: Move onto the next element and go to Step 2 (and adjust the elements #s)
Step 4: If you get this far without returning, return true

Pseudocode:

```
for (iterate through the parameter array starting with the third element (index 2))
{
    if ((list[index] – list[index – 1]) <= (list[index – 1] – list[index – 2]))
    {
        return false
    }
}
return true
```

Java code:

```
private static boolean rateIsIncreasing(ArrayList<Double> stockPrices)
{
    for (int i = 2; i < stockPrices.size(); i++) // start with the 3rd element
    {
        if ((stockPrices.get(i) - stockPrices.get(i - 1)) <=
            (stockPrices.get(i - 1) - stockPrices.get(i - 2)))
        {
            return false;
        }
    }
    return true;                // the rate is increasing
}
```

Classes and Objects (Advanced Version)

IN THIS CONCEPT

Summary: This concept is the advanced continuation of Concept 2, Classes and Objects (Basic Version). In this concept you'll find the explanations for topics such as passing parameters by reference, overloading methods and constructors, and the many uses of the word static. Full understanding of these concepts is key to scoring a 5 on the AP Computer Science A Exam.

Key Ideas

- Objects are passed by reference, while primitives are passed by value.
- Arrays are objects; therefore, they are passed by reference.
- Data encapsulation is a way of hiding user information.
- Overloaded constructors have the same name but different parameter lists.
- Overloaded methods have the same name but different parameter lists.
- Static variables are called class variables and are shared among all objects of the same class.
- Static final variables are called class constants and, once given a value, they cannot be changed during the running of the program.
- The keyword this is used to refer to the current object.
- Scope describes the region of the program in which a variable is known.

Parameters

Primitives Are Passed by Value

Values that are passed to a method or constructor are called **actual parameters** (or **arguments**), while the variables in the parameter list of the method declaration are called **formal parameters.** When an actual parameter is of a primitive type, its value cannot be changed in the method because only the *value* of the variable is sent to the method.

Said another way, the formal parameter variable becomes a copy of the actual parameter. The method cannot alter the value of the actual parameter variable. The formal parameter variable only receives the *value* of the actual parameter. The method only has the ability to change the value of the formal parameter variable, not the value of the actual parameter.

Objects Are Passed by Reference

Objects are passed by reference. What does that mean?

Simple answer: It means that when an object is passed to a method, the method has the ability to change the state of the object.

Technical answer: When an actual parameter is an object, it is the object reference that is actually the parameter. This means that when the method receives the reference and stores it in a formal parameter variable, the method now has the ability to change the state of the object *because it knows where the object lives.* The formal parameter is now an alias of the actual parameter. Essentially, you have just given away the keys to the car.

Example

Demonstrate the difference between passing by reference and passing by value. The state of the object is changed in the method, while the state of the primitive variable is unchanged.

```
public static void main(String[] args)
{
    Circle myCircle = new Circle(100);          // radius is 100
    int number = 100;                           // number is 100

    goAhead(myCircle, number);                  // send the arguments

    System.out.println(myCircle.getRadius()); // prints 50 (changed)
    System.out.println(number);                 // prints 100 (not changed)
}

public static void goAhead(Circle c, int n)
{
    c.setRadius(50);             // changes the radius of myCircle to 50
    n = 50;                      // changes n to 50, but number stays 100
}
```

```
OUTPUT
50
100
```

Arrays Are Passed by Reference

An array is an object, so this means that it will be passed by reference. If you pass an array as an actual parameter to a method or constructor, you are *actually* passing the *reference to the array*, which means that the method or constructor has the ability to change the elements in the array.

Example

Demonstrate that an array parameter can be used to modify the array. This is an example of *destruction of persistent data*, a one-point penalty on the free response. The array reference that was passed to the method allows the contents of the original array to be destroyed.

```
public static void main(String[] args)
{
    String[] myFavoriteThings = {"Memes", "Vines", "Snapchat"};

    whoops(myFavoriteThings);                    // the array is the argument

    for (String s : myFavoriteThings)
        System.out.println(s);                   // Fearless Pandas everywhere
}

public static void whoops(String[] arr)
{
    for (int i = 0; i < arr.length; i++)
        arr[i] = "Fearless Pandas";              // changes the array values
}
```

```
OUTPUT
Fearless Pandas
Fearless Pandas
Fearless Pandas
```

Destruction of Persistent Data

Recall that this is a one-point penalty on the free response. Don't ever allow a method to change the state of the argument that is passed to it (unless you are asked to do so in the specifications).

Example

Correctly demonstrate how *not* to have destruction of persistent data. In this example, the method creates a temporary array and copies each of the values of the original array over to the temporary. Then, the temporary array is assigned values while leaving the original array untouched.

```
public static void main(String[] args)
{
    String[] myFavoriteThings = {"Memes", "Vines", "Snapchat"};

    for (String s : myFavoriteThings)
        System.out.println(s);                  // the array is untouched

    String[] watchThis = yeah(myFavoriteThings);

    for (String s : watchThis)                  // the array is modified
        System.out.println(s);
}

public static String[] yeah(String[] arr)
{
    String[] temp = new String[arr.length]; // create a temporary array

    for (int i = 0; i < arr.length; i++)
        temp[i] = arr[i];                   // make a copy of the original

    for (int i = 0; i < temp.length; i++)
        temp[i] = "Fearless Pandas";        // modify the temporary array

    return temp;                            // return the temporary array
}
```

```
OUTPUT
Memes
Vines
Snapchat
Fearless Pandas
Fearless Pandas
Fearless Pandas
```

Comparing Against a `null` Reference

Recall that a nullPointerException error occurs if you attempt to perform a method on a reference variable that is null. How can you tell if a reference variable is null so you won't get that error? The answer is to perform a **null check**. Compare the reference variable using the == comparator or != comparator. Just *don't use* the equals method to perform a null check. You'll get a nullPointerException!

Example

Demonstrate how to perform a null check using the Circle class that was created in Concept 2. If the Circle reference is not null, then compute the area. If the Circle reference is null, then return -1 for the area.

```
public static void main(String[] args)
{
    Circle circle1 = new Circle(10);
    Circle circle2;                          // circle2 is null

    System.out.println(doANullCheck(circle1);   // send a valid reference
    System.out.println(doANullCheck(circle2);   // send a null reference
}

public static double doANullCheck(Circle c)
{
    if (c != null)
        return c.getArea();          // return the area if c is not null
    else
        return -1;                   // return -1 if c is null
}
```

```
OUTPUT
314.15926
-1.0
```

Checking Against `null`

Never use the equals method to determine if an object reference is null. Only use == or != to check if an object reference variable is null.

Overloaded Constructors

It is possible to have more than one constructor in a class. Constructors are always declared using the name of the class, but their parameter lists can differ. If there are two or more constructors in a class, they are all called **overloaded constructors**. The computer decides which of the overloaded constructors is the correct one to use based on the parameter list.

Overloaded constructors have the same name; however, they must differ by one of these ways:

- a different number of parameters
- the same number of parameters but at least one is a different type
- the same exact parameter types but in a different order

Example

Declare four constructors for a class called Student. The constructors have the same name as the class, but are different because their parameter lists differ.

```
public class Student
{
    private String firstName;
    private String lastName;
    private int age;

    public Student()                          // no-argument constructor
    {
    }

    public Student(String first)              // constructor has 1 parameter
    {
        firstName = first;
    }

    public Student(String first, String last)    // has 2 parameters
    {
        firstName = first;
        lastName = last;
    }

    public Student(String first, String last, int yearsOld)   // has 3
    {
        firstName = first;
        lastName = last;
        age = yearsOld;
    }
}
```

Overloaded Methods

Methods that have the same name but different parameters lists are called **overloaded methods.** The computer decides which of the overloaded methods is the correct one to use based on the parameter list.

Overloaded methods must have the same name and may have a different return type; however, they must differ by one of these ways:

- a different number of parameters
- the same number of parameters but at least one is a different type
- the same exact parameter types but in a different order

Example

Declare four overloaded methods. The methods should have the same name, but different parameter lists as described above.

```
public void doSomething(int param)             // has 1 int parameter
{
   // Do something with param
}

public void doSomething(double param)          // has 1 double parameter
{
   // Do something with param
}

public void doSomething(double a, int b)       // has 2 parameters
{
   // Do something with a and b
}

public void doSomething(int a, double b) // has 2 params; different order
{
   // Do something with a and b
}
```

Blast from the Past

Recall that the **substring** method from the String class is overloaded. There are two versions of the method, but they are different because their parameter lists are not identical.

static, static, static

static Variables (Class Variables)

There are times when a class wants to have a variable that is shared by all the objects that come from the class. If any one of the objects changes the value of this variable, then the change is reflected in the state of every object from the class. The variable is declared and assigned a value as though it were an instance variable; however, it uses the keyword **static** in its declaration. Static variables are also called **class variables** and like all instance variables, are declared private.

Example

Suppose you are hired by Starbucks to keep track of the total number of coffees that they serve. You've already decided to have a class called StarbucksStore, but how can you keep track of the total number of coffees served by all of the stores? One solution is to make a static variable, called totalCoffeesServed. Then, create a method to increment this class variable every time a coffee is served.

```
public class StarbucksStore
{
   private static int totalCoffeesServed = 0; // shared instance variable

   public void serveOneCoffee()         // method that uses shared variable
   {
      totalCoffeesServed++;          // adds one to the static variable
   }

   public int getTotalCoffeesServed()     // Accessor for static variable
   {
      return totalCoffeesServed;
   }

   /*  Other implementation not shown   */
}
```

```
public class StarbucksRunner
{
   public static void main(String[] args)
   {
      StarbucksStore store1 = new StarbucksStore();
      StarbucksStore store2 = new StarbucksStore();
      StarbucksStore store3 = new StarbucksStore();

      System.out.println(store1.getTotalCoffeesServed());
      System.out.println(store2.getTotalCoffeesServed());
      System.out.println(store3.getTotalCoffeesServed());

      store2.serveOneCoffee();   // adds one to totalCoffeesServed
      store1.serveOneCoffee();   // adds one to totalCoffeesServed
      store3.serveOneCoffee();   // adds one to totalCoffeesServed
      store3.serveOneCoffee();   // adds one to totalCoffeesServed

      System.out.println(store1.getTotalCoffeesServed());
      System.out.println(store2.getTotalCoffeesServed());
      System.out.println(store3.getTotalCoffeesServed());
   }
}
```

```
OUTPUT
0
0
0
4
4
4
```

`static` Methods

If a method is labeled static, it can only call other static methods and can only use static variables. The most common place to use a static method is in a runner class. The main method is static and so it can only call other static methods. The main method is static because there aren't any objects to initiate a call to run the program.

Example

Demonstrate how the main method can call static methods.

```
public class FunnyQuote
{
   public static void main(String[] args)
   {
      System.out.println(tellMeSomething());    // no object required
   }

   public static String tellMeSomething() // main can call this method
   {
      return "I like turtles.";
   }
}
```

```
OUTPUT
I like turtles.
```

`static final` Variables (Class Constants)

There are times when a class wants to have a **constant** that is used by all the objects of the class. As the name constant suggests, it *cannot* be reassigned a different value at any other place in the program. In fact, a compile-time error will occur if you try to give the constant a different value. Class constants are declared using the keywords **static** and **final** in their declaration and use all uppercase letters (and underscores, if necessary). These static final variables are also called **class constants.**

Example

As the programmer assigned to the Starbucks project, you decide to make a variable that holds the logo for all the stores. You choose to make this variable a constant because every store shares the logo, and none of them should be allowed to change the logo. Once the program starts, the logo cannot be changed.

```
public class StarbucksStore
{
   static final String STARBUCKS_LOGO = "Mermaid";    // class constant

   /*  Other implementation not shown   */

}
```

Data Encapsulation

Information Hiding

Protecting information by restricting its access is important in software development. Since classes are public, all the objects from the class are public. The way that we hide information or **encapsulate data** in Java is to use private access modifiers.

Encapsulated Data
It is expected on the AP Computer Science A Exam that all instance variables of a class are declared private. The only way to gain access to them is to use the accessor or mutator methods provided by the class.

What Happens if I Declare an Instance Variable `public`?

Let's suppose you just want to find out what happens if you declare an instance variable public. The result is that you can use the dot operator to access the instance variables.

Example
Declare the radius instance variable from the Circle class as public. Notice that the object has now allowed access to the public variable by anyone using the dot operator. Using the dot operator on a private variable produces a compile-time error.

```
// Pretend this code is in the Circle class
public double radius;  // public instance variable:  Don't ever do this!

// Pretend this code is in a runner class
Circle myCircle = new Circle(10); // radius is 10.0
myCircle.radius = -54;            // radius is easily changed when public
double result = myCircle.radius;  // result is -54.0
```

The Integer Class and Its `public` Fields
Remember how we accessed the largest and smallest Integer? We used the dot operator to just get the value. That's because MAX_VALUE and MIN_VALUE are public constant fields of the Integer class. You don't use an accessor method to get them.

```
int reallyBigInteger = Integer.MAX_VALUE; // MAX_VALUE is a public constant
```

Scope

The scope of a variable refers to the region of the program in which the variable is known. Attempting to use a variable outside of its scope produces a compile-time error.

Different types of variables have different scopes.

- The scope of a **local variable** is within the nearest pair of curly braces that encloses it.
- The scope of a **parameter** is the method or constructor in which it is declared.
- The scope of an **instance variable** is the class in which it is defined.

Documentation

Javadoc

Javadoc is a tool that generates an HTML-formatted document that summarizes a class (including its constructors, instance variables, methods, etc.) in a readable format. Among programmers, this document itself is often referred to as a Javadoc. The summary is called the API (application program interface) for that set of classes and is extremely useful for programmers.

Javadoc Tags

Similar to a hashtag, the **@param** is a tag that is used in Javadoc comments. When a programmer uses an @param tag, they state the name of the parameter and provide a brief description of it. Javadoc tags are automatically detected by Javadoc and appear in the API.

The **@return** is also a Javadoc tag. It is only used in return methods, and the programmer identifies the name of the variable being returned and a brief description of it.

Precondition and Postcondition

You know what happens when you assume, right? Well, a **precondition** states what you are allowed to assume about a parameter. It's a statement in the documentation comment before a constructor or method signature that describes the expectations of the parameter sent to the method. On the AP Computer Science A Exam, reading the preconditions can help you understand the nature of the parameters. A **postcondition** is what is expected of the method after it is executed.

> **Documentation Comments on the AP Computer Science A Exam**
> You won't have to write any Javadoc comments on the AP exam. You just need to know how to read them.

Example

This is the same Circle class that was created in Concept 2; however, it now includes the Javadoc tags and precondition/postcondition statements as they may appear on the AP Computer Science A Exam. It also includes an improved way to calculate the area using the Math class.

```
public class Circle                    // Class declaration
{
   private double radius;             // Instance variable

   public Circle()                     // The no-argument constructor
   {
      // The radius gets a default value of 0.0
   }

   /**
   *  A parameterized constructor for a Circle
   *  @param r The radius of the Circle
   *        Precondition:  r >= 0
   */
   public Circle(double rad)
   {
      radius = rad;  // The radius gets the value of the parameter, rad
   }

   /**
   *  Accessor method that returns the radius of the Circle object
   *  @return radius of the Circle
   */
   public double getRadius()
   {
      return radius;
   }

   /**
   *  Mutator method that reassigns the radius of the Circle object
   *  @param r The radius of the Circle
   *        Precondition:  rad > 0
   */
   public void setRadius(double rad)
   {
      radius = rad;     // The radius is changed to be the parameter value
   }

   /**
   *  Method that calculates and returns the area of the Circle
   *  @return area of the Circle
   *         Postcondition:  area >= 0
   */
   public double getArea()
   {
      return Math.PI * Math.pow(radius, 2); //A better way to compute area
   }
}  // End of Circle class
```

The Keyword this

The keyword **this** is a reference to the current object, or rather, the object whose method or constructor is being called. It is also referred to as the **implicit** parameter.

Example 1
Pass the implicit parameter **this** to a different method in a class. This usage may be tested on the exam.

```
public static void main(String[] args)
{
   MyClass m1 = new MyClass("Batman");
   m1.doSomething();
}
```

```
public class MyClass
{
   private String name;
   public MyClass(String n)
   {   name = n;   }

   public String getName()
   {   return name;   }

   public void doSomething()
   {
      nowDoSomethingWith(this);        // Pass "this" to another method
   }

   public void nowDoSomethingWith(MyClass myObject) // myObject is m1
   {
      System.out.println("I am " + myObject.getName());
   }
}
```

```
OUTPUT
I am Batman
```

Example 2

A common place to use the keyword **this** is when the instance variables have the same name as the parameters in a constructor. The keyword **this** is used on the left side of the assignment statements to assign the instance variables the values of the parameters. The left side is the instance variable; the right side is the parameter variable. Using **this** here says, "Hey, I understand that we have the same name, so these are my instance variables and not the parameters." This usage of the keyword this is not tested on the exam.

```
public class Whatever
{
   private int a;
   private int b;

   public Whatever(int a, int b)     // Parameters have the same name
   {                                 // as the instance variables.
      this.a = a;                    // Left side are the instance variables
      this.b = b;                    // Right side are the parameters
   }
}
```

IllegalArgumentException

If you pass an argument to a method and the value of the argument does not meet certain criteria required by the method, an **IllegalArgumentException** error may be thrown during run-time. Programmers may also choose to write a method that terminates with an IllegalArgumentException if it does not receive the expected input.

› Rapid Review

Passing Parameters by Value

- The parameter variables that are passed to a method or constructor are called actual parameters (or arguments).
- The parameter variables that receive the values in a method or constructor are called formal parameters.
- Passing by value means that only the value of the variable is passed.
- Primitive variables are passed by value.
- It is impossible to change the value of actual parameters that are passed by value.

Passing Parameters by Reference

- Passing by reference means that the address of where the object is stored is passed to the method or constructor.
- Objects are passed by reference (including arrays).
- The state of the object can be modified when it is passed by reference.
- It is possible to change the value of actual parameters that are passed by reference.

Overloaded

- Overloaded constructors may have (1) a different number of parameters, (2) the same number of parameters but of a different type, or (3) the same exact parameter types but in a different order.
- Overloaded methods have the same name; however, their method parameter lists are different in some way.
- Overloaded methods may have (1) a different number of parameters, (2) the same number of parameters but of a different type, (3) the same exact parameter types but in a different order.

`static`

- Static variables are also known as class variables.
- A class variable is a variable that is shared by all instances of a class.
- Changes made to a class variable by any object from the class are reflected in the state of each object for all objects from the class.
- Static final variables are also called class constants.
- Class constants are declared using both the keywords **static** and **final**.
- Constants, by naming convention, use uppercase letters and underscores.
- Constants cannot be modified during run-time.
- If a method is declared static, it can only call other static methods and can reference static variables.

Scope

- The scope of a variable refers to the area of the program in which it is known.
- The scope of a local variable is in the block of code in which it is defined.
- The scope of an instance variable is the class in which it is defined.
- The scope of a parameter is the method or constructor in which it is defined.

Documentation

- Documentation tags are used by Javadocs.
- The @param tag is used to describe a parameter.
- The @return tag is used to describe what is being returned by a method.
- A precondition describes what can be expected of the values that the parameters receive.
- A postcondition describes the end result criteria for a method.

Miscellaneous

- Data encapsulation is the act of hiding the values of the instance variables from other classes.
- Declaring instance variables as private encapsulates them.
- Declaring instance variables as public allows instant access to them using the dot operator.
- Use the keyword **this** when you want to refer to the object itself.
- An IllegalArgumentException error occurs when a parameter that is passed to a method fails to meet criteria set up by the programmer.

› Review Questions

Basic Level

Questions 1–2 refer to the following information.

Consider the following method:

```
public int calculate(int a, int b)
{
    a = a - b;
    int c = b;
    b = a * a;
    a = b - (a + c);
    return a;
}
```

1. The following code segment appears in another method in the same class.

```
int var = 9;
int count = 2;
System.out.println(calculate(var, count));
```

What is printed as a result of executing the code segment?

(A) `2`
(B) `9`
(C) `23`
(D) `40`
(E) `49`

2. The following code segment appears in another method in the same class.

```
int a = 9;
int b = 2;
int c = calculate(a, b);
System.out.println(a + " " + b + " " + c);
```

What is printed as a result of executing the code segment?

(A) `40 49 40`
(B) `9 2 40`
(C) `40 49 49`
(D) Run-time error: `a` and `b` cannot contain two values at once.
(E) Compile-time error. Duplicate variable.

3. The following method performs the same calculation as the method in questions 1 and 2, but now the int values are contained in an array.

```
public static int calculate(int[] arr)
{
    arr[0] = arr[0] - arr[1];
    int c = arr[1];
    arr[1] = arr[0] * arr[0];
    arr[0] = arr[1] - (arr[0] + c);
    return arr[0];
}
```

The following code segment appears in another method in the same class.

```
int[] myArr = new int[2];
myArr[0] = 9;
myArr[1] = 2;
int c = calculate(myArr);
System.out.println(myArr[0] + " " + myArr[1] + " " + c);
```

What is printed as a result of executing the code segment?
(A) `40 49 40`
(B) `9 2 40`
(C) `40 49 2`
(D) `9 2 49`
(E) Nothing will be printed. Program will terminate with an error.

4. Consider the following calculation of the area of a circle.

```
double area = Math.PI * Math.pow(radius, 2);
```

`Math.PI` can best be describes as:
(A) A method in the `Math` class.
(B) A method in the class containing the given line of code.
(C) A public static final double in the `Math` class.
(D) A private static final double in the `Math` class.
(E) A private static final double in the class containing the given line of code.

5. A programmer is designing a `BankAccount` class. In addition to the usual information needed for a bank account, she would like to have a variable that counts how many `BankAccount` objects have been instantiated. She asks you how to do this. You tell her:

(A) It cannot be done, since each object has its own variable space.
(B) She should use a class constant.
(C) She should call a mutator method to change the `numAccounts` variable held in each object.
(D) She should create a static variable and increment it in the constructor as each object is created.
(E) Java automatically keeps a tally. She should reference the variable maintained by Java.

6. Having multiple methods in a class with the same name but with a different number or different types of parameters is called:

(A) abstraction
(B) method overloading
(C) encapsulation
(D) method visibility
(E) parameterization

Advanced Level

7. Consider the following method.

```
public int mystery(int a, int b, int c)
{
    if (a < b && a < c)
    {
        return a;
    }
    if (b < c && b < a)
    {
        return b;
    }
    if (c < a && c < b)
    {
        return c;
    }
}
```

Which statement about this method is true?

(A) mystery returns the smallest of the three integers a, b, and c.
(B) mystery always returns the value of a.
(C) mystery always returns both values a and c.
(D) mystery sometimes returns all three values a, b, and c.
(E) mystery will not compile, because the return statement appears to be unreachable under some conditions.

8. The method from the problem above is re-written as shown.

```
public int mystery2(int a, int b, int c)
{
    if (a < b && a < c)
    {
        return a;
    }
    if (b < c && b < a)
    {
        return b;
    }
    return c;
}
```

Now, which statement about the method is true?

(A) mystery2 returns the smallest of the three integers a, b, and c.
(B) mystery2 always returns the value of c.
(C) mystery2 returns either the values of both a and c or both b and c.
(D) mystery2 sometimes returns all three values a, b, and c.
(E) mystery2 will not compile, because the return statement appears to be unreachable under some conditions.

Questions 9–10 refer to the following information.

Consider the following class declaration.

```
public class AnotherClass
{
    private static int val = 31;
    private String data;

    public AnotherClass(String s)
    {
        data = s;
        val /= 2;
    }

    public void join(String s)
    {   data = data + s;   }

    public void setData(String s)
    {   data = s;   }

    public String getData()
    {   return data;   }

    public int getVal()
    {   return val;    }
}
```

9. The following code segment appears in the main method of another class.

```
AnotherClass word = new AnotherClass("Hello");
word.join("Hello");
AnotherClass sentence = new AnotherClass(word.getData());
sentence.join("Hi");
word.setData(sentence.getData());
word.join("Hello");
sentence.join("Hi");
sentence.setData("Hi");
System.out.println(word.getVal());
```

What is printed as a result of executing the code segment?

(A) `30`
(B) `15`
(C) `7.75`
(D) `7.5`
(E) `7`

10. The following code segment appears in the main method of another class.

```
AnotherClass[] acArray = new AnotherClass[10];
acArray[2] = new AnotherClass("two");
acArray[3] = new AnotherClass("three");
for (int i = 0; i < acArray.length; i++)
{
    /* missing condition */
    {
        System.out.println(acArray[i].getData());
    }
}
```

Which of the following statements should be used to replace `/* missing condition */` so that the code segment will not terminate with a `NullPointerException`?

(A) `if (acArray[i] != null)`
(B) `if (acArray.get(i) != null)`
(C) `if (acArray.getData() != null)`
(D) `if (acArray.getData().equals("two") || acArray.getData().equals("three"))`
(E) Since not all of acArray's entries contain an AnotherClass object, there is no way to loop through the array without a `NullPointerException`.

11. Consider the following method declaration.

```
public int workaround(int a, double b)
```

Which of these method declarations would correctly overload the given method?

I. `public double workaround(int a, double b)`
II. `public int workaround(double b, int a)`
III. `public int workaround(String s, double b)`
IV. `public double workaround(int a, double b, String s)`
V. `public int workaround(int b, double a)`

(A) I only
(B) IV only
(C) II and III only
(D) II, III, and IV only
(E) All of the above

12. Free-Response Practice: The `Height` Class

The purpose of the `Height` class is to hold a measurement in feet and inches, facilitate additions to that measurement, and keep the measurement in simplest form (inches < 12). A `Height` object can be instantiated with a measurement in feet and inches, or in total inches.

Write the entire `Height` class that includes the following:

a. Instance variables `int feet` for the number of feet and `int inches` for the number of inches. These instance variables must be updated anytime they are changed so that they hold values in simplest form. That is, inches must *always* be less than 12 (see method described in part C below).
b. Two constructors, one with two parameters representing feet and inches, and one with a single parameter representing inches. The parameters do not necessarily represent the measurement simplest form.
c. A method called `simplify()` that recalculates the number of feet and inches, if necessary, so that the number of inches is less than 12.
d. An `add(int inches)` method that takes the number of inches to add and updates the appropriate instance variables.
e. An `add(Height ht)` method that takes a parameter that is a Height object and adds that object's inches and feet to this object's instance variables, updating if necessary.
f. Accessor (getter) methods for the instance variables `inches` and `feet`.

› Answers and Explanations

Bullets mark each step in the process of arriving at the correct solution.

1. The answer is D.

- When the method call calculate(var, count) is executed, the arguments 9 and 2 are passed by value. That is, their values are copied into the parameters a and b, so a = 9 and b = 2 at the beginning of the method.
- a = 9 − 2 = 7
- int c = 2
- b = 7 * 7 = 49
- a = 49 – (7 + 2) = 40 and that is what will be returned.

2. The answer is B.

- When the method call calculate(a, b) is executed, the arguments 9 and 2 are passed by value, that is their values are copied into the parameters a and b. The parameters a and b are completely different variables from the a and b in the calling method and any changes made to a and b inside the method do not affect a and b in the calling method in any way.
- We know from problem #1 that calculate(9, 2) will return 40.
- 9 2 40 will be printed.

3. The answer is A.

    - This time, when the method call calculate(myArr) is executed, the argument myArr is passed by reference. The *reference* to myArr is copied into the parameter arr, so now arr and myArr both point to the same object. Any changes made to arr in the method are also made to myArr, since it references the same object as arr. When the method returns, the variable arr no longer exists, but the changes made using its name can still be seen by accessing the array with the variable myArr.
    - Refer to the explanation for problem #1 to see the calculations (only the names have changed).
    - 40 49 40 will be printed.

4. The answer is C.

    - All instance variables in the classes we write must be declared private. In fact, the AP exam readers may deduct points if you do not declare your variables private. However, some built-in Java classes provide us with useful constants by making their fields public.
    - We access these fields by using the name of the class followed by a dot, followed by the name of the field. We can tell that it is a constant because its name is written entirely in capital letters. Some other public constants that we use are Integer.MAX_VALUE and Integer.MIN_VALUE.
    - Note that we can tell this is not a method call, because the name is not followed by (). The Math.pow(radius, 2) is a method call to a static method of the Math class.

5. The answer is D.

    - Option A is incorrect. It correctly states that each object has its own variable space, and each object's variables will be different. However, it is not correct about it being impossible.
    - Option B is incorrect. It says to use a constant, which doesn't make sense because she wants to be able to update its value as objects are created.
    - Option C is incorrect. There is no one mutator method call that will access the instance variables of each object of the class.
    - Option D is correct. A static variable or class variable is a variable held by the class and accessible to every object of the class.
    - Option E is simply not true.

6. The answer is B.

    - Method overloading allows us to use the same method name with different parameters.

7. The answer is E.

    - The intent of the mystery method is to find and return the smallest of the three integers. We know that all possible conditions are covered by the if statements and one of them has to be true, but the compiler does not know that. The compiler just sees that all of the return statements are in if clauses, and so it thinks that if all the conditions are false, the method will not return a value. The compiler generates an error message telling us that the method must return a value.

8. The answer is A.

    - The first time a return statement is reached, the method returns to the caller. No further code in the method will be executed, so no further returns will be made.

9. The answer is E.

- There are a lot of calls here, but only two that will affect the value of val, since the only time val is changed is in the constructor.
- It's important to notice that val is a static variable. That means that the class maintains one copy of the variable, rather than the objects each having their own copy. Since there is only one shared copy, changes made to val by any object will be seen by all objects.
- The first time the constructor is called is when the word "object" is instantiated. val = 31 when the constructor is called and this call to the constructor changes its value to 15 (integer division).
- The second time the constructor is called is when the sentence object is instantiated. This time val = 15 when the constructor is called and this call to the constructor changes its value to 7 (integer division).

10. The answer is A.

- acArray is an array of objects. Initially, all the entries in the array are null. That is, they do not reference anything. If a program tries to use one of these entries as if it were an object, the program will terminate with a NullPointerException.
- To avoid the NullPointerException, we must not attempt to use a null entry as if it contained a reference to an object.
- Option B is incorrect. It uses ArrayList syntax, not array syntax, so it will not even compile.
- Options C and D are incorrect. They attempt to call getData() on a null entry and so will terminate with a NullPointerException.
- Option A is correct. It checks to see whether the array entry is null before allowing the System.out.println statement to treat it as an object.

11. The answer is D.

- Options I and V are incorrect. An overloaded method must have a different parameter list so that the compiler knows which version to use.
- Options II, III, and IV are valid declarations for overloaded methods. They can all be distinguished by their parameter lists.

12. The `Height` Class

```
public class Height
{
    // Part a - remember to simplify when these variables are changed
    private int feet;
    private int inches;

    // Part b - overloaded constructors
    // notice the calls to simplify.
    public Height(int i)
    {
        inches = i;
        simplify();
    }

    public Height(int f, int i)
    {
        feet = f;
        inches = i;
        simplify();
    }

    // Part c - This method puts feet and inches in simplest form
    // see note below for an alternate solution
    public void simplify()
    {
        int totalInches = 12 * feet + inches;
        feet = totalInches / 12;
        inches = totalInches % 12;
    }

    // Parts d and e - overloaded add methods
    // Notice the call to the Height object len passed as a parameter
    public void add(Height ht)
    {
        feet += ht.getFeet();
        inches += ht.getInches();
        simplify();
    }
    public void add(int in)
    {
        inches += in;
        simplify();
    }

    //Part f - getters for the instance variables inches and feet
    public int getInches()
    {
        return inches;
    }
    public int getFeet()
    {
        return feet;
    }
}
```

Note: There are many ways to write the `simplify` method. If you aren't sure if yours works, type it into your IDE. Here's one alternate solution:

```
while (inches > 12)
{
    inches -= 12;
    feet++;
}
```

Inheritance and Polymorphism

IN THIS CONCEPT

Summary: In order to take advantage of everything an object-oriented language has to offer, you must learn about its ability to handle inheritance. This concept explains how superclasses and subclasses are related within a class hierarchy. It also explains what polymorphism is and how objects within a class hierarchy may act independently of each other.

Key Ideas

- A class hierarchy describes the relationship between classes.
- Inheritance allows objects to use features that are not defined in their own class.
- A subclass extends a superclass (a child class extends a parent class).
- Child classes do not have direct access to their parent's private instance variables.
- A child class can override the parent class methods by simply redefining the method.
- Polymorphism is what happens when objects from child classes are allowed to act in a different way than objects from their parent classes.
- The Object class is the parent of all classes (the mother of all classes).

Inheritance

Sorry, this inheritance is not about a large sum of money that your crazy, rich uncle has left you. It is, however, a feature of Java that is immensely valuable. It allows objects from one class to inherit the features of another class.

Class Hierarchy

One of the most powerful features of Java is the ability to create a **class hierarchy**. A hierarchy allows a class to inherit the attributes of another class, making the code **reusable**. This means that the classes on the lower end of the hierarchy inherit features (like methods and instance variables) without having to define them in their own class.

In biology, animals are ranked with classifications such as species, genus, family, class, kingdom, and so on. Any given rank **inherits** the traits and features from the rank above it.

Example

Demonstrate the hierarchy of cats, dogs, horses, and elephants. Note that all German shepherds are dogs, and all dogs are mammals, so all German shepherds are mammals.

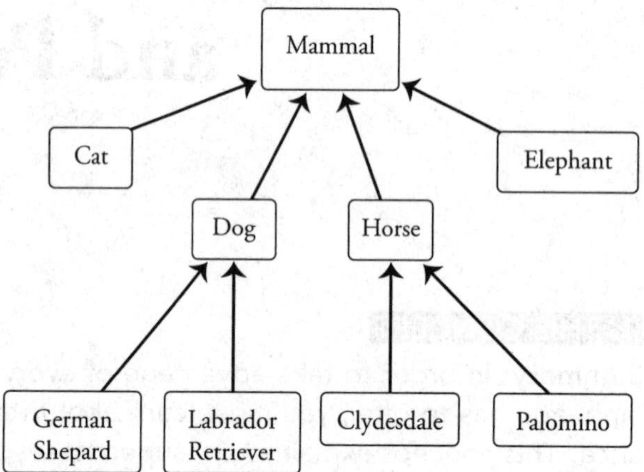

Parent Versus Child Classes (Superclass Versus Subclass)

So what does this have to do with computer programming? Well, in Java, a class has the ability to extend another class. That is, a **child** class (**subclass**) has the ability to extend a **parent** class (**superclass**). When a child class extends a parent class, it inherits the instance variables and methods of the parent class. This means that even though the child class doesn't define these instance variables or methods on its own, it has them. This is a powerful way to reuse code without having to rewrite it.

Note: The methods from a parent class are inherited and can be used by the child class. However, the child class gets its *own set* of instance variables that are separate from the parent class. The instance variables for the parent and the instance variables for the child are two different sets of instance variables. The child class can only access the private instance variables of the parent by using the accessor and mutator methods from the parent class.

Something for Nothing

Child classes inherit the methods of their parent class. This means that the child objects automatically get these methods without having to define them for themselves.

The Keyword `extends`

The words **superclass** and **parent** are interchangeable as are **subclass** and **child**. We say, "a child class **extends** a parent class" or "a subclass is derived from a superclass." The phrase **is-a** refers to class hierarchy. So when we refer to classes such as Dog or Mammal, we say "a Dog is-a Mammal" or even "a GermanShepherd is-a Mammal."

A class can extend *at most one other class*; however, there is no limit to how many classes can exist in a class hierarchy.

The Keyword `extends`
The keyword **extends** is used to identify that inheritance is taking place. A child class can only extend one parent class.

Example

Simulate class hierarchy and inheritance using the Mammal hierarchy. Each subclass inherits the traits of its superclass and can add additional traits of its own.

```
public class Mammal
{
    // All mammals are warm blooded
    // All mammals have lungs to breath air
}
```

```
public class Dog extends Mammal
{
    // The Dog Class is a child class of the Mammal Class
    // A Dog is-a Mammal
    // All Dogs are warm blooded (inherited)
    // All Dogs have lungs to breath air (inherited)
    // Dogs are domesticated (unique trait of a Dog)
}
```

```
public class LabradorRetriever extends Dog
{
    // The LabradorRetriever Class is a direct child class of the Dog Class
    // A LabradorRetriever is-a Dog
    // A LabradorRetriver is-a Mammal
    // LabradorRetriever's are warm blooded (inherited)
    // LabradorRetriever's have lungs to breath air (inherited)
    // LabradorRetrievers are domesticated (inherited)
    // LabradorRetrievers are cute (unique trait of LabradorRetrievers)
}
```

Be Your Own Class
Child classes are allowed to define additional methods of their own. The total number of methods that the child class has is a combination of both the parent class and child class methods.

The Keyword super

When you make an object from a child class, the child class constructor automatically calls the no-argument constructor of the parent class using the **super()** call (some integrated development environments, or IDEs, display this instruction). However, if you want to call a parent constructor with arguments from the child class constructor, you need to put that in your code explicitly as the first line of code in the child class constructor using the **super(arguments)** instruction. The child is then making a call to a parameterized constructor of the parent class.

Example

Demonstrate a child class that has two constructors. The no-parameter constructor makes a call to the parent's no-parameter constructor. The parameterized constructor makes a call to the parent's parameterized constructor. Note: The super() call *must* be the first line of code in each of the child's constructors.

```
public class Mammal
{
    boolean vertebrate;                  // instance variable for a Mammal
    boolean milkProducer;                // instance variable for a Mammal
    String hairColor;                    // instance variable for a Mammal

    public Mammal()                      // No parameter Mammal constructor
    {
        vertebrate = true;               // mammals are vertebrates
        milkProducer = true;             // mammals produce milk
    }

    public Mammal(String color)
    {
        vertebrate = true;
        milkProducer = true;
        hairColor = color;               // Assign the hair color
    }

    /**  Assume all accessor methods are declared  */
}
```

```
public class Dog extends Mammal
{
    String name;               // every Dog has a name

    public Dog()
    {
        super();               // This call is made with or without typing it
    }

    public Dog(String hairColor, String nameOfDog)
    {
        super(hairColor);      // the call to the parent must come first!

        name = nameOfDog;      // assign the Dog's name in the Dog constructor
    }

    /**  Assume all accessor methods are declared  */
}
```

```
public static void main(String[] args)
{
    Dog myDog1 = new Dog();
    System.out.println(myDog1.getName());           // will not have a name
    System.out.println(myDog1.getHairColor());      // will not have hair color
    System.out.println(myDog1.isVertebrate());      // this is true
    System.out.println(myDog1.isMilkProducer());    // this is true
    System.out.println();

    Dog myDog2 = new Dog("Brown", "Bella");
    System.out.println(myDog2.getName());           // the name is Bella
    System.out.println(myDog2.getHairColor());      // the color is Brown
    System.out.println(myDog2.isVertebrate());      // this is true
    System.out.println(myDog2.isMilkProducer());    // this is true
}
```

```
OUTPUT
null
null
true
true

Bella
Brown
true
true
```

Data Encapsulation Within Class Hierarchy

A subclass can only access or modify the private instance variables of its parent class *by using the accessor and modifier methods* of the parent class.

Children Don't Have Direct Access to Their Parent's `private` Instance Variables
Parent and child classes may both contain instance variables. These instance variables should be declared private. This is data encapsulation. Children have to use the accessor and modifier methods of the parent class to access their parent's private instance variables.

Polymorphism

"Sometimes children want to act in a different way from their parents."

Overriding a Method of the Parent Class

In Java, a child class is allowed to **override** a method of a parent class. This means that even though the child inherited a certain way to do something from its parent, it can do it in a different way if it wants to. This is an example of **polymorphism**.

The way to make a child override a parent method is to redefine a method (the signature must be exactly the same) and change what is done in the method. When an object from the child class calls the method, it will perform the action that is found in the child class rather than the method of the parent class.

Example

All objects from the FamilyMember class drink from a cup and eat with a fork. However, objects from the Baby class drink from a cup and eat with their hands. All objects from the SporkUser class drink from a cup and eat with a spork. In this case, the eat method of the FamilyMember class is overridden by both the Baby class and the SporkUser class. Therefore, even though the name of the method is the same, the objects from Baby class and SporkUser class eat in a different way than objects from the FamilyMember class. The decision of when to use the overridden method by all objects that extend a superclass is made when the program is compiled and this process is called **static binding**.

```
public class FamilyMember // This is the superclass
{
   public String drink()
   {
      return "cup";       // the way that all FamilyMember objects drink
   }

   public String eat()
   {
      return "fork";      // the default way that all FamilyMember's eat
   }
}
```

```
public class Baby extends FamilyMember
{
   public String eat()        // Baby is overriding the eat() method
   {
      return "hands";         // all Baby objects eat with their hands
   }
}
```

```
public class SporkUser extends FamilyMember
{
   public String eat()   // SporkUser is overriding the eat() method
   {
      return "spork";    // all SporkUser objects eat with a spork
   }
}
```

```
public static void main(String[] args)
{
    FamilyMember mom = new FamilyMember();
    Baby junior = new Baby();
    SporkUser auntSue = new SporkUser();

    System.out.println(mom.drink());          // mom drinks with a cup
    System.out.println(junior.drink());       // junior drinks with a cup
    System.out.println(auntSue.drink());      // auntSue drinks with a cup

    System.out.println(mom.eat());            // mom eats with a fork
    System.out.println(junior.eat());         // junior eats with his hands
    System.out.println(auntSue.eat());        // auntSue eats with a spork
}
```

```
OUTPUT
cup
cup
cup
fork
hands
spork
```

Polymorphism
Java uses the term **polymorphism** to describe how different child classes of the same parent class can act differently. This is accomplished by having the child class override the methods of the parent class.

Dynamic Binding

A problem arises when we want to make a list of family members. *To make an array or ArrayList of people in the family, they all have to be of the same type!* To solve this problem, we make the reference variables all of the *parent* type.

Example

Create a FamilyMember ArrayList to demonstrate dynamic binding. Make the reference variables all the same type (the name of the parent class) and make the objects from the child classes. Allow each object to eat and drink the way they were designed to eat and drink. The decision of how each FamilyMember object eats is decided while the program is running (known as **run-time**) and this process is called **dynamic binding**. It's pretty cool!

```
public static void main(String[] args)
{
    FamilyMember mom = new FamilyMember();
    FamilyMember junior = new Baby();
    FamilyMember auntSue = new SporkUser();

    // Make a list of family members
    List<FamilyMember> family = new ArrayList<FamilyMember>();

    family.add(mom);
    family.add(junior);
    family.add(auntSue);

    // Demonstrate polymorphism by allowing everyone to eat their way
    for (FamilyMember member : family)
    {
        System.out.println(member.drink()); // everyone drinks with a cup
        System.out.println(member.eat());   // everyone eats their way
        System.out.println();
    }
}
```

```
OUTPUT
cup
fork

cup
hands

cup
spork
```

Reference Variables and Hierarchy

The reference variable *can* be the parent of a child object.

```
FamilyMember somebody = new Baby();    // Legal
```

The reference variable *cannot* be a child of a parent object.

```
Baby somebody = new FamilyMember();    // Illegal:  Type Mismatch error
```

Using super from a Child Method

Even if a child class overrides a method from the parent class, an object from a child class can call its parent's method if it wants to; however, it must do it from the overriding method. By using the keyword **super**, a child can make a call to the parent method that it has already overridden. We say, "The child can invoke the parent's method if it wants to."

Example

Revise the Baby class so that if a baby is older than three years, it eats with either its hands or a fork; otherwise it eats with its hands. Use the keyword **super** to make a call to the parent's eat method. Also, add an instance variable for the age.

```
public class Baby extends FamilyMember
{
   private int age;          // every Baby has an age

   public Baby(int myAge)
   {
      age = myAge;           // assign the age to the Baby
   }

   public String eat()
   {
      if (age > 3)           // older baby may eat with hands or a fork
         return "hands or a " + super.eat();
      else
         return "hands";
   }
}
```

```
public static void main(String[] args)
{
   Baby youngBaby = new Baby(2);           // youngBaby is 2 years old
   Baby olderBaby = new Baby(4);           // olderBaby is 4 years old

   System.out.println(youngBaby.eat());   // youngBaby uses only hands
   System.out.println(olderBaby.eat());   // olderBaby uses hands or a fork
}
```

```
OUTPUT
hands
hands or a fork
```

Beware When Using a Parent Reference Variable for a Child Object

A child object has the ability to perform all the public methods of its parent class. However, a parent does not have the ability to do what the child does if the child action is original. Also, if a child object has a parent reference variable, the child *does not* have the ability to perform any of the unique methods that are defined in its own class.

Example

Demonstrate that a parent reference variable cannot perform any of the methods that are unique to a child class. A FamilyMember object does not know how to throwTantrum. Only a Baby object knows how to throwTantrum. Note: Even though babyCousin is a Baby object, the FamilyMember reference variable prevents it from being able to throwTantrum.

```
public class Baby extends FamilyMember
{
    public String throwTantrum()      // A Baby's impressive new ability
    {
        return "scream loudly";
    }

    /*  all other implementation is not shown  */
}
```

```
public static void main(String[] args)
{
    Baby junior = new Baby();                 // reference is a Baby
    FamilyMember babyCousin = new Baby(); // reference is a FamilyMember

    System.out.println(junior.throwTantrum());       // legal
    System.out.println(babyCousin.throwTantrum());   // illegal
}
```

```
OUTPUT
COMPILE TIME ERROR:  the method, throwTantrum is undefined for FamilyMember
```

Downcasting

The problem described in the previous example can be solved using a technique called **downcasting**. This is similar to—but not exactly the same as—casting an int to a double. A parent object reference variable is cast to the object type of the object that it references.

Downcasting

You can downcast a parent reference variable to a child object as long as the object it is referring to is that child.

```
FamilyMember somebody = new Baby();  // Parent reference and child object
((Baby) somebody).uniqueMethodForBaby(); // Correct way to downcast
```

Example 1

Demonstrate downcasting by allowing the parent reference variable to perform a method that is unique to a child class. The parent reference variable is cast as a child and must refer to the child object. Pay close attention to the parentheses when casting.

```
public static void main(String[] args)
{
   Baby junior = new Baby();                   // reference is a Baby
   FamilyMember babyCousin = new Baby(); // reference is a FamilyMember

   System.out.println(junior.throwTantrum());              // legal
   System.out.println(((Baby) babyCousin).throwTantrum());  // legal
}
```

```
OUTPUT
scream loudly
scream loudly
```

Example 2

You *are not allowed* to cast a parent reference object to a child if the object is a parent. In this example, mom is a FamilyMember reference variable that refers to a FamilyMember object. Attempting to cast mom as a Baby results in an error.

```
FamilyMember mom = new FamilyMember();
((Baby) mom).eat(); // ClassCastException:FamilyMember cannot be cast to Baby
```

The `Object` Class

The **Object class** is the mother of all classes. Every class, even the ones you write, is a descendent of the Object class. The extends is *implied*, so you don't have to write, "extends Object", in your own classes. The Object class does not have any instance variables, but it does have several methods that every descendent inherits. Two of these methods, the **toString** and the **equals**, are tested on the AP Computer Science A Exam. You must be able to implement the toString method; however, you only need to know how to use the equals method (like we do when we compare String objects).

Overriding the `toString` Method

The **toString** method is inherited from the Object class and was created to describe the object in some way, typically by referencing its instance variables. The Object class's toString method doesn't give you anything valuable (it returns the hex value of the object's address). Most of the time, when creating your own classes, you will override the toString method so it returns a String that describes the object in a meaningful way.

Example

Override the toString method for the Circle class (used in Concept 2) so the method gives a description of the Circle object that calls it. The String that is returned should include the radius of the circle.

```
public class Circle
{
   /*  All other implementation not shown  */

   public String toString()              // Override the toString method
   {
      return "Circle with a radius of " + radius;
   }
}
```

```
public static void main(String[] args)
{
   Circle circle = new Circle(5);            // Circle with a radius of 5
   System.out.println(circle.toString()); // call the toString() method
}
```

```
OUTPUT
Circle with a radius of 5.0
```

Overriding the `toString` Method
All classes extend the Object class. When you design your own classes, *be sure to override the toString() method* of the Object class so that it describes objects from your class in a unique way using its own instance variables.

Casting an `Object`

If a method's parameter list contains an Object reference variable, then you will have to downcast the reference before using it. You must make sure that the object that you are casting **is-a** object from the class.

Example
Demonstrate how to handle an Object reference parameter. In this example, when the FamilyMember object, uncleDon, gets passed to the timeToEat method, he is sent as a FamilyMember reference variable. When he is received by the timeToEat method, he is renamed as hungryMember and is now an Object reference. Finally, he is cast to a FamilyMember and is then able to eat.

```
public static void main(String[] args)
{
    FamilyMember uncleDon = new FamilyMember();
    timeToEat(uncleDon);
}

public static void timeToEat(Object hungryMember)
{
    ((FamilyMember) hungryMember).eat();  // casting Object to FamilyMember

    // Even though uncleDon is an Object reference, he can eat like a
    // FamilyMember because he was downcast to a FamilyMember
}
```

```
OUTPUT
fork
```

❯ Rapid Review

Hierarchy and Inheritance

- One major advantage of an object-oriented language is the ability to create a hierarchy of classes that allows objects from one class to inherit the features from another class.
- A class hierarchy is formed when one class extends another class.
- A class that extends another class is called the child or subclass.
- The class that is extended is called the parent or superclass.
- A class can only extend one other class.
- A class hierarchy can extend as many times as you want. That is, a child can extend a parent, then the parent can extend another parent, and so on.
- When a class hierarchy gets extended to more than just one parent and one child, the lower ends of the hierarchy inherit the features of each of the parent classes above it.
- The phrase "is-a" helps describe what parent classes a child class belongs to.
- If you want a child class to modify a parent class's method, then you override the parent class's method. To do this, you simply redefine the method in the child class (using the exact same method declaration and replace the code with what you want the child to do).
- The word "super" can be used to refer to either a parent's constructor or methods.
- To purposefully call the no-argument super constructor, you must put super() as the first line of code in the child's constructor. Other instructions are placed after the super() call.
- To call the parameterized super constructor, you must put super(arguments) as the first line of code in the child's parameterized constructor. Other instructions are placed after the super() call.
- If you write a method that overrides the parent class's method, you can still call the original parent's method if you want to. Example: super.parentMethod().
- A reference variable can be a parent class type if the object being created is its child class type.
- A reference variable cannot be a child class type if the object being created is its parent class type.

- A reference variable that is of a parent data type can be cast to a child data type as long as the object that it is referencing is a child object of that type.

Polymorphism

- Polymorphism is when more than one child class extends the same parent class and each child object is allowed to act in a different way than its parent and even its siblings.
- If several child classes extend the same parent class and each child class overrides the same method of the parent class in its own unique way, then each child will act differently for that method.
- Static binding is the process of deciding which method the child reference variable will perform during compile-time.
- Dynamic binding is the process of deciding which method the parent reference will perform during run-time.
- The Object class is the mother of all classes in Java since it is the root of the class hierarchy and every class has Object as a superclass.
- The Object class contains two methods that are tested on the AP exam: equals and toString.
- The public boolean equals(Object other) method returns true if the object reference variable that calls it is referencing the same exact object as other. It returns false if the object reference variables are not referencing the same object.
- It is common to override the equals method when you want to determine if two objects from a class are the same.
- The public String toString() method returns a String representation of the object.
- It is very common to override the toString() method when generating your own classes.
- Downcasting is casting a reference variable of a parent class to one of its child classes.
- Downcasting is commonly used when an Object is in a parameter list or ArrayList.
- A major problem with downcasting is that the compiler will not catch any errors, and if the cast is made incorrectly, a run-time error will be thrown.

› Review Questions

Basic Level

1. Assuming all classes are defined in a manner appropriate to their names, which of the following statements will cause a compile-time error?

(A) `Object obj = new Lunch();`
(B) `Lunch lunch = new Lunch();`
(C) `Lunch sandwich = new Sandwich();`
(D) `Lunch lunch = new Object();`
(E) `Sandwich sandwich = new Sandwich();`

Questions 2–4 refer to the following classes.

```
public class Student
{
   public int gradYear;
   private double gpa;

    /*  other implementation not shown  */
}

public class UnderClassman extends Student
{
   public int homeRoomNum;
   private String counselor;

    /*  other implementation not shown  */
}

public class Freshman extends UnderClassman
{
   private int middleSchoolCode;

    /*  other implementation not shown  */
}
```

2. Which of the following is true with respect to the classes defined above?

I. `Freshman` is a subclass of `UnderClassman` and `UnderClassman` is a subclass of `Student`
II. `Student` is a superclass of both `UnderClassman` and `Freshman`
III. `UnderClassman` is a superclass of `Freshman` and a subclass of `Student`

(A) I only
(B) II only
(C) I and II only
(D) II and III only
(E) I, II, and III

3. Which of the following lists of instance data contains only variables that are directly accessible to a `Freshman` class object?

(A) `middleSchoolCode, homeRoomNum, gpa, gradYear`
(B) `middleSchoolCode, counselor, homeRoomNum`
(C) `homeRoomNum, counselor, gradYear`
(D) `middleSchoolCode, counselor, gpa`
(E) `middleSchoolCode, homeRoomNum, gradYear`

4. Assume that all three classes contain parameterized constructors with the following declarations and that all variables are logically named.

```
public Student (int gradYear, double gpa)

public UnderClassman (int gradYear, double gpa, int homeRoomNum, String counselor)

public Freshman (int gradYear, double gpa, int homeRoomNum, String counselor,
        int middleSchoolCode)
```

Which of the following calls to `super` would appear in the constructor for the `Freshman` class?

(A) `super(gradYear, gpa, homeRoomNum, counselor, middleSchoolCode);`
(B) `super(gradYear, gpa, homeRoomNum, counselor);`
(C) `super(homeRoomNum, counselor);`
(D) `super(gradYear, gpa);`
(E) `super();`

5. Consider the following method in the `Student` class.

```
public void attendClass(int roomNumber, String subject)
{
   /* implementation not shown */
}
```

A `Freshman` object must `attendClass` like a `Student` object, but also have the teacher initial his/her homework planner (using the `initialPlanner` method, not shown).

Which of the following shows a possible implementation of the `Freshman` `attendClass` method?

(A)
```
public void attendClass extends Student(int roomNumber, String subject)
{
    super.attendClass(int roomNumber, String subject);
    initialPlanner();
}
```

(B)
```
public void attendClass extends Student(int roomNumber, String subject)
{
    super.attendClass();
    initialPlanner();
}
```

(C)
```
public void attendClass(int roomNumber, String  subject)
{
    super(int roomNumber, String subject);
    initialPlanner();
}
```

(D)
```
public void attendClass(int roomNumber, String  subject)
{
    super(roomNumber, subject);
    initialPlanner();
}
```

(E)
```
public void attendClass(int roomNumber, String  subject)
{
    super.roomNumber = roomNumber;
    super.subject = subject;
    initialPlanner();
}
```

Advanced Level

Questions 6–7 refer to the following classes.

```
public class Club
{
    private ArrayList<String> members;

    public Club() {  }

    public Club(ArrayList<String> theMembers)
    {  members = theMembers;  }

    /* Additional implementation not shown */
}

public class SchoolClub extends Club
{
    private String advisor;

    public SchoolClub(String theAdvisor, ArrayList<String> theMembers)
    {
        /* missing code */
    }

    /* Additional implementation not shown */
}
```

Assume `ArrayList<String> members_1` and `ArrayList<String> members_2` have been properly instantiated and initialized with a non-zero number of appropriate elements.

6. Which of the following declarations is NOT valid?

(A) `Club[] clubs = new SchoolClub[7];`

(B) `Club[] clubs = {new Club(members_1), new Club()};`

(C) `SchoolClub[] clubs = {new SchoolClub("Mr. Johnson", members_1),`
`new Club(members_2)};`

(D) `SchoolClub[] clubs = {new SchoolClub("Ms. Paymer", members_1),`
`new SchoolClub("Mr. Johnson", members_2)};`

(E) All of the above are valid.

7. Which of the following could replace `/* missing code */` in the `SchoolClub` constructor to ensure that all instance variables are initialized correctly?

(A)
```
super(theMembers);
advisor = theAdvisor;
```

(B)
```
super(new Club(theMembers));
advisor = theAdvisor;
```

(C)
```
this.SchoolClub = new Club(members);
advisor = theAdvisor;
```

(D)
```
advisor = theAdvisor;
```

(E)
```
advisor = theAdvisor;
super(theMembers);
```

Questions 8–9 refer to the following classes.

```
public class Present
{
    private String contents;

    public Present(String theContents)
    {   contents = theContents;   }

    public String getContents()
    {   return contents;   }

    public void setContents(String theContents)
    {   contents = theContents;   }

    public String toString()
    {   return "contains " + getContents();   }
}

public class KidsPresent extends Present
{
    private int age;

    public KidsPresent(String contents, int theAge)
    {
        super(contents);
        age = theAge;
    }

    public int getAge()
    {   return age;   }

    public String toString()
    {   return super.toString() + " for a child age " + getAge();   }
}
```

8. Consider the following code segment.

```
Present p = new KidsPresent("kazoo", 4);
System.out.println(p);
```

What is printed as a result of executing the code segment?

(A) `contains kazoo`
(B) `kazoo`
(C) `contains kazoo for a child age 4`
(D) Nothing is printed. Compile-time error: illegal cast
(E) Nothing is printed. Run-time error: illegal cast

9. Consider the following code segment.

```
p.setContents("blocks");
System.out.println(p);
```

What is printed as a result of executing the code segment?

(A) `contains blocks`
(B) `contains blocks for a child age 4`
(C) `contains kazoo for a child age 4`
(D) Nothing is printed. Compile-time error: there is no `setContents` method in the `KidsPresent` class
(E) Nothing is printed. Run-time error: there is no `setContents` method in the `KidsPresent` class

10. Free-Response Practice: Point3D Class

Consider the following class:

```
public class Point
{
    private int x;
    private int y;

    public Point(int newX, int newY)
    {
        x = newX;
        y = newY;
    }

    public int getX()
    {  return x;  }

    public int getY()
    {  return y;  }
}
```

Write the class `Point3D` that extends this class to represent a three-dimensional point with x, y, and z coordinates. Be sure to use appropriate inheritance. Do not duplicate code.

You will need to write:

a. A constructor that takes three int values representing the x, y, and z coordinates

b. Appropriate mutator and accessor methods (setters and getters)

› Answers and Explanations

Bullets mark each step in the process of arriving at the correct solution.

1. The answer is D.

- When you instantiate an object, you tell the compiler what type it is (left side of the equal sign) and then you call the constructor for a specific type (right side of the equal sign). Here's where you can put *is-a* to good use. You can construct an object as long as it *is-a* version of the type you have declared. You can say, "Make me a sandwich, and I'll say thanks for lunch," and you will always be right, but you can't say, "Make me lunch and I'll say thanks for the sandwich." What if I made soup?
- Option A is correct, since Lunch *is-a* Object.
- Options B and E are correct since Lunch *is-a* Lunch, and Sandwich *is-a* Sandwich. (Don't confuse the variable name sandwich with the type Lunch. It's a bad programming choice, since a reader expects that a variable named sandwich is an object of the class Sandwich, but it's legal.)
- Option C is correct since Sandwich *is-a* Lunch.
- Option D is incorrect. Object is not a Lunch. It's not reversible. The one on the right has to have an *is-a* relationship to the one on the left in that order. Option D will not compile.

2. The answer is E.

- Superclass refers to grandparents as well as parents, all the way up the inheritance line, and subclass refers to grandchildren as well as children, all the way down the inheritance line, so all three relationships are true.

3. The answer is E.

- A Freshman object will have direct access to its own public and private variables, and the public variables of its superclasses.
- Please note that it is considered a violation of encapsulation to declare instance variables public, and points may be taken off on the AP exam if you do so!

4. The answer is B.

- Since the super call must be the first thing in the constructor, the Freshman constructor will hang on to the parameter middleSchoolCode and use the super call to pass the remaining parameters to the constructor for Underclassman. It will then set its own instance variable to the middleSchoolCode parameter. (It's worth noting that the Underclassman constructor will set its own two variables and pass the remaining variables to its super constructor.)

5. The answer is D.

- Options A and B are incorrect. They show an incorrect usage of "extends". The keyword extends belongs on the class declaration, not the method declaration.
- Option C is incorrect. It puts data types next to the arguments (actual parameters) in the super method call. Types only appear next to the formal parameters in the method declaration.
- Option D is correct. It has fixed both of the above errors. The method with the same name in the superclass is called using the super.methodName notation, and it is passed the two parameters it is expecting. Then the initialPlanner method is called.
- Option E is incorrect. It attempts to assign the parameters directly to their corresponding variables in the Student class. The variables in the parent class should be private, so the child doesn't know if they exist, and can't access them even if they do. Also, the child class shouldn't assume that assigning variables is all the parent class method does. Surely attendClass involves more than that!

6. The answer is C.

- Option A is valid. It instantiates an array of Club objects. The right side of the instantiation specifies that there will be two entries, but it does not instantiate those entries. This is a valid statement, but note that the entries in this array are null.
- Option B is valid. It uses an initialization list to instantiate the array of Club objects. Club has overloaded constructors. There is a constructor that takes an ArrayList<String> as an argument, so the first entry is correct, and there is a no-argument constructor, so the second entry is correct.
- Option C is not valid. It uses an initialization list to instantiate an array of SchoolClub objects, but the list includes not only a SchoolClub object, but also a Club object. We could make a Club[] and include SchoolClub objects, but not the other way around. Remember the *is-a* designation. A Club is not a SchoolClub.
- Option D is valid. It instantiates an array of SchoolClub objects by giving an initialization list of correctly instantiated SchoolClub objects.

7. The answer is A.

- The first statement in the constructor of an extended class is a super call to the constructor of the parent class. This may be an implicit call to the no-argument constructor (not explicitly written by you) or it may be an explicit call to any of the superclass's constructors.
- Option B is incorrect. It calls the super constructor but passes a parameter of the wrong type.
- Option C is incorrect. this.SchoolClub has no meaning.
- Option D is incorrect. This option is syntactically correct. The no-argument constructor will be called implicitly, but the members ArrayList instance variable will not be set.
- Option E is incorrect. The super call must be the first statement in the constructor.

8. The answer is C.

- First of all, remember that when an object is printed, Java looks for that object's toString method to find out how to print the object. (If there is no toString method, the memory address of the object is printed, which is not what you want!)
- Variable p is of type Present. That's what the compiler knows. But when we instantiate variable p, we make is a KidsPresent. That's what the runtime environment knows. So when the runtime environment looks for a toString method, it looks in the KidsPresent class. The KidsPresent toString prints both the contents and the appropriate age.

9. The answer is B.

- Since p is declared as a Present, and the Present class has a setContents method, the compiler is fine with this sequence.
- When setContents is called, the run-time environment will look first in the KidsPresent class and it will not find a setContents method. But it doesn't give up! Next it looks in the parent class, which is Present and there it is. So the run-time environment happily uses the Present class's setContents method and successfully sets the Present class's contents variable to "blocks".
- When the KidsPresent toString is called, the first thing it does is call super.toString(), which accesses the value of the contents variable. Since that variable now equals "blocks", the code segment works as you might hope it would and prints "contains blocks for a child age 4".

10. On the AP exam, you are often asked to extend a class, as you were in this question. Check your answer against my version below. A few things to notice:

- The constructor takes x, y, and z but then immediately passes x and y to the Point class constructor through the super(x, y) call.
- x and y, their setters and getters are in the Point class; they don't need to be duplicated here. Only variable z, setZ and getZ need to be added.

```
public class Point3D extends Point
{
    private int z;

    // Part a
    public Point3D(int myX, int myY, int myZ)
    {
        super(myX, myY);
        z = myZ;
    }

    // Part b
    public void setZ(int myZ)
    {
        z = myZ;
    }

    public int getZ()
    {
        return z;
    }
}
```

The abstract class and the interface

IN THIS CONCEPT

Summary: This concept contains two advanced features of the Java language: abstract classes and interfaces. These two concepts are similar in some respects and different in others. They are both used extensively in high-level software designs.

Key Ideas

✪ Abstract classes and interfaces are similar in that:
- They both contain method(s) without implementation.
- They cannot be instantiated.
- They work well with polymorphism.
- You can make reference variables that refer to them.

✪ Abstract classes and interfaces are different in that:
- An abstract class may contain instance variables.
- An abstract class may contain methods that do provide implementation.
- An abstract class must be *extended* by another class to complete the implementation.
- An interface cannot contain instance variables.
- An interface cannot contain methods that provide implementation.
- An interface must be *implemented* by another class to complete the implementation.

The abstract class

Definition of an abstract class

I had a high school coach who told his athletes what to do at practice during the week. However, on the weekends, he told us to work out but he didn't tell us what kind of workout to do. So, how we worked out varied from person to person. Some kids played basketball; some worked out in the gym; others played video games. He told us that we had to work out but didn't tell us how. My coach's instructions were like an abstract method in an abstract class.

An **abstract class** is similar to the Java classes that we have learned about so far, but it is also different in several ways. An abstract class must have the keyword **abstract** in its class declaration. It does not contain any constructors but should contain **abstract methods**. Abstract methods are methods that have the keyword abstract in the method declaration but *do not contain any code*. The fancy way of saying this is, "abstract methods do not contain any implementation." Only the method declaration (followed by a semicolon) is stated; the method does not have a method body.

The fact that an abstract class does not contain any constructors poses a problem that we've never seen before. How can you possibly create an object from a class that doesn't contain a constructor? The answer is, you can't.

So What's the Point of an abstract class?

The answer lies in design. In the class hierarchy that we've studied so far, a subclass may override a superclass's method if it wants to. In an abstract class, the subclass *must override the methods that are labeled abstract.* Typically, you choose to create an abstract class when you do not want to define a generic way to carry out a method, therefore leaving the job of implementing the method up to the child class. In the example above, the coach gave specific directions during the week and the athletes had to follow the directions. However, on the weekends, the coach gave very vague directions and the athletes got to carry out the directions any way they wanted to. An abstract class is used in software design when you want a parent class to contain some implemented methods but you also want it to contain some unimplemented methods. The abstract class requires any subclass of the abstract class to carry out the implementation of the abstract methods.

Declaring an abstract class

The keyword abstract must be included in the **class declaration**. Also, the methods that do not contain any implementation must have the keyword abstract in their declarations.

General Form for Declaring an abstract class with One Abstract Method

```
public abstract class NameOfAbstractClass
{
   accessModifier abstract returnType nameOfAbstractMethod();    // No body
}
```

General Form for Declaring a Class That `extends` an `abstract class`

```
public class NameOfSubClass extends NameOfAbstractClass()
{
    accessModifier returnType nameOfAbstractMethod()
    {
        // code that carries out the abstract method
    }
}
```

Some of my methods have implementations
and some of them don't. Since I don't really
want to make a generic way to do something
(that you are going to override anyway),
I'm going to make those methods abstract.
You figure out how to implement them.
abstract class

Who Has to Implement the Code for an `abstract class`?

Any class that extends an abstract class must make sure that all abstract methods are implemented.

However, it is possible for one abstract class to extend another abstract class. If this is the case, then the first subclass that is *not* an abstract class must make sure that all abstract methods are implemented.

Example

Suppose you have been hired by a video game company to design a game in which characters in the game move around from one place to another. Each character will extend a superclass called GameCharacter. Each character will move in its own individual way (some fly, some swim, some teleport, etc.); there is no generic way that they all move. The runner class will call a move method that will allow all of the characters to move.

Solution: Use an Abstract Class

Choose to make the *GameCharacter* superclass an abstract class that includes an abstract method called *move*. The move method does not have a method body, so there is no generic way for each character to move. When each subclass extends the abstract class, they will be forced to implement the move method. Then, you can make a collection of all of the characters and whenever you tell each character to move, they will move in the way that they know how. Polymorphism at work!

```
public abstract class GameCharacter       // abstract class
{
   public String sayHello()  // non-abstract method in an abstract class
   {
      return "Hello";        // Every GameCharacter object can say Hello
   }

   public abstract String move();         // no generic way to move
}
```

```
public class Bird extends GameCharacter    // subclass of abstract class
{
   public String move()
   {
      return "I fly";   // The Bird's implementation of the move method
   }
}
```

```
public class Fish extends GameCharacter      // subclass of abstract class
{
   public String move()
   {
      return "I swim"; // The Fish's implementation of the move method
   }
}
```

```
public static void main(String[] args)
{
   // notice that the object references are the abstract class
   // and the objects are the subclasses
   GameCharacter character1 = new Bird();
   GameCharacter character2 = new Fish();

   // The ArrayList is a list of the abstract class references
   List<GameCharacter> team = new ArrayList<GameCharacter>();

   team.add(character1);
   team.add(character2);

   for (GameCharacter character : team)
   {
      System.out.println(character.move());  // each moves its own way
   }
}
```

```
OUTPUT
I fly
I swim
```

An `abstract class` Cannot Be Instantiated

You cannot make an object using an abstract class.

```
GameCharacter character = new GameCharacter();
// Error:  GameCharacter cannot be instantiated because it is Abstract
```

You can, however, make an object reference variable of the abstract class type.

```
GameCharacter character = new Bird();        // Legal declaration
```

Summary: `abstract class`

- An abstract class
  - must include the keyword abstract in the class declaration
  - should include at least one method declaration that has the keyword abstract in its method signature and does not contain implementation
  - cannot be instantiated
  - can include instance variables and implemented methods
- A subclass that extends the abstract class must implement the abstract methods from the abstract class (unless the subclass is also an abstract class!).
- A subclass can only extend one abstract class.
- An abstract class works really well for designs that include polymorphism.

The `interface`

Definition of an `interface`

The **interface** is a Java feature that is similar in some respects to the abstract class. The main difference between the two is that an interface *only* contains method declarations (without any implementation). The classes that contain the implementation for the interface use the keyword **implements** in their class declaration.

An interface is *not a class* but rather a **reference type**. Even though an interface is not a class, it can be a reference variable for an object of the class that implements it. You cannot make an object from an interface; however, you can make a reference variable of the interface type. Finally, unlike extending, *a class can implement more than one interface.*

An Interpretation of the English Word *Interface*

If you think of the use of the word *interface* in the English language, it may help to explain what a Java interface is. For example, a video service like Netflix streams movies to its users. Based on the device that the user is using, the *interface* for Netflix may look different. The Netflix screen on a laptop may look different than the Netflix screen on a smartphone, even though the movie is streaming to it in the same way! Each device has to have different code that takes in the video and displays it so the user can see it the way it should be seen on that particular device.

In this diagram, Netflix streams a movie and tells the devices that they have to figure out how to display it. The software in the devices must implement the code that actually displays the movie since the Netflix interface doesn't supply it.

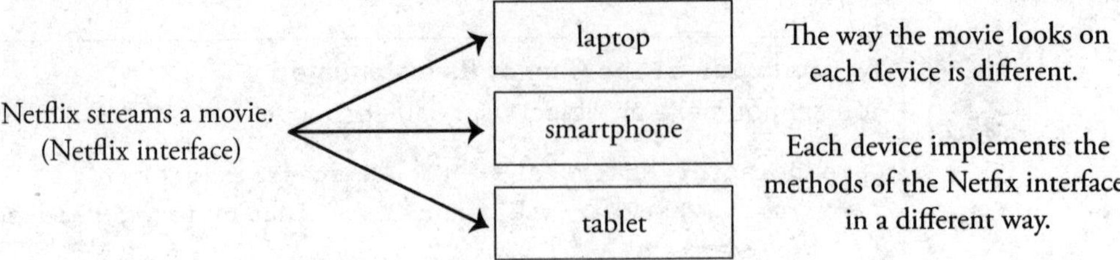
Netflix streams a movie.
(Netflix interface)
laptop
smartphone
tablet
The way the movie looks on each device is different.
Each device implements the methods of the Netfix interface in a different way.

Declaring an `interface`

The interface declaration contains the keyword interface and *does not contain* the keyword class. The class that implements the interface is responsible for carrying out the code for the methods defined in the interface.

General Form for Declaring an `interface` and a Subclass That Implements It

```
public interface nameOfInterface
{
   returnType methodName();   // Only the method declaration is provided
}
public class SomeClass implements nameOfInterface
{
   public returnType methodName()
   {
      // Code that carries out method is placed here
   }
}
```

Hey, you signed a contract with me.
You must write the code for
every one of my method declarations!
interface

Example

Suppose you have been hired by Netflix to write code for devices to play their streamed movies. You decide to use an interface design structure. The interface will have three methods: play, pause, and rewind. Since the hardware is different for different devices, each device has to have particular software that needs to be written specifically for the device. The important thing to remember is that Netflix simply streams the video and doesn't tell the devices how to handle the play, pause, and rewind features. Each device must implement the methods for playing, pausing, and rewinding in the way that is specific to it. Design an interface and two classes that implement the interface, one for a smartphone and one for a laptop.

```
public interface NetflixPlayer
{
   String play();        // No generic way to play for every device
   String pause();       // No generic way to pause for every device
   String rewind();      // No generic way to rewind for every device
}
```

```
public class SmartPhone implements NetflixPlayer
{
   /* Actual implementation not shown!  */

   public String play()
   {
      return "Pressed play on a SmartPhone.";
   }

   public String pause()
   {
      return "Pressed pause on a SmartPhone.";
   }

   public String rewind()
   {
      return "Pressed rewind on a SmartPhone.";
   }
}
```

```
public class Laptop implements NetflixPlayer
{
   /* Actual implementation not shown!  */

   public String play()
   {
      return "Pressed play on a Laptop.";
   }

   public String pause()
   {
      return "Pressed pause on a Laptop.";
   }

   public String rewind()
   {
      return "Pressed rewind on a Laptop.";
   }
}
```

```
public static void main(String[] args)
{
    NetflixPlayer device1 = new SmartPhone();
    NetflixPlayer device2 = new Laptop();

    //  ArrayList devices is a list of interface references
    List<NetflixPlayer> devices = new ArrayList<NetflixPlayer>();

    devices.add(device1);
    devices.add(device2);

    for (NetflixPlayer device : devices)
    {
        System.out.println(device.play());      // each device can play
        System.out.println(device.pause());     // each device can pause
        System.out.println(device.rewind());    // each device can rewind
        System.out.println();
    }
}
```

```
OUTPUT
Pressed play on a SmartPhone.
Pressed pause on a SmartPhone.
Pressed rewind on a SmartPhone.

Pressed play on a Laptop.
Pressed pause on a Laptop.
Pressed rewind on a Laptop.
```

A Class Can `implement` More Than One `interface`

A class can only extend one other class; however, a class can implement as many interfaces as it wants to. It just has to implement the code for all the methods in every interface.

Example

Write three different interfaces and have one class implement all three.

```
public interface Interface1
{
    void doSomething();          // Only the method signature is provided
}
```

```
public interface Interface2
{
    void doSomethingElse();     // Only the method signature is provided
}
```

```
public interface Interface3
{
    void doAnotherThing();      // Only the method signature is provided
}
```

```
public class SomeClass implements Interface1, Interface2, Interface3
{
    public void doSomething()
    {
        // Code that carries out the method in Interface1
    }

    public void doSomethingElse()
    {
        // Code that carries out the method in Interface2
    }

    public void doAnotherThing()
    {
        // Code that carries out the method in Interface3
    }
}
```

Summary: `interface`

- An interface
  - must include the keyword interface in its declaration
  - can only contain method declarations (and not the code)
  - cannot include any instance variables or constructors
  - cannot be instantiated (you cannot make an object from an interface)
  - cannot include any implemented methods
- Any subclass that implements an interface must contain the code for the methods of the interface.
- A class can implement more than one interface.
- An interface works really well for designs that include polymorphism.

The `List interface`

The **List interface** is a popular interface in Java because it is used by many complex data structures. As the name suggests, it defines methods for handling items that are stored in some kind of list. Examples of classes that (ultimately) implement the List interface are ArrayList, LinkedList, and Vector. Each of these complex data structures has its own way of performing actions like getting, setting, adding, and removing items from a list. Therefore, the List interface defines these methods. However it does not include the code for carrying out the actions—that is the responsibility of the classes (or subclasses) that implement the interface.

The ArrayList class is the only class that implements the List interface that you are required to know on the AP Computer Science A Exam.

List vs. ArrayList

On the exam, you will see the following two ways to instantiate an ArrayList:

```
ArrayList<String> myWords = new ArrayList<String>();
```

and

```
List<String> myWords = new ArrayList<String>();
```

Don't be confused! These two techniques for making an ArrayList are essentially the same. They both instantiate a new ArrayList object; however, in the first example, the reference variable, myWords, can only hold the address of an ArrayList object, but in the second example, myWords can hold the address of any object from any class that implements the List interface.

› Rapid Review

The `abstract class`

- An abstract class is similar to the classes we've studied so far, with a few exceptions.
- An abstract class is allowed to have unimplemented methods. These methods are declared abstract.
- Abstract classes cannot be instantiated. They must be extended.
- If a class extends an abstract class, then it must implement all the abstract methods from the abstract class.
- If a subclass of an abstract class is also an abstract class, then the first subclass in the hierarchy that is not abstract must implement any remaining abstract methods that have not yet been implemented.
- Abstract classes can have any number of methods or instance variables.
- Technically, an abstract class does not have to have any abstract methods, but if this is the case, then you should just define the class without labeling it abstract.
- You can create a reference variable of an abstract class type; however, you are not allowed to create objects from an abstract class.
- You can make an array or an ArrayList of reference variables whose type is an abstract class.

The `interface`

- An interface contains only method declarations. None of the methods in an interface contain any code.
- Interfaces cannot have any instance variables, constructors, or implemented methods.
- An interface is a contract with the class that implements it. By using the keyword **implements**, the class signs this contract; the class is obligated to implement every method that is defined in the interface.
- A class can implement more than one interface. If a class implements more than one interface, it must contain the implementation for the methods from all of the interfaces.
- You can create a reference variable of an interface type; however, you are not allowed to create new objects from an interface.
- A class that implements an interface can be instantiated.
- You can make an array or an ArrayList of reference variables whose type is an interface.

› Review Questions

Basic Level

1. Consider the following declarations.

```
public abstract class Abs1 { /* implementation not shown */ }

public abstract class Abs2 { /* implementation not shown */ }

public interface Iface1 { /* implementation not shown */ }

public interface Iface2 { /* implementation not shown */ }
```

Which of the following is a valid class declaration?

(A) `public class TryThis extends Iface1`
(B) `public class TryThis implements Abs1`
(C) `public class TryThis extends Abs1, Abs2`
(D) `public class TryThis extends Iface1 implements Abs1, Abs2`
(E) `public class TryThis extends Abs1 implements Iface1, Iface2`

2. Consider the following interface and classes.

```
public interface Decorator()
{
   void consultWithClient();
   void buyDecorations();
   void hangDecorations();
}

public class WeddingDecorator implements Decorator
{
   /* implementation not shown */
}

public class KidsPartyDecorator implements Decorator
{
   /* implementation not shown */
}
```

Which of the following is true of `WeddingDecorator` and `KidsPartyDecorator`?

(A) They have identical copies of the three methods defined in the `Decorator` interface.
(B) They have exactly three methods, as defined in the `Decorator` interface.
(C) They may have multiple methods, which may or may not include the three methods defined in the `Decorator` interface.
(D) They may have multiple methods, but the three methods defined in the `Decorator` interface must be among them.
(E) There is no way to tell anything about their implementation from the information given.

Advanced Level

Questions 3–4 refer to the following classes.

```
public abstract class Alpha
{
   private int value = 4;

   public Alpha()
   {  }

   public abstract String display();

   public void setValue(int n)
   {  value = n;  }

   public int getValue()
   {  return value;  }
}

public class Beta extends Alpha
{
   private int num;

   public Beta(int myNum)
   {
      super();
      num = myNum;
      super.setValue(myNum - 3);
   }

   public String display()
   {
      return "value: " + getValue() + "  num: " + num;
   }
}

public class Gamma extends Alpha
{
   private int num;

   public Gamma(int myNum)
   {
      super();
      num = myNum;
   }

   public String display()
   {
      return "num: " + num + "  value: " + getValue();
   }
}
```

3. Which of the following is true of the class hierarchy?

(A) `Alpha` as written could be defined as an interface, and `Beta` and `Gamma` could implement the interface with the same results.
(B) `Alpha` contains an abstract method that `Beta` and `Gamma` must implement, but they may implement them in different ways.
(C) `Alpha` contains an abstract method that `Beta` and `Gamma` must implement in exactly the same way.
(D) `Alpha` is the parent of `Beta` and `Beta` is the parent of `Gamma`.
(E) These classes will not compile. They are written incorrectly.

4. Consider the following code segment in the main method of another class.

```
Alpha bObject  = new Beta(5);
Alpha gObject  = new Gamma(6);
System.out.println(bObject.display());
System.out.println(gObject.display());
```

What is printed as a result of executing the code segment?

(A)
```
value: 2   num: 5
num: 6   value: 3
```
(B)
```
value: 2   num: 5
value: 3   num: 6
```
(C)
```
value: 2   num: 5
value: 4   num: 6
```
(D)
```
value: 2   num: 5
num: 6   value: 4
```
(E)
```
value: 2   num: 5
value: 2   num: 5
```

5. Consider the following interface.

```
public interface TwoDShape
{
    double getArea();
    double getPerimeter();
}
```

Which of these classes correctly implements the interface?

I.
```
public class Square implements TwoDShape
{
    int sideLength;

    public Square(int sLen)
    {
        sideLength = sLen;
    }

    public int getArea()
    { return sideLength * sideLength; }

    public int getPerimeter()
    { return 4 * sideLength; }
}
```

II.
```
public class Square implements TwoDShape
{
    int sideLength;

    public Square(int sLen)
    {
        sideLength = sLen;
    }

    public double getArea(int sLen)
    { return sLen * sLen; }

    public double getPerimeter()
    { return 4 * sideLength; }
}
```

III.
```
public class Square implements TwoDShape
{
    int sideLength;

    public Square(int sLen)
    {
        sideLength = sLen;
    }

    public double getArea(int sLen)
    { return sLen * sLen; }

    public double getArea()
    { return sideLength * sideLength; }

    public double getPerimeter()
    { return 4 * sideLength; }
}
```

(A) I only
(B) II only
(C) III only
(D) I and II only
(E) II and III only

6. Free-Response Practice: Music Venue

A certain band plays at several venues throughout the year. Some venues are stadiums and some are clubs. When the band plays at a club, the ticket prices are all the same. When the band plays at a stadium, there are premium seats and general seats, each with a different price. The ticket prices for the stadium shows are decided ahead of time and cannot be changed: premium seats are $200 and general seats are $75. All venues have a name and maximum seating capacity. A club also has a genre.

Consider the following abstract class `Venue` and its subclasses, `Stadium` and `Club`.

a. Implement the abstract method `getRevenue` for the `Stadium` class.

b. Implement the abstract method `getRevenue` for the `Club` class.

```
public abstract class Venue
{
    private String name;
    private int capacity;

    public Venue(String myName, int myCapacity)
    {
        name = myName;
        capacity = myCapacity;
    }

    /**
     * @return the revenue generated
     */
    public abstract double getRevenue();

    /* other methods not shown */
}

public class Stadium extends Venue
{
    private final int  $_PREMIUM_SEAT = 200;
    private final int $_GENERAL_SEAT = 75;
    private final int MAX_PREMIUM_SEATS;
    private final int MAX_GENERAL_SEATS;
    private int premiumSeatsSold, generalSeatsSold;

    /**
     * All seats that are not premium seats are general seating.
     * @param name the name of the Stadium
     * @param capacity the seating capacity of the Stadium
     * @param premiumSeats the number of premium seats in the Stadium
     */
    public Stadium(String name, int capacity, int premiumSeats)
    {
        super(name, capacity);
        MAX_PREMIUM_SEATS = premiumSeats;
        MAX_GENERAL_SEATS = capacity - premiumSeats;
    }
```

```
    /**
     * A Stadium calculates revenue by multiplying the premium or
     * general seat price by the number of seats sold for each.
     * @return the revenue generated
     */
    public double getRevenue()
    {
        /* to be implemented in part a */
    }

    /* other methods not shown */
}

public class Club extends Venue
{
    private String genre;
    private final double $_SEAT;
    private int ticketsSold;

    public Club(String name, int capacity, String musicType, double ticketPrice)
    {
        super(name, capacity);
        $_SEAT = ticketPrice;
        genre = musicType;
    }

    /**
     * A Club calculates revenue by multiplying the ticket price
     * by the number of tickets sold.
     * @return the revenue generated
     */
    public double getRevenue()
    {
        /* to be implemented in part b */
    }

    /* other method not shown */
}
```

› Answers and Explanations

Bullets mark each step in the process of arriving at the correct solution.

1. The answer is E.
   - You *extend* an abstract class, and you can only extend one.
   - You *implement* an interface, and you can implement as many as you want.
2. The answer is D.
   - A class that implements an interface must implement every method defined in the interface, but how the class implements the method is entirely up to each individual class. WeddingPartyDecorator and KidsPartyDecorator must implement the three methods defined in the interface, but the methods may do entirely different things.
   - Classes that implement an interface may have as many other methods as they want in addition to the methods required by the interface.

3. The answer is B.

    - Option A is incorrect. Alpha cannot be defined as an interface, because it contains a constructor and methods that have been implemented.
    - Option C is incorrect. When extending an abstract class, all methods must be implemented, but the details of the implementation are entirely up to each child class.
    - Option D is incorrect. Beta and Gamma are both subclasses (or children) of Alpha.
    - Option E is incorrect. The classes are written correctly.

4. The answer is D.

    - bObject and gObject both have reference variables that are of the type Alpha. This will be OK with the compiler because all of the methods that they call exist in the Alpha class.
    - bObject is instantiated as an instance of the Beta class and gObject is instantiated as an instance of the Gamma class. At run-time, Java will treat them as Beta or Gamma objects, respectively.
    - Both the Beta class and the Gamma class define a display method, but those methods are different. When bObject.display() is executed, the Beta version of display is executed. When gObject.display() is executed, the Gamma version of display is executed.
    - When bObject is instantiated, it subtracts 3 from Alpha's instance variable, but it is not a static variable. The different objects have their own copies of Alpha's variables, just like different objects have their own copies of their own variables. When getValue() is called, it retrieves their individual copies of value. In the case of aObject, that's 2, and in the case of gObject, that's 4.

5. The answer is C.

    - All three versions have methods with the right names, but:
        - Option I is incorrect. The methods that implement the interface must have the exact method declaration given in the interface. In this option, the return types are different.
        - Option II is incorrect. The methods that implement the interface must have the exact method declaration given in the interface. In this option, the getArea method takes a parameter.
        - Option III is correct. Like option II, option III has a getArea method that takes a parameter, but in option III, getArea is overloaded. The required version of getArea is also implemented.

6. On the AP Exam, you are often asked to implement an interface or extend an abstract class like you were in this question. Check your answer against my version below. A few things to notice:

   - Final variables (constants) do not have to be declared with their value, but once a value has been assigned, it can never be changed. These variables do not have to be declared final, but doing so is good programming practice.
   - $ and _ are the only symbols that can be used in identifiers.
   - You don't have to calculate the result and assign it to a variable, then return that variable. You can just return the calculation. This is especially common in methods that return a boolean. You will see return statements like `return (x > 5);` instead of `if (x > 5) return true; else return false;`

Part a: The implementation in the Stadium class:

```
/**
 * A Stadium calculates revenue by multiplying the premium or
 * general seat price by the number of seats of each sold.
 * @return the revenue generated
 */
public double getRevenue()
{
    return $_PREMIUM_SEAT * premiumSeatsSold + $_GENERAL_SEAT * generalSeatsSold;
}
```

Part b: The implementation in the Club class:

```
/**
 * A Club calculates revenue by multiplying the ticket price
 * by the number of tickets sold.
 * @return the revenue generated
 */
public double getRevenue()
{
    return $_SEAT * ticketsSold;
}
```

Recursion

IN THIS CONCEPT

Summary: This concept defines and explains recursion. It also shows how to read and trace a recursive method.

Key Ideas

- A recursive method is a method that calls itself.
- Recursion is not the same as nested for loops or while loops.
- A recursive method must have a base case that tells the method to stop calling itself.
- You will need to know how to trace recursive methods on the AP Computer Science A Exam. You will not have to write recursive methods.

Recursion Versus Looping

The for loop and the while loop allow the programmer to repeat a set of instructions. **Recursion** is the process of repeating instructions too, but in a **self-similar** way. I know that's kind of weird. Let me explain by using the following example.

Example 1

Imagine you are standing in a really long line and you want to know how many people are in line in front of you, but you can't see everyone. If you ask the person in front of you how many people are in front of him, and he turns to the person in front of him and asks the person in front of him how many people are in front of him, and so on, eventually this chain reaction will reach the first person in line. This person will turn to the person behind him and say, "No one." Then something fun happens. The second person in line will turn to the third person and say, "One." The third person will turn to the fourth person and say, "Two," and so on, until finally the person in front of you will respond with the number of people who are in front of him. This creative solution to the problem is an example of how a recursive algorithm works.

Recursion is a programming technique that can be used to solve a problem in which a solution to the problem can be found by making subsequently smaller iterations of the same problem. When used correctly, it can be an extremely elegant solution to a problem.

One method can call another method, right? What if the method called itself? If you stop and think about that, after the method was **invoked** the first time, the program would never end because the method would continue to call itself forever (think of the movie *Inception*).

Example 2

Here's an example of a really bad recursive method.

```
public static void IAmABadRecursiveMethod()
{
    IAmABadRecursiveMethod();    // This method never stops calling itself
}
```

However, if the method had some kind of trigger that told it when to stop calling itself, then it would be a good recursive method.

General Form for a Recursive Method

```
public static returnType goodRecursiveMethod(parameterList)
{
    if ( /* base case is true */)
        return something;
    else
        return goodRecursiveMethod(argumentList);
}
```

Note: All recursive methods must have a base case.

The Base Case

The **base case** of a recursive method is a comparison between a parameter of the method and a predefined value strategically chosen by the programmer. The base case comparison determines whether or not the method should call itself again. Referring back to the example in the introduction of this concept, the response by the first person in line of "no one" is the base case. It is the answer that reverses the recursive process.

Each time the recursive method calls itself and the base case is not satisfied, the method temporarily sets that call aside before making a new call to itself. This continues until the base case is satisfied. When there is finally a call to the method that makes the base case true (the base case is satisfied), the method stops calling itself and starts the process of returning a value for each of the previous method calls. Since the calls are placed on a **stack**, the returns are executed in the reverse order of how they were placed. This order is known as last-in, first-out (LIFO), which means that the last recursive call made is the first one to return a value.

Furthermore, each successive call of the method should bring the value of the parameter *closer and closer to making the base case true.* This means that with each call to the method, the value of the parameter should be *closing in* on making the base case true rather than moving further away from it being true.

If a recursive method does not contain a valid base case comparison, then it may continue to call itself into oblivion. That would be bad.

What do I mean by bad? Remember that every time a recursive method is called, it has to set that call aside temporarily until the base case is satisfied. This process is called *putting the call on the stack* and the computer stores this stack in its RAM. If the base case is *never* satisfied, then the method calls pile up and the computer eventually runs out of memory. This will cause a **stack overflow error** and it occurs when you have an **infinite recursion.**

The Base Case
The base case is a comparison between a parameter of the method and some specific predefined value chosen by the programmer. When the base case is true, the method stops calling itself and starts the process of returning values. Every recursive method needs a valid base case or else it will continue to call itself indefinitely (infinite recursion) and cause an out-of-memory error, called a stack overflow error.

Beginner Example: The Factorial Recursive Method

A very popular problem to solve using recursion is the factorial calculation. The factorial of a number uses the notation **n!** and is calculated by multiplying the integer n by each of the integers that precede it all the way down to 1. Factorial is useful for computing combinations and permutations when doing probability and statistics.

Example
Find the answer to 4! and 10!

4! = 4 * 3 * 2 * 1 = 24. Therefore 4! is 24.
10! = 10 * 9 * 8 * 7 * 6 * 5 * 4 * 3 * 2 * 1 = 3628800. Therefore, 10! is 3628800.

Why the Factorial Calculation Is a Good Recursive Example

The definition of a factorial can be written recursively. This means that the factorial of the number, n, can use the factorial of the previous number, n − 1, in its computation.

$$n! = n * (n - 1)! \qquad \text{where } 1! \text{ is defined as } 1$$

Example

Write 4! recursively using math.

When you start to compute 4! using the definition of factorial, you see that 4! is 4 times 3! This means that in order to compute 4!, you need to figure out what 3! is. But 3! is 3 times 2! so we have to figure out what 2! is. And 2! is 2 times 1!, and 1! is 1.

Do you see how we had to put our calculations on hold until we got to 1!? Once we got a final answer of 1 for 1!, we were able to compute 4! by working backward.

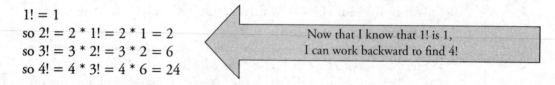

1! = 1
so 2! = 2 * 1! = 2 * 1 = 2
so 3! = 3 * 2! = 3 * 2 = 6
so 4! = 4 * 3! = 4 * 6 = 24

Now that I know that 1! is 1,
I can work backward to find 4!

Implementation of the Factorial Recursive Method

```
public class Factorial
{
    public static void main(String[] args)
    {
        System.out.println(factorial(5));     // find factorial(5)
    }

    /**
     * This is a recursive method that computes the
     * factorial of a number.
     *
     * @param n the number we are finding the factorial of
     * @return the answer to the factorial of n
     *
     * PRECONDITION:  n >= 1
     * POSTCONDITION:  The result of the recursive call
     *                 is the factorial of the original argument
     */
    public static int factorial(int n)
    {
        if (n == 1)                            // base case comparison
            return 1;
        else
            return n * factorial(n - 1);       // move closer to base case
    }
}
```

```
OUTPUT
120
```

> **Good News, Bad News**
> On the AP Exam, you will never be asked to write your own original recursive method. However, you will be asked to hand-trace recursive methods and state the results.

Hand-Tracing the Factorial Recursive Method

This is kind of tricky, so read it over very carefully. When tracing a recursive method call, write out each call to the method including its parameter. When the base case is reached, replace the result from each method call with its result, one at a time. This process will force you to calculate the return values in the reverse order from the way that the calls were made.

Goal: Compute the value of factorial(5) using the process of recursion.

Step 1: This problem is similar to the "find the number of people in line" example. The answer to a factorial requires knowledge of the factorial preceding it.

Suppose I ask the number 5, "Hey what's your factorial?" 5 responds with, "I don't know. I need to ask 4 what its factorial is before I can find out my own." Then 4 asks 3 what its factorial is. Then 3 asks 2 what its factorial is. Then 2 asks 1 what its factorial is. Then 1 says, "1".

Step 2: After 1 says, "1", the process reverses itself and each number answers the question that was asked of it. The answer of "1" is the base case.

1 says, "My factorial is 1."

2 then says, "My factorial is 2 because 2 * 1 is 2."

3 then says, "My factorial is 6 because 3 * 2 is 6."

4 then says, "My factorial is 24 because 4 * 6 is 24."

And finally, 5 turns to me and says, "My factorial is 120, because 5 * 24 is 120."

factorial(1) = 1
factorial(2) = 2 * factorial(1) = 2 * 1 = 2
factorial(3) = 3 * factorial(2) = 3 * 2 = 6
factorial(4) = 4 * factorial(3) = 4 * 6 = 24
factorial(5) = 5 * factorial(4) = 5 * 24 = 120

Conclusion: factorial(5) = 120

A Visual Representation of Computing a Recursive Value

Example

Explain how to compute the value of factorial(5) for the visual learner. As you trace through this problem, think of the original example of the people in the long line.

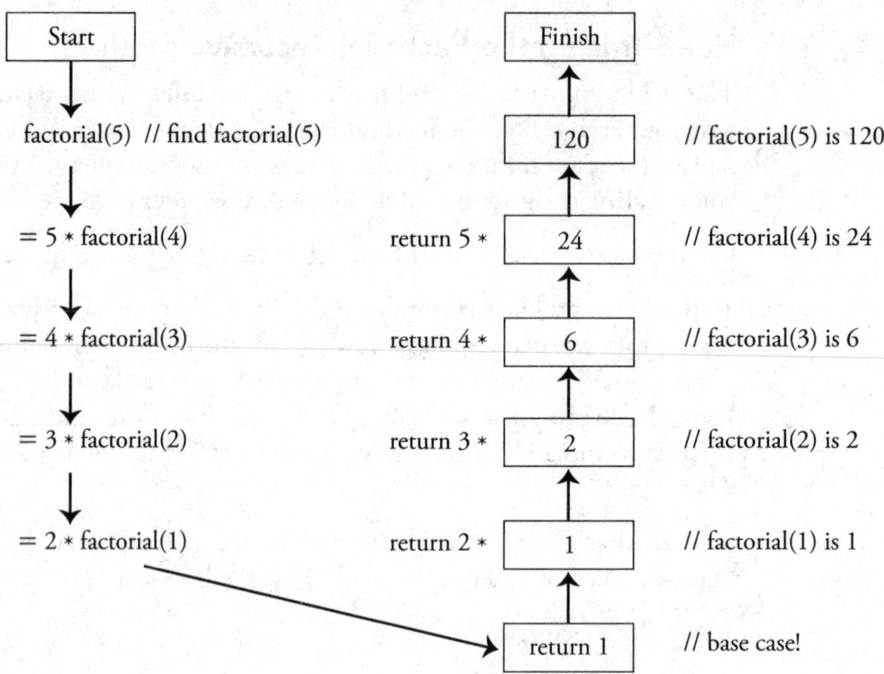

Advanced Example: The Fibonacci Recursive Method

On the AP Exam, you will need to be able to hand-trace recursive method calls that have either multiple parameters or make multiple calls within the return statement. The following example shows a recursive method that has multiple calls to the recursive method within its return statement.

The Fibonacci Sequence

This popular sequence is often covered in math class as it has some interesting applications. In the Fibonacci sequence, the first two terms are 1, and then each term starting with the third term is equal to the sum of the two previous terms.

1, 1, 2, 3, 5, 8, 13, 21, 34, 55 . . .

The Java code that is provided below is a recursive method that returns the value of the nth term in a Fibonacci sequence. **Example:** fibonacci(6) is 8. The value of the sixth term is 8.

```
public class AdvancedRecursiveProblem
{
    public static void main(String[] args)
    {
        System.out.println(fibonacci(6));           // find fibonacci(6)
    }

    /**
     * This is a recursive method that has multiple base cases and
     * uses two recursive calls within the return statement.
     *
     * @param n The fibonacci term to be calculated
     * @return The nth term of the fibonacci sequence
    */
    public static int fibonacci(int n)
    {
        if (n == 1 || n == 2)
            return 1;
        else
            return fibonacci(n - 1) + fibonacci(n - 2);
    }
}
```

```
OUTPUT
8
```

Hand-Tracing the Fibonacci Recursive Method

This is harder to follow than the factorial recursive method because the return statement includes two calls to the recursive method. Also, there are two base cases.

Goal: Compute the value of fibonacci(6).

Step 1: In a manner similar to the previous example, I ask 6, "What's your fibonacci?" 6 responds by saying, "I don't know right now, I need to ask both 5 and 4 what their fibonaccis are." In response, 5 says, "I need to ask both 4 and 3 what their fibonaccis are." And, 4 says, "I need to ask both 3 and 2 what their fibonaccis are." This continues until 2 and 1 are asked what their fibonaccis are.

Step 2: Ultimately 2 and 1 will respond with "1" and "1" and each number can then answer the question that was asked of them. The 2 and 1 are the base cases for this problem.

fibonacci(1) = 1
fibonacci(2) = 1
fibonacci(3) = fibonacci(2) + fibonacci(1) = 1 + 1 = 2
fibonacci(4) = fibonacci(3) + fibonacci(2) = 2 + 1 = 3
fibonacci(5) = fibonacci(4) + fibonacci(3) = 3 + 2 = 5
fibonacci(6) = fibonacci(5) + fibonacci(4) = 5 + 3 = 8

Conclusion: fibonacci(6) is 8

› Rapid Review

- Recursive methods are methods that call themselves.
- Choose recursion rather than looping when the problem you are trying to solve can be solved by repeatedly shrinking the problem down to a single, final simple problem.
- Recursive methods usually have a return type (rather than void) and must take a parameter.
- The re-calling of the method should bring you closer to the base case.
- Many problems can be solved without recursion; however, if done properly, recursion can be an extremely elegant way to solve a problem.
- Once the recursive process has started, the only way to end it is to satisfy the base case.
- The base case is a comparison that is done in the method. When this comparison becomes true, the method returns a value that begins the return process.
- If you write a recursive method and forget to put in a base case, or your subsequent calls to the method do not get you closer to the base case, you will probably cause infinite recursion, which will result in a stack overflow error.
- To trace a recursive call, write out each call of the method with the value of the parameter. When the base case is reached, replace each method call with its result. This will happen in reverse order.

› Review Questions

Basic Level

1. Consider the following recursive method.

```
public int puzzle(int num)
{
    if (num <= 1)
        return 1;
    else
        return num + puzzle(num / 2);
}
```

What value is returned when `puzzle(10)` is called?

(A) `18`
(B) `15`
(C) `11`
(D) `1`
(E) Nothing is returned. Infinite recursion causes a stack overflow error.

2. Consider the following recursive method.

```
private int mystery(int k)
{
    if (k == 0)
    {
        return 1;
    }
    return 2 * k + mystery(k - 2);
}
```

What value is returned when `mystery(11)` is called?

(A) `1`
(B) `49`
(C) `73`
(D) `665280`
(E) Nothing is returned. Infinite recursion causes a stack overflow error.

3. Consider the following recursive method.

```
public int enigma(int n)
{
    if (n < 3)
        return 2;
    if (n < 5)
        return 2 + enigma(n - 1);
    return 3 + enigma(n - 2);
}
```

What value is returned when `enigma(9)` is called?

(A) 7
(B) 10
(C) 13
(D) 15
(E) 16

4. Consider the following recursive method.

```
public void printStars(int n)
{
    if (n == 1)
    {
        System.out.println("*");
    }
    else
    {
        System.out.print("*");
        printStars(n - 1);
    }
}
```

What will be printed when `printStars(15)` is called?

(A) ***************
(B) *
(C) Nothing is printed. Method will not compile. Recursive methods cannot print.
(D) Nothing is printed. Method returns successfully.
(E) Many stars are printed. Infinite recursion causes a stack overflow error.

5. Consider the following method.

```
public String weird(String s)
{
    if (s.length() >= 10)
    {
        return s;
    }
    return weird(s + s.substring(s.indexOf("lo")));
}
```

What value is returned when `weird("Hello")` is called?

(A) "Hello"
(B) "Hellolo"
(C) "Hellolololo"
(D) "Hellolololololololo"
(E) Nothing is returned. Infinite recursion causes a stack overflow error.

Advanced Level

Questions 6–7 refer to the following methods.

```
public int factors(int number)
{
    return factors(number, number - 1, 0);
}

public int factors(int number, int check, int count)
{
    if (number % check == 0)
    {
        count++;
    }
    check--;
    if (check == 1)
    {
        return count;
    }
    return factors(number, check, count);
}
```

6. What value is returned when `factors(10)` is called?

(A) 0
(B) 1
(C) 2
(D) 3
(E) 4

7. Which of the following statements will execute successfully?

I. `int answer = factors(0);`
II. `int answer = factors(2);`
III. `int answer = factors(12, 2, 5);`

(A) I only
(B) II only
(C) III only
(D) I and II only
(E) I, II, and III

8. Consider the following recursive method.

```
public int function(int x, int y)
{
    if (x <= y)
        return x + y;
    else
        return function(x - y, y + 1) + 2;
}
```

What value is returned when `function(24, 3)` is called?

(A) 13
(B) 21
(C) 25
(D) 27
(E) Nothing is returned. Infinite recursion causes a stack overflow error.

9. Consider this recursive problem.

In elementary school, we learned that if you add up the digits of a number and get a sum of 9, then that number is divisible by 9. If your answer is less than 9, then that number is not divisible by 9. For example,

- 81 → 8 + 1 = 9, divisible by 9.
- 71 → 7 + 1 = 8, not divisible by 9.

The trouble is, if you add up the digits of a big number, you get a sum that is greater than 9, for example, 999 → 9 + 9 + 9 = 27. We can fix this by using the same trick on that sum: 27 → 2 + 7 = 9, divisible by 9.

Here's another example:

- 457829 → 4 + 5 + 7 + 8 + 2 + 9 = 35, keep going,
- 35 → 3 + 5 = 8, not divisible by 9.

If your number is big enough, you may have to repeat the process three, four, five, or even more times, but eventually, you will get a sum that is <= 9 and that will let you determine whether your number is divisible by 9.

Your client code will call method `divisible`, passing a number and expecting a `boolean` result, and `divisible` will call the recursive method `sumUp` to do the repeated addition.

Consider the class below that implements a solution to the problem.

```
public class Div9
{
    public boolean divisible (int dividend)
    {
        return sumUp(dividend)  == 9;
    }

    public int sumUp(int dividend)
    {
        int sum = 0;

        // This loop adds the digits of dividend into the sum variable
        while (dividend > 0)
        {
            sum += dividend % 10;
            dividend = dividend /10;
        }

        if ( /* condition */ )
            return sum;
        else
            return sumUp( /* argument */ );
    }
}
```

Which of the following can be used to replace `/* condition */` and `/* argument */` so that `sumUp` will work as intended?

(A) condition: `sum <= 9` argument: `sum`
(B) condition: `sum = 9` argument: `dividend`
(C) condition: `sum <= 9` argument: `dividend`
(D) condition: `sum < 9` argument: `sum`
(E) condition: `sum < 9` argument: `dividend / sum`

› Answers and Explanations

Bullets mark each step in the process of arriving at the correct solution.

1. The answer is A.

- Let's trace the calls. The parts in *italics* were filled in on the way back up. That is, the calls in the plain type were written top to bottom until the base case returned 1. Then the answer were filled in *bottom to top.*

  puzzle(10) = num + puzzle(num/2) = 10 + puzzle(5) = *10 + 8 =* ***18***
  puzzle(5) = num + puzzle(num/2) = 5 + puzzle(2) = *5 + 3 = 8*
  puzzle(2) = num + puzzle(num/2) = 2 + puzzle(1) = *2 + 1 = 3*
  puzzle(1) = base case! return 1

- Don't forget to do integer division.

2. The answer is E.

- Let's trace the calls just like we did in problem 1.

  mystery(11) = 2 * 11 + mystery(9)
  mystery(9) = 2 * 9 + mystery(7)
  mystery(7) = 2 * 7 + mystery(5)
  mystery(5) = 2 * 5 + mystery(3)
  mystery(3) = 2 * 3 + mystery(1)
  mystery(1) = 2 * 1 + mystery(-1)

  Uh oh—the parameter skipped right over 0, and it's only going to get smaller. This example will recurse until there is a stack overflow error.

- You might be confused by the fact that sometimes people leave out else when writing code like this. Let's take a look at the code again:

```
if (k == 0)
    return 1;
return 2 * k + mystery(k - 2);
```

  Why doesn't this have to say *else* return 2 * k + mystery(k – 2)? The purpose of an else clause is to tell the flow of control to skip that section when the if part is executed. So *either* the if clause or the else clause is executed. In this case, the if clause contains a return. Once a return is executed, nothing further in the method will be read and control returns to the calling method. We don't have to tell it to skip the else clause, because it has already gone off to execute code in another method. It doesn't matter whether you choose to write your code like this or to include the else, but don't be confused if you see it on the exam.

3. The answer is C.

- Here we go again!

  enigma(9) = 3 + enigma(7) = *3 + 10 =* ***13***
  enigma(7) = 3 + enigma(5) = *3 + 7 = 10*
  enigma(5) = 3 + enigma(3) = *3 + 4 = 7*
  ...here comes one of the base cases!
  enigma(3) = 2 + enigma(2) = *2 + 2 = 4*
  ...the other base case!
  enigma(2) = 2...and up we go!

4. The answer is A.

- We don't really want to trace 15 calls of the printStars method, so let's see if we can just figure it after a few calls.

  printStars(15): prints one * and calls printStars(14)
  printStars(14): prints one * and calls printStars(13)
  . . .
  . . .
  printStars(1): prints one * and returns (remember, void methods also return, they just don't return a value).

- Without tracing every single call, we can see the pattern. 15 stars will be printed.
- This recursive method is a little different because it is not a return method. That's allowed, but it is not very common.

5. The answer is C.

- This time with Strings!

  weird("Hello") = weird("Hello" + "lo") = weird("Hellolo") = ***"Hellolololo"***
  weird("Hellolo") = weird("Hellolo" + "lo") = weird("Hellololo") = *"Hellolololo"*
  weird("Hellololo") = weird("Hellololo" + "lo") = weird("Hellolololo") = *"Hellolololo"*
  . . . that has a length >10, so we just start returning s.

- This one is a little different because there is no computation on the "return trip."

6. The answer is C.

- Instead of tracing the code, this time we are going to reason out what the method is doing.
- Notice that factors is an overloaded method. Our first call has one int parameter, but the remaining calls all have three. Our first call, factors(10), will result in the call factors(10, 9, 0).
- Looking at the factors method with three parameters, we can see that each time through, we subtract one from check, and the base case is check == 1. We start with check = 9 and call the method eight more times before returning count. Count begins at 0 and will be incremented when number % check == 0. Since number is 10, and check will equal all the numbers between 1 and 9, that will happen twice, at 10 % 5 and at 10 % 1. When the return statement is reached, count = 2.

7. The answer is C.

- Option I is incorrect. factors(0) will call factors(0, -1, 0). Since check is decremented each time, we will move further and further from the base case of 1. Infinite recursion will result in a stack overflow exception.
- Option II is incorrect. factors(2) will call factors(2, 1, 0). Since check is decremented to 0 before checking for the base case, once again we have infinite recursion resulting in a stack overflow error.
- Option III is correct. factors(12, 2, 5) will increment count (12 % 2 == 0), decrement check to 1, and then check for the base case and return count.

8. The answer is B.

- Let's trace.

  function(24, 3) = function(21, 4) + 2 = *19 + 2 =* ***21***
  function(21, 4) = function(17, 5) + 2 = *17 + 2 = 19*
  function(17, 5) = function(12, 6) + 2 = *15 + 2 = 17*
  function(12, 6) = function(6, 7) + 2 = *13 + 2 = 15*
  function(6, 7) = 6 + 7 = 13 (base case!)

9. The answer is A.

- The if statement we are completing represents the base case and the recursive call.
- We need to keep going if sum > 9, so the base case, the case that says we are done, occurs at sum <= 9. That is the *condition*. If sum <= 9, all we need to do is return sum.
- If sum >= 9, we need to add the digits up again, but the question is, the digits of *what*? If you go back to the examples in the description, 999, for example, the while loop will add 9 + 9 + 9 and put the result in sum, which now = 27. Then the next step is to add 2 + 7. Since 27 is held in the variable sum, that's what we need to pass to the next round of recursion. The *argument* is sum.
- You don't need to understand the while loop to answer the question, but it is a pretty cool loop, so let's explain it here anyway.
  - dividend % 10 gives you the last digit of dividend
  - dividend / 10 gets rid of the last digit of dividend (because it is integer division)
  - Here's an example, using the number is 365.

  365 / 10 = 36 remainder 5 (mod will give us 5, add it to sum)
  36 / 10 = 3 remainder 6 (mod will give us 6, add it to sum)
  3 / 10 = 0 remainder 3 (mod will give us 3, add it to sum)
  0 will cause the loop to terminate.

Algorithms (Advanced Version)

IN THIS CONCEPT

Summary: In the previous Algorithms section (Concept 6), you learned basic algorithms for swapping, copying, searching, accumulating, and finding the highest or lowest value in a list. What if the information being processed is buried in some complex data structure that is also buried in a class hierarchy? Or what if you are asked to solve a problem that you've never seen before? This concept focuses on algorithms that require you to access information that is more in depth and models how to write algorithms for different situations.

Key Ideas

- The process of developing an original algorithm requires concentration and analytical thought.
- Accessing data that is buried within a class hierarchy requires the programmer to use the methods that belong to the class.

The Accumulate Advanced Algorithm

You have been hired by the high school baseball team to write a program that calculates statistics for the players on the team. You decide to have a class called BaseballPlayer. Every BaseballPlayer has a name, a numberOfHits, and a numberOfAtBats. The BaseballRunner has an array of BaseballPlayer objects called roster. Your goal is to find the team batting average.

```
public class BaseballPlayer
{
    private String name;
    private int hits;
    private int atBats;

    public String getName()
    {   return name;   }

    public int getHits()
    {   return hits;   }

    public int getAtBats()
    {   return atBats;   }

    public double getBattingAverage()
    {   /*  implementation not shown  */   }

    /*  Additional implementation not shown   */
}
```

```
public class BaseballRunner
{
    public static void main(String[] args)
    {
        private BaseballPlayer[] roster;       // array of players

        /*  Additional implementation not shown   */
    }

    public static double findTeamAverage()
    {
        /*  implementation to be completed in this example  */
    }
}
```

General Problem: Given a roster of baseball players, find the team's batting average.

Refined Problem: You are given a list of baseball players. Every baseball player knows his number of hits and number of at-bats. The team average is found by first computing the sum of the at-bats and the sum of the hits and then dividing the total hits by the total at-bats. Write a method that computes the team batting average.

Final Problem: Write a method called findTeamAverage that has one parameter: a BaseballPlayer array. The method should return a double that is the team's batting average. Compute the team average by dividing the total hits by the total at-bats. Make sure a player exists before processing him. Also, perform a check against zero before computing the team average. If the team total at-bats is 0, return a team batting average of 0.0. Finally, adapt your algorithm to work with an ArrayList of BaseballPlayer objects.

Algorithm:
Step 1: Create a variable called totalHits and set it equal to 0
Step 2: Create a variable called totalAtBats and set it equal to 0
Step 3: Look at each player on the roster
Step 4: Get the number of hits for the player and add it to totalHits
Step 5: Get the number of atBats for the player and add it to totalAtBats
Step 6: Continue until you reach the end of the list
Step 7: Compute the teamBattingAverage by dividing the totalHits by the totalAtBats
Step 8: Return the teamBattingAverage

Pseudocode:

```
set totalHits = 0
set totalAtBats = 0
for (iterate through all the players in the roster)
{
    if (player exists)
    {
        totalHits = totalHits + player's hits
        totalAtBats = totalAtBats + player's at-bats
    }
}
if (totalAtBats = 0)
{
    return 0
}
else
{
    return totalHits / totalAtBats
}
```

Java Code 1: Java code using an array (`for` loop)

```
public static double findTeamAverage(BaseballPlayer[] arr)
{
   int totalHits = 0;                              // initialize totals
   int totalBats = 0;

   for (int i = 0; i < arr.length; i++)
   {
      if (arr[i] != null)                          // make sure player exists
      {
         totalHits += arr[i].getHits();            // add hits to total
         totalAtBats += arr[i].getAtBats();        // add at-bats to total
      }
   }

   if (totalAtBats == 0)                           // make sure total is not 0
      return 0;
   else
      return (double)totalHits / totalAtBats;     // cast average to a double
}
```

Java Code 2: Java code using an array (`for-each` loop)

```
public static double findTeamAverage(BaseballPlayer[] arr)
{
   int totalHits = 0;
   int totalBats = 0;

   for (BaseballPlayer player : arr)
   {
      if (player != null)
      {
         totalHits += player.getHits();
         totalAtBats += player.getAtBats();
      }
   }

   if (totalAtBats == 0)
      return 0;
   else
      return (double)totalHits / totalAtBats;
}
```

Java Code 3: Java code using an `ArrayList` (`for` loop)

```
public static double findTeamAverage(ArrayList<BaseballPlayer> arr)
{
    int totalHits = 0;
    int totalAtBats = 0;

    for (int i = 0; i < arr.size(); i++)
    {
        if (arr.get(i) != null)
        {
            totalHits += arr.get(i).getHits();
            totalAtBats += arr.get(i).getAtBats();
        }
    }

    if (totalAtBats == 0)
        return 0;
    else
        return (double)totalHits / totalAtBats;
}
```

Java Code 4: Java code using an `ArrayList` (`for-each` loop)

```
public static double findTeamAverage(ArrayList<BaseballPlayer> arr)
{
    int totalHits = 0;
    int totalAtBats = 0;

    for (BaseballPlayer player : arr)
    {
        if (player != null)
        {
            totalHits += player.getHits();
            totalAtBats += player.getAtBats();
        }
    }

    if (totalAtBats == 0)
        return 0;
    else
        return (double)totalHits / totalAtBats;
}
```

```
public class BaseballRunner
{
    public static void main(String[] args)
    {
        BaseballPlayer[] roster = new BaseballPlayer[9];

        roster[0] = new BaseballPlayer("Joe", 1, 4);
        roster[1] = new BaseballPlayer("Sam", 2, 4);
        roster[2] = new BaseballPlayer("Kyle", 4, 4);
        roster[3] = new BaseballPlayer("Chris", 3, 4);
        roster[4] = new BaseballPlayer("Tommy", 1, 3);
        roster[5] = new BaseballPlayer("David", 2, 3);
        roster[6] = new BaseballPlayer("Jordan", 1, 3);
        roster[7] = new BaseballPlayer("Brian", 1, 3);
        roster[8] = new BaseballPlayer("John", 2, 3);

        System.out.println("Team average:  " + findTeamAverage(roster));
    }

    public static double findTeamAverage(BaseballPlayer[] arr)
    {
        /* implementation is either Code 1 or Code 2 described above */
    }
}
```

```
OUTPUT
Team average:  0.5483870967741935
```

The Find-Highest Advanced Algorithm

General Problem: Given a roster of baseball players, find the name of the player that has the highest batting average.

Refined Problem: You are given a list of baseball players. Every baseball player has a name and knows how to calculate his batting average. Write a method that gets the batting average for every player in the list, and find the name of the player that has the highest batting average.

Final Problem: Write a method called findBestPlayer that has one parameter: a *BaseballPlayer* array. The method should return a String that is the name of the player that has the highest batting average. Make sure a player exists before processing him. If two players have the same high average, only select the first player. Also, adapt your algorithm to work with an ArrayList of BaseballPlayer objects.

Note: We will use the BaseballPlayer class from the previous example.

Algorithm:

Step 1: Create a variable called highestAverage and assign it a really small number

Step 2: Create a variable called bestPlayer and assign it the empty String

Step 3: Look at each of the players in the roster

Step 4: If the batting average of the player is greater than highestAverage, then set the highestAverage to be that player's average and set bestPlayer to be the name of the player

Step 5: Continue until you reach the end of the list

Step 6: Return the name of bestPlayer

Pseudocode:

```
set highestAverage = the smallest available number in Java
set bestPlayer = ""
for (iterate through all the players in the list)
{
    if (player exists)
    {
        if (current player's batting average > highestAverage)
        {
            set highestAverage = current player's batting average
            set bestPlayer = name of current player
        }
    }
}
return bestPlayer
```

Java Code 1: Java code using an array (`for` loop)

```
public static String findBestPlayer(BaseballPlayer[] arr)
{
    double highestAverage = Integer.MIN_VALUE;
    String bestPlayer = "";

    for (int i = 0; i < arr.length; i++)
    {
        if (arr[i] != null)
        {
            if (arr[i].getBattingAverage() > highestAverage)
            {
                highestAverage = arr[i].getBattingAverage();
                bestPlayer = arr[i].getName();
            }
        }
    }

    return bestPlayer;
}
```

Java Code 2: Java code using an array (`for-each` loop)

```
public static String findBestPlayer(BaseballPlayer[] arr)
{
    double highestAverage = Integer.MIN_VALUE;
    String bestPlayer = "";

    for (BaseballPlayer player : arr)
    {
        if (player != null)
        {
            if (player.getBattingAverage() > highestAverage)
            {
                highestAverage = player.getBattingAverage();
                bestPlayer = player.getName();
            }
        }
    }

    return bestPlayer;
}
```

Java Code 3: Java code using an `ArrayList` (`for` loop)

```
public static String findBestPlayer(ArrayList<BaseballPlayer> arr)
{
    double highestAverage = Integer.MIN_VALUE;
    String bestPlayer = "";

    for (int i = 0; i < arr.size(); i++)
    {
        if (arr.get(i) != null)
        {
            if (arr.get(i).getBattingAverage() > highestAverage)
            {
                highestAverage = arr.get(i).getBattingAverage();
                bestPlayer = arr.get(i).getName();
            }
        }
    }

    return bestPlayer;
}
```

Java Code 4: Java code using an `ArrayList` (`for-each` loop)

```
public static String findBestPlayer(ArrayList<BaseballPlayer> arr)
{
   double highestAverage = Integer.MIN_VALUE;
   String bestPlayer = "";

   for (BaseballPlayer player : arr)
   {
      if (player != null)
      {
         if (player.getBattingAverage() > highestAverage)
         {
            highestAverage = player.getBattingAverage();
            bestPlayer = player.getName();
         }
      }
   }

   return bestPlayer;
}
```

```
Example: Runner Program

public class BaseballRunner
{
   public static void main(String[] args)
   {
      BaseballPlayer[] roster = new BaseballPlayer[9];

      roster[0] = new BaseballPlayer("Joe", 1, 4);
      roster[1] = new BaseballPlayer("Sam", 2, 4);
      roster[2] = new BaseballPlayer("Kyle", 4, 4);
      roster[3] = new BaseballPlayer("Chris", 3, 4);
      roster[4] = new BaseballPlayer("Tommy", 1, 3);
      roster[5] = new BaseballPlayer("David", 2, 3);
      roster[6] = new BaseballPlayer("Jordan", 1, 3);
      roster[7] = new BaseballPlayer("Brian", 1, 3);
      roster[8] = new BaseballPlayer("John", 2, 3);

      System.out.println("Best Player:  " + findBestPlayer(roster));
   }

   public static double findBestPlayer(BaseballPlayer[] arr)
   {
      /*  implementation is either Code 1 or Code 2 describe above  */
   }
}
```

```
OUTPUT
Best Player:  Kyle
```

The Connect-Four Advanced Algorithm

Connect Four is a two-player game played on a grid of six rows and seven columns. Players take turns inserting disks into the grid in hopes of connecting four disks in a row to win the game. You have been hired to work on the app version of the game, and it is your responsibility to determine whether a player has won the game.

General Problem: Determine if a player has won the game of Connect Four.

Refined Problem: The Connect Four app version uses a 2-D array that is 6 × 7. After a player makes a move, determine if he or she has won by checking all possible ways to win. If four of the same color disks are found in a line, then the game is over. Decide who won the game (player one or player two). A way to win can be vertical, horizontal, diagonally downward, or diagonally upward.

Final Problem: Write a method called checkForWinner that has one parameter: a 2-D array of int. The 2-D array consists of 1s and 2s to represent a move by player one or player two. Cells that do not contain a move have a value of 0. This method checks all four directions to win (vertical, horizontal, diagonally downward, and diagonally upward). The method should return the number of the player who won or return 0 if nobody has won.

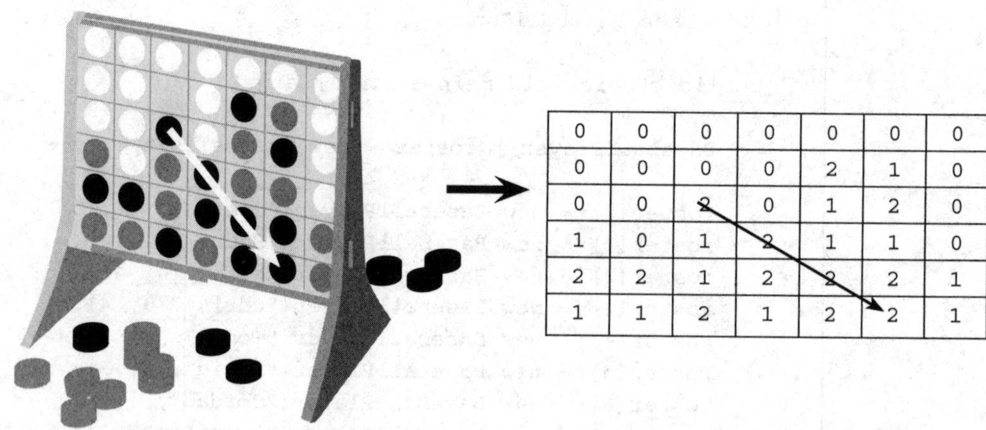

0 0 0 0 0 0 0
0 0 0 0 2 1 0
0 0 2 0 1 2 0
1 0 1 2 1 1 0
2 2 1 2 2 2 1
1 1 2 1 2 2 1

	0	1	2	3	4	5	6
0	0	0	0	0	0	0	0
1	0	0	0	0	2	1	0
2	0	0	2	0	1	2	0
3					1	1	0
4	2	2	1	2	2	2	1
5	1	1	2	1	2	2	1

Algorithm (for vertical win):

Note: There are a total of 21 ways to win vertically. Start with the top left cell and move in Row-Major order.

Step 1: Start with row 0 and end with row 2 (row 2 is four less than the number of rows)
Step 2: Start with column 0 and end with column 6 (use every column)
Step 3: If current cell value is not equal to 0 and the current cell = cell below it = cell below it = cell below it, then a player has won
Step 4: Continue until you reach the end of row 2
Step 5: Return the player who won or zero

Pseudocode (for vertical win):

```
for (start with row 0 and end with row 2)
{
    for (start with column 0 and end with column 6)
    {
        if (current cell != 0 && current cell = cell below it = cell below it = cell below it)
        {
            return current cell value
        }
    }
}
return 0 (no winning player is found)
```

	0	1	2	3	4	5	6
0	0	0	0	0	0	0	0
1	0	0	0	0	2	1	0
2	0	0	2	0	1	2	0
3	1	0	1	2	1	1	0
4	2	2	1	2	2	2	1
5	1	1	2	1	2	2	1

Algorithm (for diagonal downward win):

Note: There are a total of 12 ways to win diagonally downward. Start with the top left cell and go in Row-Major order.

Step 1: Start with row 0 and end with row 2 (row 2 is four less than the number of rows)
Step 2: Start with column 0 and end with column 3 (column 3 is four less than the number of columns)
Step 3: If current cell value is not equal to 0 and the cell below and to the right = cell below and to the right = cell below and to the right, then a player has won
Step 4: Continue until you reach the end of row 2
Step 5: Return the player who won or zero

Pseudocode (for diagonal downward win):

```
for (start with row 0 and end with row 2)
{
  for (start with column 0 and end with column 3)
  {
    if (current cell != 0 && current cell = cell below/right = cell below/right
        = cell below/right)
    {
        return current cell value
    }
  }
}
return 0
```

> **Note:** I have provided the algorithm and pseudocode for only the vertical win and the diagonal downward win. The code for these, as well as the horizontal win and the diagonal upward win, are provided in the Java code below.

Java Code: Java code using a 2-D array and nested `for` loops (includes all four ways to win):

```
public static int checkForWinner(int[][] board)
{
    // Check for Vertical Win
    for (int r = 0; r <= board.length - 4; r++)
    {
        for (int c = 0; c < board[0].length; c++)
        {
            if (board[r][c] != 0 &&
                board[r][c] == board[r+1][c] &&
                board[r+1][c] == board[r+2][c] &&
                board[r+2][c] == board[r+3][c])
            return board[r][c];
        }
    }

    // Check for Diagonal Downward Win
    for (int r = 0; r <= board.length - 4; r++)
    {
        for (int c = 0; c <= board[0].length - 4; c++)
        {
            if (board[r][c] != 0 &&
                board[r][c] == board[r+1][c+1] &&
                board[r+1][c+1] == board[r+2][c+2] &&
                board[r+2][c+2] == board[r+3][c+3])
            return board[r][c];
        }
    }
```

```
    // Check for Horizontal Win
    for (int r = 0; r < board.length; r++)
    {
       for (int c = 0; c <= board[0].length - 4; c++)
       {
          if (board[r][c] != 0 &&
               board[r][c] == board[r][c+1] &&
               board[r][c+1] == board[r][c+2] &&
               board[r][c+2] == board[r][c+3])
          return board[r][c];
       }
    }

    // Check for Diagonal Upward Win
    for (int r = 3; r < board.length; r++)
    {
       for (int c = 0; c <= board[0].length - 4; c++)
       {
          if (board[r][c] != 0 &&
               board[r][c] == board[r-1][c+1] &&
               board[r-1][c+1] == board[r-2][c+2] &&
               board[r-2][c+2] == board[r-3][c+3])
          return board[r][c];
       }
    }

    // No winner was found after checking all 4 ways to win
    return 0;
}
```

```
public class ConnectFourRunner
{
   public static void main(String[] args)
   {
      int[][] board = {{0,0,0,0,0,0,0},
                       {0,0,0,0,2,1,0},
                       {0,0,2,0,1,2,0},
                       {1,0,1,2,1,1,0},
                       {2,2,1,2,2,2,1},
                       {1,1,2,1,2,2,1} };

      System.out.println("Winner:  " + checkForWinner(board));
   }

   public static int checkForWinner(int[][] arr)
   {
      /*  implementation is described above  */
   }
}
```

```
OUTPUT
Winner:  2
```

The Twitter-Sentiment-Analysis Advanced Algorithm

Twitter has become very popular and Twitter Sentiment Analysis has grown in popularity. The idea behind Twitter Sentiment is to pull emotions and/or draw conclusions from the tweets. A large collections of tweets can be used to make general conclusions of how people feel about something. The important words in the tweet are analyzed against a library of words to give a tweet a sentiment score.

One of the first steps in processing a tweet is to remove all of the words that don't add any real value to the tweet. These words are called **stop words,** and this step is typically part of a phase called **preprocessing**. For this problem, your job is to remove all the stop words from the tweet. There are many different ways to find meaningless words, but for our example, the stop words will be all words that are three characters long or less.

General Problem: Given a tweet, remove all of the meaningless words.

Refined Problem: You are given a tweet as a list of words. Find all of the words in the list that are greater than three characters and add them to a new list. Leave the original tweet unchanged. The new list will contain all the words that are not stop words.

Final Problem: Write a method called removeStopWords that has one parameter: an ArrayList of Strings. The method should return a new ArrayList of Strings that contains only the words from the original list that are greater than three characters long. Do not modify the original ArrayList. Also, modify the code to work on a String array.

Solution 1: Using an `ArrayList`

Algorithm 1:

Step 1: Create a list called longWords
Step 2: Look at each word in the original list
Step 3: If the word is greater than three characters, add that word to longWords
Step 4: Continue until you reach the end of the original list
Step 5: Return longWords

Pseudocode 1:

```
create a list called longWords
for (every word in the original list)
{
    if (word length > 3)
    {
        add the word to longWords
    }
}
return longWords
```

Java Code 1: Using an `ArrayList` (`for-each` loop)

```
public static ArrayList<String> removeStopWords(ArrayList<String> words)
{
   ArrayList<String> longWords = new ArrayList<String>();
   for (String word : words)
   {
      if (word.length() > 3)
      {
         longWords.add(word);
      }
   }
   return longWords;
}
```

Solution 2: Using an Array

This solution requires more work than the ArrayList solution. Since we don't know how big to make the array, we need to count how many words are greater than three characters before creating it.

Algorithm 2:

Step 1: Create a counter and set it to zero
Step 2: Look at each word in the original list
Step 3: If the length of the word is greater than three characters, add one to the counter
Step 4: Continue until you reach the end of the list
Step 5: Create an array called longWords that has a length of counter
Step 6: Create a variable that will be used as an index for longWords and set it to zero
Step 7: Look at each word in the original list
Step 8: If the length of the word is greater than three characters, add the word to longWords and also add one to the index that is used for longWords
Step 9: Continue until you reach the end of the list
Step 10: Return the longWords array

Pseudocode 2:

```
set counter = 0
for (iterate through every word in the original list)
{
   if (word length > 3)
   {
      counter = counter + 1
   }
}
create an array called longWords whose length is the same as counter
set index = 0
for (iterate through every word in the original list)
{
   if (word length > 3)
   {
      add the word to longWords using index
      set index = index + 1
   }
}
return longWords
```

Java Code 2: Using an array (using a `for-each` loop and a `for` loop)

```
public static String[] removeStopWords(String[] words)
{
    int count = 0;
    for (String word : words)
    {
        if (word.length() > 3)
        {
            count++;
        }
    }
    String[] longWords = new String[count];
    index = 0;
    for (int i = 0; i < words.length; i++)
    {
        if (words[i].length() > 3)
        {
            longWords[index] = words[i];
            index++;
        }
    }
    return longWords;
}
```

Example

Using an ArrayList:

```
public class TweetSentimentRunner
{
    public static void main(String[] args)
    {
        ArrayList<String> tweet = new ArrayList<String>();

        tweet.add("If");
        tweet.add("only");
        tweet.add("Bradley's");
        tweet.add("arm");
        tweet.add("was");
        tweet.add("longer");
        tweet.add("best");
        tweet.add("photo");
        tweet.add("ever");

        ArrayList<String> processedTweet = removeStopWords(tweet);

        System.out.println(processedTweet);       // print the ArrayList
    }

    public static ArrayList<String> removeStopWords(ArrayList<String> arr)
    {
        /*  implementation is described above Java code 1 */
    }
}
```

```
OUTPUT
[only, Bradley's, longer, best, photo, ever]
```

❯ Rapid Review

- The process of developing an original algorithm requires concentration and analytical thought.
- Accessing data buried within a class hierarchy requires the programmer to use the methods that belong to the class.

❯ Review Questions

1. Count the empty lockers.

A high school has a specific number of lockers so students can store their belongings. The office staff wants to know how many of the lockers are empty so they can assign these lockers to new students. Consider the following classes and complete the implementation for the `countEmptyLockers` method.

```
public class Locker
{
   private boolean inUse;

   public boolean isInUse()
   {    return inUse;    }

   public void setInUse(boolean isInUse)
   {    inUse = isInUse;     }
}

public class School
{
   private static Locker[] lockerList;

   /*   Additional implementation not shown      */

   /**
    * This method counts the number of empty lockers in the list of lockers.
    * @param lockers the list of Locker objects
    * @return the number of empty lockers
    * PRECONDITION:  No object in the list lockers is null
    *                (every locker has a true/false value)
    */
   private static int countEmptyLockers(Locker[] lockers)
   {
      /* to be implemented */
   }
}
```

2. Find the total revenue for the school lunch program.

A high school lunch program serves lunch to students. When the student checks out, the lunch person enters the cost of the meal. You have been hired to write the method that calculates the total amount of money received for every lunch order for all the days of the school year.

The following classes are used in the software for the lunch program. Complete the implementation for the method `getTotalOfOrders`.

```
public class Order
{
    private double amount;

    public Order(double cost)
    {
        amount = cost;
    }

    public double getAmount()
    {   return amount;   }

    public void setAmount(double cost)
    {   amount = cost;   }
}

public class Day
{
    private List<Order> ordersForTheDay;

    public Day()
    {
        ordersForTheDay = new ArrayList<Order>();
    }

    public void addOrder(Order order)
    {   ordersForTheDay.add(order);   }

    public List<Order> getOrdersForTheDay()
    {   return ordersForTheDay;   }

    public void setDaysOrders(List<Order> daysOrders)
    {   ordersForTheDay = daysOrders;   }
}

public class LunchRunner
{
    public static void main(String[] args)
    {
        ArrayList<Day> allDays = new ArrayList<Day>();

        allDays = /* data entered by lunch workers */

        /*  Additional implementation not shown    */
    }

    /**
     * This method finds the total of all the orders for the list of Days.
     * @param allDays  the list of Days
     * @return the total dollar amount of all the Orders in each Day
     */
    public static double getTotalOfOrders(ArrayList<Day> allDays)
    {
        /* to be implemented  */
    }
}
```

❯ Answers and Explanations

1. Count the empty lockers.

Algorithm:
Step 1: Create a variable called total and set it to zero
Step 2: Look at each of the lockers in the list
Step 3: If the locker is not in use, increment the total
Step 4: Continue until you reach the end of the list
Step 5: Return the total

Pseudocode:
```
set total = 0
for (iterate through all the lockers in the list)
{
    if (the locker is not in use)
    {
        total = total + 1
    }
}
return total
```

Java Code: Java code using an array (`for-each` loop)
```
private static int countEmptyLockers(Locker[] lockers)
{
    int total = 0;
    for (Locker locker : lockers)
    {
        if (!locker.isInUse())
        {
            total++;
        }
    }
    return total;
}
```

2. Find the total revenue for the school lunch program.

Algorithm:
Step 1: Create a variable called total and set it to zero
Step 2: Look at each Day in the list of Days
Step 3: Look at each Order in the list of Orders for each Day
Step 4: Add the current order amount to the total
Step 5: Continue until you reach the end of the list of Orders
Step 6: Continue until you reach the end of the list of Days
Step 7: Return the total

Pseudocode:

```
set total = 0
for (iterate through all the Days in the list of Days)
{
    for (iterate through all the Orders in the list of Orders)
    {
        total = total + current order
    }
}
return total
```

Java Code: Java code using an ArrayList (2 `for-each` loops)

```
public static double getTotalOfOrders(ArrayList<Day> allDays)
{
    double total = 0;
    for (Day day : allDays)
        for (Order order : day.getOrdersForTheDay())
            total += order.getAmount();

    return total;
}
```

Sorting Algorithms and the Binary Search

IN THIS CONCEPT

Summary: Welcome to the Information Age. There has never before been a time in human history that so much information was produced and recorded. So, what's a person to do with all that information? Search it? Sort it? Find patterns that can help people? In this concept, you will learn how to implement algorithms that allow you to search and sort data, regardless of how many pieces of data there are.

Key Ideas

- The Insertion Sort, Selection Sort, and Merge Sort are sorting routines.
- The Binary Search is an efficient way to search for a target in a sorted list.

Background on Sorting Data

The idea of sorting is quite easy for humans to understand. In contrast, teaching a computer to sort items all by itself can be a challenge. Through the years, hundreds of sorting algorithms have been developed, and they have all been critiqued for their efficiency and memory usage. The AP Computer Science A Exam expects you to be able to read and analyze code that uses three main sorts: the Insertion Sort, the Selection Sort, and the Merge Sort.

The most important thing to remember is different sorting algorithms are good for different things. Some are easy to code, some use the CPU efficiently, some use memory efficiently, some are good at adding an element to a pre-sorted list, some don't care whether the list is pre-sorted or not, and so on. Of our three algorithms, Merge Sort is the fastest, but it uses the most memory.

The Swap Algorithm

Recall the swapping algorithm that you learned in Concept 6: Algorithms (Basic Version). It is used in some of the sorting algorithms in this concept.

Sorting Algorithms on the AP Computer Science A Exam
You will not have to write the full code for any of the sorting routines described in this concept. You should, however, understand how each works.

Insertion Sort

The **Insertion Sort** algorithm is similar to the natural way that people sort a list of numbers if they are given the numbers one at a time. For example, if you gave a person a series of number cards one at a time, the person could easily sort the numbers "on the fly" by inserting the card where it belonged as soon as the card was received. Insertion Sort is considered to be a relatively simple sort and works best on very small data sets.

To write the code that can automate this process on a list of numbers requires an algorithm that does a lot of comparing. Let's do an example of how Insertion Sort sorts a list of numbers from smallest to largest.

Insertion Sort	1st number	2nd number	3rd number	4th number	5th number	6th number
Original list	67	23	12	54	35	18
After 1st pass	23	67	12	54	35	18
After 2nd pass	12	23	67	54	35	18
After 3rd pass	12	23	54	67	35	18
After 4th pass	12	23	35	54	67	18
After 5th pass	12	18	23	35	54	67

This algorithm uses a temporary variable that I will call temp. The first step is to put the number that is in the second position into the temporary variable (temp). Compare temp with the number in the first position. If temp is less than the number in the first position, then move the number that is in the first position to the second position and put temp in the first position. Now, the first pass is completed and the first two numbers are sorted from smallest to largest.

Next, put the number in the third position in temp. Compare temp to the number in the second position in the list. If temp is less than the second number, move the second number into the third position and compare temp to the number in the first position. If temp is less than the number in the first position, move the number in the first position to the second position, and move temp to the first position. Now, the second pass is completed and the first three numbers in the list are sorted. Continue this process until the last number is compared against all of the other numbers and inserted where it belongs.

I'll Pass

A **pass** in programming is one iteration of a process. Each of these sorts makes a series of passes before it completes the sort. An efficient algorithm uses the smallest number of passes that it can.

Implementation

The following class contains a method that sorts an array of integers from smallest to greatest using the Insertion Sort algorithm.

```
public class InsertionSort
{
    public static void main(String[] args)
    {
        int[] myArray = {67, 23, 12, 54, 35, 18};
        insertionSort(myArray);
        for (int i : myArray)
        {
            System.out.print(i + "\t");
        }
    }

    /**
     * This method sorts an int array using Insertion Sort.
     *
     * @param element the array containing the items to be sorted
     *
     * Postcondition:  arr contains the original elements and
     *                 elements are sorted in ascending order
     */
    public static void insertionSort(int[] arr)
    {
        for (int j = 1; j < arr.length; j++)
        {
            int temp = arr[j];
            int index = j;
            while (index > 0 && temp < arr[index - 1])
            {
                arr[index] = arr[index - 1];
                index--;
            }
            arr[index] = temp;
        }
    }
}
```

OUTPUT

```
12    18    23    35    54    67
```

Selection Sort

The **Selection Sort** algorithm forms a sorted list by repeatedly finding and selecting the smallest item in a list and putting it in its proper place.

Selection Sort	1st number	2nd number	3rd number	4th number	5th number	6th number
Original list	67	23	12	54	35	18
After 1st pass	**12**	23	67	54	35	18
After 2nd pass	**12**	**18**	67	54	35	23
After 3rd pass	**12**	**18**	**23**	54	35	67
After 4th pass	**12**	**18**	**23**	**35**	54	67
After 5th pass	**12**	**18**	**23**	**35**	**54**	67
After 6th pass	**12**	**18**	**23**	**35**	**54**	**67**

To sort a list of numbers from smallest to largest using Selection Sort, search the entire list for the smallest item, select it, and swap it with the first item in the list (the two numbers change positions). This completes the first pass. Next, search for the smallest item in the remaining list (not including the first item), select it, and swap it with the item in the second position. This completes the second pass. Then, search for the smallest item in the remaining list (not including the first or second items), select it, and swap it with the item in the third position. Repeat this process until the last item in the list becomes (automatically) the largest item in the list.

Implementation

The following class contains a method that sorts an array of integers from smallest to greatest using the Selection Sort algorithm.

```
public class SelectionSort
{
    public static void main(String[] args)
    {
        int[] myArray = {67, 23, 12, 54, 35, 18};
        selectionSort(myArray);
        for (int i : myArray)
        {
            System.out.print(i + "\t");
        }
    }
```

```
    /**
     * This method sorts an int array using Selection Sort.
     *
     * @param arr  the array to be sorted
     *
     * Postcondition:    arr contains the original elements and
     *                   elements are sorted in ascending order
     */
    public static void selectionSort(int[] arr)
    {
        for (int j = 0; j < arr.length - 1; j++)
        {
            int index = j;
            for (int k = j + 1; k < arr.length; k++)
            {
                if (arr[k] < arr[index])
                    index = k;
            }

            int temp = arr[j];
            arr[j] = arr[index];
            arr[index] = temp;
        }
    }
}
```

```
OUTPUT
12    18    23    35    54    67
```

Merge Sort

The **Merge Sort** is called a "divide and conquer" algorithm and uses recursion. The Merge Sort algorithm repeatedly divides the numbers into two groups until it can't do it anymore (that's where the recursion comes in). Next, the algorithm merges the smaller groups together and sorts them as it joins them. This process repeats until all the groups form one sorted group.

Merge Sort is a relatively fast sort and is much more efficient on large data sets than Insertion Sort or Selection Sort. Although it seems like Merge Sort is the best of these three sorts, its downside is that it requires more memory to execute.

Example

Sort a group of numbers using the Merge Sort algorithm.

Goal: Sort these numbers from smallest to largest: 8, 5, 2, 6, 9, 1, 3, 4

Step 1: Split the group of numbers into two groups: 8, 5, 2, 6 and 9, 1, 3, 4
Step 2: Split each of these groups into two groups: 8, 5 and 2, 6 and 9, 1 and 3, 4
Step 3: Repeat until you can't anymore: 8 and 5 and 2 and 6 and 9 and 1 and 3 and 4
Step 4: Combine two groups, sorting as you group them: 5, 8 and 2, 6 and 1, 9 and 3, 4
Step 5: Combine two groups, sorting as you group them: 2, 5, 6, 8 and 1, 3, 4, 9
Step 6: Repeat the previous step until you have one sorted group: 1, 2, 3, 4, 5, 6, 8, 9

Implementation

The following class contains three interdependent methods that sort a list of integers from smallest to greatest using the Merge Sort algorithm. The method setUpMerge is a recursive method.

```
public class MergeSort
{
    private static int[] myArray;
    private static int[] tempArray;

    public static void main(String[] args)
    {
        int[] inputArray = {8, 5, 2, 6, 9, 1, 3, 4};
        mergeSort(inputArray);
        for (int i: inputArray)
        {
            System.out.print(i + "\t");
        }
    }

    /**
     * This method starts the process of the Merge Sort.
     *
     * @param arr   the array to be sorted
     */
    private static void mergeSort(int arr[])
    {
        myArray = arr;
        int length = arr.length;
        tempArray = new int[length];
        setUpMerge(0, length - 1);
    }

    /**
     * This method is called recursively to divide the array
     * into each of the groups to be sorted.
     *
     * @param lower   the lower index of the subgroup
     * @param higher  the higher index of the subgroup
     */
     private static void setUpMerge(int lower, int higher)
     {
         if (lower < higher)
         {
             int middle = lower + (higher - lower) / 2;
             setUpMerge(lower, middle);
             setUpMerge(middle + 1, higher);
             doTheMerge(lower, middle, higher);
         }
     }
```

```
    /**
    * This method merges the elements in two subgroups
    * ([start, middle] and [middle + 1, end]) in ascending order.
    * @param lower the index of the lower element
    * @param middle the index of the  middle element
    * @param higher the index of the higher element
    */
    private static void doTheMerge(int lower, int middle, int higher)
    {
        for (int i = lower; i <= higher; i++)
        {
            tempArray[i] = myArray[i];
        }
        int i = lower;
        int j = middle + 1;
        int k = lower;
        while (i <= middle && j <= higher)
        {
            if (tempArray[i] <= tempArray[j])
            {
                myArray[k] = tempArray[i];
                i++;
            }
            else
            {
                myArray[k] = tempArray[j];
                j++;
            }
            k++;
        }
        while (i <= middle)
        {
            myArray[k] = tempArray[i];
            k++;
            i++;
        }
    }
}
```

```
OUTPUT
1   2   3   4   5   6   7   8   9
```

Binary Search

So far, the only algorithm we know to search for a target in a list is the Sequential Search. It starts at one end of a list and works toward the other end, comparing each element to the target. Although it works, it's not very efficient. The **Binary Search** algorithm is the most efficient way to search for a target item *provided the list is already sorted*. It repeatedly eliminates half of the search list with each pass by guessing in the middle of the existing list and determining if the guess was too high or too low. The secret of the Binary Search algorithm is that the data must be sorted prior to doing the search.

Sorted Data
The Binary Search only works on sorted data. Performing a Binary Search on unsorted data produces invalid results.

Example
The "Guess My Number" game consists of someone saying something like, "I'm thinking of a number between 1 and 100. Try to guess it. I will tell you if your guess is too high or too low."

Solution 1: Terrible way to win the "Guess My Number" game

First guess:	1	Response: Too Low
Second guess:	2	Response: Too Low
Third guess:	3	Response: Too Low
and so on		
and so on		

Solution 2: Fastest way to win the "Guess My Number" game (the Binary Search algorithm)

Guess halfway between 1 and 100	First Guess: 50	Response: Too Low
Guess halfway between 51 and 100	Second Guess: 75	Response: Too High
Guess halfway between 51 and 74	Third Guess: 62	Response: Too Low
Guess halfway between 63 and 74	Fourth Guess: 68	Response: Too Low
Guess halfway between 69 and 74	Fifth Guess: 71	Response: Too High
Guess halfway between 69 and 70	Sixth Guess: 69	Response: YOU GOT IT!

Visual description of the "Guess My Number" game using the Binary Search algorithm:

A list of sorted numbers from 1 to 100

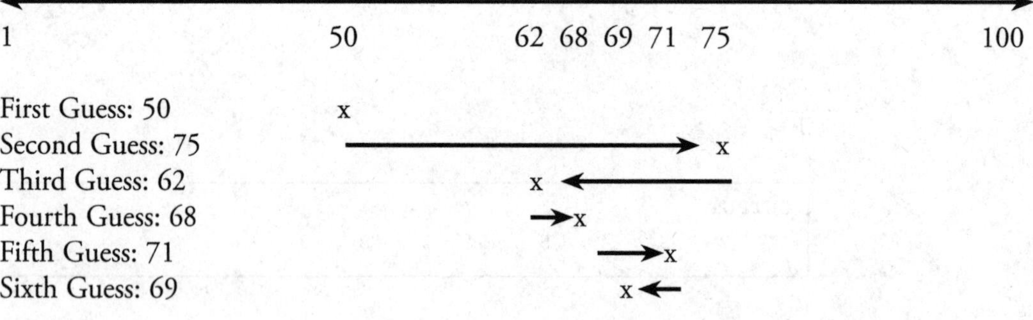

Implementation

The following class contains a method that searches an array of integers for a target value using the Binary Search algorithm. Notice that the main method makes a method call to sort the array prior to calling the binarySearch method.

```
public class BinarySearch
{
   public static void main(String[] args)
   {
      int[] myArray = {23, 146, 57, 467, 69, 36, 184, 492, 100};
      int searchTarget = 57;
      insertionSort(myArray);            // Any sorting method can be used

      System.out.println(binarySearch(searchTarget, myArray));
   }

   /**
    * This method performs a Binary Search on a sorted array of integers.
    *
    * @param target the value that you are searching for
    * @param data the array you are searching
    * @return the result of the search. True if found, False if not found.
    *
    * Precondition:  The data is already sorted from smallest to largest
    */
   public static boolean binarySearch(int target, int[] data)
   {
      int low = 0;
      int high = data.length;

      while(high >= low)
      {
         int middle = (low + high) / 2;
         if (data[middle] == target)
         {
            return true;
         }
         if (data[middle] < target)
         {
           low = middle + 1;
         }
         if (data[middle] > target)
         {
            high = middle - 1;
         }
      }
      return false;
   }
}
```

```
OUTPUT
true
```

› Rapid Review

Insertion Sort

- Insertion Sort uses an algorithm that repeatedly compares the next number in the list to the previously sorted numbers in the list and inserts it where it belongs.

Selection Sort

- Selection Sort uses an algorithm that repeatedly selects the smallest number in a list and swaps it with the current element in the list.

Merge Sort

- Merge Sort uses an algorithm that repeatedly divides the data into two groups until it can divide no longer. The algorithm joins the smaller groups together and sorts them as it joins them. This process repeats until all the groups are joined to form one sorted group.
- Merge Sort uses a divide and conquer algorithm and recursion.
- Merge Sort is more efficient than Selection Sort and Insertion Sort.
- Merge Sort requires more internal memory (RAM) than Selection Sort and Insertion Sort.

Binary Search

- A Binary Search algorithm is the most efficient way to search for a target value in a sorted list.
- A Binary Search algorithm requires that the list be sorted prior to searching.
- A Binary Search algorithm eliminates half of the search list with each pass.

› Review Questions

Basic Level

Questions 1–4 refer to the following information.

Array `arr` has been defined and initialized as follows

```
int[] arr = {5, 3, 8, 1, 6, 4, 2, 7};
```

1. Which of the following shows the elements of the array in the correct order after the first pass through the outer loop of the Insertion Sort algorithm?

(A) 1 5 3 8 6 4 2 7
(B) 1 3 8 5 6 4 2 7
(C) 5 3 7 1 6 4 2 8
(D) 3 5 8 1 6 4 2 7
(E) 1 2 3 4 5 6 7 8

2. Which of the following shows the elements of the array in the correct order after the fourth pass through the outer loop of the Insertion Sort algorithm?

(A) 1 3 5 6 8 4 2 7
(B) 1 2 3 4 6 5 8 7
(C) 1 2 3 4 5 6 7 8
(D) 3 5 8 1 6 4 2 7
(E) 1 2 3 4 5 6 8 7

3. Which of the following shows the elements of the array in the correct order after the first pass through the outer loop of the Selection Sort algorithm?

(A) 1 2 3 5 6 4 8 7
(B) 1 3 8 5 6 4 2 7
(C) 1 3 5 8 6 4 2 7
(D) 1 5 3 8 6 4 2 7
(E) 5 3 1 6 4 2 7 8

4. Which of the following shows the elements of the array in the correct order after the fourth pass through the outer loop of the Selection Sort algorithm?

(A) 3 1 4 2 5 6 7 8
(B) 3 1 4 2 5 8 6 7
(C) 1 2 3 4 5 6 7 8
(D) 1 2 3 4 5 8 6 7
(E) 1 2 3 4 6 5 8 7

5. Array `arr2` has been defined and initialized as follows.

```
int[] arr2 = {1, 2, 3, 4, 5, 6, 7, 8, 9, 10, 11};
```

If the array is being searched for the number 4, which sequence of numbers is still in the area to be searched after two passes through the `while` loop of the Binary Search algorithm?

(A) `1 2 3 4 5`
(B) `2 3 4`
(C) `4 5 6`
(D) `4 5`
(E) The number has been found in the second pass. Nothing is left to be searched.

Advanced Level

6. Consider the following method.

```
public static int mystery (int[] array, int a)
{
    int low = 0;
    int high = array.length - 1;
    while (low <= high)
    {
        int mid = (high + low) / 2;
        if (a < array[mid])
            high = mid - 1;
        else if (a > array[mid])
            low = mid + 1;
        else
            return mid;
    }
    return -1;
}
```

The algorithm implemented by the method can best be described as:

(A) Insertion Sort
(B) Selection Sort
(C) Binary Search
(D) Merge Sort
(E) Sequential Sort

7. Consider the following code segment that implements the Insertion Sort algorithm.

```
public void insertionSort(int[] arr)
{
    for (int i = 1; i < arr.length; i++)
    {
        int key = arr[i];
        int j = i - 1;
        while (j >= 0 && /* condition */)
        {
            arr[j + 1] = arr[j];
            j--;
        }
        arr[j + 1] = key;
    }
}
```

Which of the following can be used to replace `/* condition */` so that `insertionSort` will work as intended?

(A) `arr[i] > key`
(B) `arr[j] > key`
(C) `arr[i + 1] > key`
(D) `arr[j + 1] > key`
(E) `arr[i - 1] > key`

❯ Answers and Explanations

Bullets mark each step in the process of arriving at the correct solution.

1. The answer is D.

- Our array starts as:

5	3	8	1	6	4	2	7

- The 5 is bigger than everything to its left (which is nothing), so let it be. Our algorithm starts with the 3.
- Take out the 3 and store it in a variable: tempKey = 3. Now we have a gap.

5		8	1	6	4	2	7

- Is 3 < 5? Yes, so move the 5 to the right.

	5	8	1	6	4	2	7

- There's nothing more to check, so pop the 3 into the open space the 5 left behind.

3	5	8	1	6	4	2	7

- That's the order after the first pass through the loop (question 1).

2. The answer is A.

- We start where we left off at the end of question 1.

3	5	8	1	6	4	2	7

- Now for the second pass. The 3 and 5 are in order; tempKey = 8, leaving a gap.

3	5		1	6	4	2	7

- Since 5 < 8, we put the 8 back into its original spot.

3	5	8	1	6	4	2	7

- Now for the third pass: 3, 5, 8 are sorted; tempKey = 1, leaving a gap.

3	5	8		6	4	2	7

- 1 < 8, move 8 into the space 1 left behind. 1 < 5, move 5 over. 1 < 3, move 3 over, and put 1 in the first spot.

1	3	5	8	6	4	2	7

- Now for the fourth pass: 1, 3, 5, 8 are sorted; tempKey = 6, leaving a gap.

1	3	5	8		4	2	7

- 6 < 8, move the 8 into the space 6 left behind. 5 < 6, so we've found the spot for 6 and we pop it in.

1	3	5	6	8	4	2	7

- Giving us the answer to question 2.
- Note: One thing to look for in the answers when trying to recognize Insertion Sort—the end of the array doesn't change until it is ready to be sorted. So the 4 2 7 in our example are in the same order that they were in at the beginning.

3. The answer is B.

- Our array starts as:

5	3	8	1	6	4	2	7

- We scan the array to find the smallest element: 1.
- Now swap the 1 and the 5.

1	3	8	5	6	4	2	7

- That's the order after the first pass through the loop (question 3).

4. The answer is E.

- We start where we left off at the end of question 3.
- Now for the second pass. Leave the 1 alone and scan for the smallest remaining item: 2.

1	3	8	5	6	4	2	7

- The 3 is in the spot the 2 needs to go into, so swap the 3 and the 2.

1	2	8	5	6	4	3	7

- The third pass swaps the 8 and the 3.

1	2	3	5	6	4	8	7

- The fourth pass swaps the 5 and the 4.

1	2	3	4	6	5	8	7

- This gives us the answer to question 4.

5. The answer is D.

- The Binary Search algorithm looks at the element in the middle of the array, sees if it is the right answer, and then decides if the target item is higher or lower than that element. At that point it knows which half of the array the target item is in. Then it looks at the middle element of that half, and so on.
- Here's our array:

1	2	3	4	5	6	7	8	9	10	11

- First pass: The middle element is 6. We are looking for 4. 4 < 6, so eliminate the right half of the array (and the 6). Now we are considering:

1	2	3	4	5	6	~~7~~	~~8~~	~~9~~	~~10~~	~~11~~

- Second pass: The middle element is 3. 4 > 3, so we eliminate the left half of the section we are considering (and the 3). Now we are considering:

~~1~~	~~2~~	~~3~~	4	5	6	~~7~~	~~8~~	~~9~~	~~10~~	~~11~~

- You can see that we will find the 4 on our next round, but the answer to the question is 4 5.
- Good to know: In this example, there always *was* a middle element when we needed one. If there is an even number of elements, the middle will be halfway in between two elements. The algorithm will just decide whether to round up or down in those cases (usually down, as integer division makes that easy).

6. The answer is C.

- Binary Search works like this:
  - We look at the midpoint of a list, compare it to the element to be found (let's call it the key) and decide if our key is >, <, or = that midpoint element.
  - If our key = midpoint element, we have found our key in the list.
  - If our key > midpoint element, we want to reset the list so that we will search just the top half of the current list.
  - If our key < midpoint element, we want to reset the list so that we will search just the bottom half of the current list.
- Look at the given code. We can see those comparisons, and we can see the high and low ends of the list being changed to match the answers to those comparisons. This code implements Binary Search.

7. The answer is B.

   - There are many small alterations possible when writing any algorithm. Do you go from 1 to length or from 0 to length -1, for example. This algorithm makes those kinds of alterations to the version given in the text. But in all cases, the general approach is:
     - The outer loop goes through the array. Everything to the left of the outer loop index (i in this case) is sorted.
     - Take the next element and put it into a temporary variable (key in this case).
     - Look at the element to the left of the element you are working on. If it is larger than the element to be sorted, move it over. Keep doing that until you find an element that is smaller than the element to be sorted (this is the condition we need).
     - Now you've found where our element needs to go. Exit inner loop and pop the element into place.
   - So the condition we are looking for reflects *keep doing that until we find an element that is smaller than the element to be sorted,* or, phrasing it like a while loop ("as long as" instead of "until"), *keep going as long as the element we are looking at is larger than our key.* In code, we want to keep going as long as arr[j] > key, so that's our condition.

Seeing the Big Picture: Design

IN THIS CONCEPT

Summary: If you take a bird's eye view of the idea of software development, you can see that it is important for programmers to follow guidelines. The rules that guide programmers help make sure that programs are easily maintainable.

Key Ideas

- Program specifications define what a program is supposed to accomplish.
- Someone other than the software developer often writes program specifications.
- Software engineering is a field that studies how software is designed, developed, and maintained.
- Top-down and bottom-up are two approaches to designing class hierarchy.
- Procedural abstraction helps make software more modular.
- Software developers use testing techniques to verify that their program is working correctly.

The Software Development Cycle

Program Specifications

When you are in computer science class, your teacher probably gives you assignments that have **program specifications**. This means that in order to receive full credit for the assignment, you have to follow the directions and make sure that your program does what it is supposed to do. When you write your own programs for your own purpose, you make your own specifications.

In the real world of programming, it is common for someone from a design team to give the software developers the program specifications. Specifications are descriptions of what the program should look like and how it should operate. Someone other than the actual programmer writes these **specs**. You may have heard that it is important to be a good communicator. Well, great programmers have good communication skills so that they can talk back and forth with the person who writes the specs to clarify any questions they may have.

Working for Others
Programmers must learn how to write code from specifications. On the Free-Response Questions section of the AP Computer Science A Exam, you have to write code that matches the specifications that are described in the questions.

Software Engineering

Software engineering is the study of designing, developing, and maintaining software.

Over the years, many different software development models have been designed that help developers create quality software. Here is a brief list.

- Prototyping—an approximation of a final system is built, tested, and reworked until it is acceptable. The complete system is then developed from this prototype.
- Incremental development—The software is designed, implemented, and tested a little bit at a time until the product is finished.
- Rapid application development—the user is actively involved in the evaluation of the product and modifications are made immediately as problems are found. Radical changes in the system are likely at any moment in the process.
- Agile software development—the developers offer frequent releases of the software to the customer and new requirements are generated by the users. Short development cycles are common in this type of development.
- Waterfall model—the progress of development is sequential and flows downward like a waterfall. Steps include conception, initiation, analysis, design, construction, testing, implementation, and maintenance.

Fun Fact: *Margaret Hamilton, a computer scientist who helped develop the on-board flight software for the Apollo space program, coined the phrase "software engineering." Her code prevented an abort of a moon landing!*

So What's the Best Way?

It depends on the situation. It varies from problem to problem, situation to situation, and even company to company. My belief is that you should view the models of software engineering as tools in a toolbox in which the technique that you apply depends on the project you are working on.

Designing Class Hierarchy

Top-Down Versus Bottom-Up Design

Top-down and **bottom-up** are two ways of approaching class hierarchy design. Top-down is also referred to as **functional decomposition**. The two designs are different only in their approach to the problem. Top-down design starts with the big picture, whereas a bottom-up design starts with the details. They each work toward the other, but where they begin is different.

Top-Down Versus Bottom-Up in Object-Oriented Programming

This concept can be applied to designing a complex class hierarchy. If you start your design by thinking of what would be at the top of the class hierarchy, then you are doing a top-down design. If you start your design by thinking of what would be at the bottom of the class hierarchy, then you are doing a bottom-up design.

For example, suppose you have been given the opportunity to design a video game for the NFL (National Football League). If you approach the design from a bottom-up perspective, you would say, "Let's design the Player class first. We'll determine all the instance variables and methods for the Player. Then we'll move on to the Team class. After we get those done, we'll figure out where to go from there." This approach starts with the lowest level object that could be in your game first (the Player), and then works *up* from there.

If you approach the design from a top-down perspective, you would say, "Let's design the League class first. We'll identify what every League has and what it can do. Then, we'll move onto the Conference class. After we get those done, we'll figure out where to go from there." This approach starts with the highest possible object first (the League), and then works *down* from there.

> **Top-Down Versus Bottom-Up**
> Top-down design begins with the parent classes, while bottom-up starts with the child classes. On the AP Computer Science A Exam, you will need to be able to recognize a top-down versus a bottom-up design.

Procedural Abstraction

When a program is broken down into pieces and each piece has a responsibility, then the program is following **procedural abstraction**. For example, if you write a quest game and have separate methods for hunting, gathering, and collecting, then you are following procedural abstraction. If your hunt method *also* gathers, collects, and slays dragons while eating a peanut butter and jelly sandwich, then you are not following procedural abstraction.

Fun Fact: *The word procedure comes from early procedural programming languages like Fortran and Pascal. Procedures are similar to methods in Java. Each procedure has a purpose and does its job when you ask it to.*

Procedural Abstraction
Every method in Java should have a specific job. Complex tasks should be broken down into smaller tasks.

Testing

You may have heard of video game testers whose job it is to find bugs in the software. Talk about a dream job, right? Well, professional programmers test their work to make sure that it does what it's supposed to do before they release it. There are a few standard ways that developers test code.

Unit Testing

Unit tests are typically short methods that test whether or not a specific method is performing its task correctly. Suppose you have a method that performs some kind of mathematical calculation. To test if this method is working correctly, you would write another method, called a **unit test**. The unit test calls the method you want to test to see if it is producing correct answers. To accomplish this, you figure out what the answer is *before* you make the call to the method and compare the method's answer with your answer.

Here are the steps to creating a unit test for a method:

1. Create a method that performs some kind of calculation.
2. Think of an appropriate input for the method and calculate the answer.
3. Create a unit test. The unit test must contain a call to the original method.
4. Run the unit test with your input and compare the result to the answer you calculated.
5. Repeat this process for every possible type of input that you can think of. For example, when creating a unit test method that does a mathematical calculation, you may want to test it with a positive number, a negative number, and zero.
6. If the method produces correct values for every one of your inputs, the method *passes the unit test.*

Example
Create a unit test method for the getArea method of the Circle class from Concept 2. Pass it a value for the radius as well as the known answer for its area to determine if the method is doing its calculation correctly.

```
public class UnitTestForCircle
{
    public static void main(String[] args)
    {
        testGetArea(new Circle(10), 314.15926); // Supply the correct answers
        testGetArea(new Circle(0), 0.0);
    }

    public static void testGetArea(Circle c, double correctAnswer)
    {
        // Use a tolerance to determine closeness of answer
        double tolerance = 0.0001;

        if (Math.abs(c.getArea() - correctAnswer) <= tolerance)
        {
            System.out.println(c.getRadius() + " passed the test");
        }
        else
        {
            System.out.println(c.getRadius() + " failed the test");
        }
    }
}  // End of Circle class
```

```
public class Circle
{
    /*  Additional implementation not shown */

    public double getArea()
    {
        return Math.PI * Math.pow(radius, 2);
    }
}  // End of Circle class
```

```
OUTPUT
10.0 passed the test
0.0 passed the test
```

Integration Testing

Integration testing is used to make sure that connections to outside resources are working correctly. For example, if you are writing software that connects to Snapchat, then you would write a method that tests whether the connection is made correctly.

› Rapid Review

- Program specifications define what a program is supposed to accomplish.
- Someone other than the software developer often writes program specifications.
- A good software developer has the ability to communicate with others.
- Software engineering is a field that studies how software is designed, developed, and maintained.
- There isn't one correct way to design a computer program. How you write the program depends on the application.
- Top-down and bottom-up are two approaches to designing class hierarchy.
- Top-down design starts with the parent classes and works toward the child classes.
- Bottom-up design starts with the child classes and works toward the parent classes.
- Procedural abstraction helps make software more modular.
- When applying procedural abstraction to a complex task, the developer breaks the complex task down into smaller parts and implements these tasks individually.
- Software developers use testing techniques to verify that their program is working correctly.
- Unit testing is the process of testing a method to make sure that it is working correctly.
- Integration testing is the process of testing connections to outside resources to make sure that they are working correctly.

❯ Review Questions

1. You are designing a program to simulate the students at your school. Starting with which one of these classes would best demonstrate top-down design?

(A) The student body
(B) The senior class
(C) The girls in the senior class
(D) The girls taking programming in the senior class
(E) Mary Zhang, a girl in the programming class

2. If you were designing a unit test for the `isEven` method, which of these methods would you prefer to be working with and why? Note that the two methods both work and return the same result.

Option 1:
```
public boolean isEven(int num)
{
    int digit = num % 10;
    if (digit >= 5)
        if (digit == 6 || digit == 8)
            return true;
        else
            return false;
    else if (digit == 0 || digit == 2 || digit == 4)
        return true;
    else
        return false;
}
```

Option 2:
```
public boolean isEven(int num)
{
    return (num % 2 == 0);
}
```

(A) Option 1 because it doesn't use modulus, so it's easier to understand.
(B) Option 1 because it is longer.
(C) Option 2 because there is only one path through the code, so fewer test cases are needed.
(D) Neither option requires testing because the intent of the method is clear from the name.
(E) It doesn't matter. Testing them would be the same.

3. What is the term used to describe breaking a large program into smaller, well-defined sections that are easier to code and maintain?

(A) Encapsulation
(B) Procedural abstraction
(C) Inheritance
(D) Static design
(E) Polymorphism

› Answers and Explanations

1. The answer is A.

In top-down design, you start by designing the broadest category, and then work your way down. The broadest category here is the student body.

2. The answer is C.

Unit tests have to test every path through the code. The fewer the paths, the shorter the test, the more sure the tester is that the code will work in every case.

3. The answer is B.

This process is called procedural abstraction.

Building Your Test-Taking Confidence

AP Computer Science A Practice Exam 1

AP Computer Science A Practice Exam 2

AP Computer Science A: Practice Exam 1

Multiple-Choice Questions

ANSWER SHEET

1	A	B	C	D	E	21	A	B	C	D	E
2	A	B	C	D	E	22	A	B	C	D	E
3	A	B	C	D	E	23	A	B	C	D	E
4	A	B	C	D	E	24	A	B	C	D	E
5	A	B	C	D	E	25	A	B	C	D	E
6	A	B	C	D	E	26	A	B	C	D	E
7	A	B	C	D	E	27	A	B	C	D	E
8	A	B	C	D	E	28	A	B	C	D	E
9	A	B	C	D	E	29	A	B	C	D	E
10	A	B	C	D	E	30	A	B	C	D	E
11	A	B	C	D	E	31	A	B	C	D	E
12	A	B	C	D	E	32	A	B	C	D	E
13	A	B	C	D	E	33	A	B	C	D	E
14	A	B	C	D	E	34	A	B	C	D	E
15	A	B	C	D	E	35	A	B	C	D	E
16	A	B	C	D	E	36	A	B	C	D	E
17	A	B	C	D	E	37	A	B	C	D	E
18	A	B	C	D	E	38	A	B	C	D	E
19	A	B	C	D	E	39	A	B	C	D	E
20	A	B	C	D	E	40	A	B	C	D	E

AP Computer Science A: Practice Exam 1

Part I (Multiple Choice)

Time: 90 minutes
Number of questions: 40
Percent of total score: 50

Directions: Choose the best answer for each problem. Some problems take longer than others. Consider how much time you have left before spending too much time on any one problem.

Notes:

- You may assume all import statements have been included where they are needed.
- You may assume that the parameters in method calls are not null.
- You may assume that declarations of variables and methods appear within the context of an enclosing class.

1. Consider the following code segment.

```
int myValue = 17;
int multiplier = 3;
int answer = myValue % multiplier + myValue / multiplier;
answer = answer * multiplier;
System.out.println(answer);
```

What is printed as a result of executing the code segment?

(A) `9`
(B) `10`
(C) `7`
(D) `30`
(E) `21`

2. Assume that `a`, `b`, and `c` have been declared and correctly initialized with `int` values. Consider the following expression.

```
boolean bool = !(a < b || b <= c) && !(a < c || b >= a);
```

Under what conditions does `bool` evaluate to true?

(A) `a = 1, b = 2, c = 3`
(B) `a = 3, b = 2, c = 1`
(C) `a = 3, b = 1, c = 2`
(D) All conditions; `bool` is always `true`.
(E) No conditions; `bool` is always `false`.

3. Consider the following code segment.

```
int[] myArray = {2, 3, 4, 1, 7, 6, 8};
int index = 0;
while (myArray[index] < 7)
{
    myArray[index] += 3;
    index++;
}
```

What values are stored in `myArray` after executing the code segment?

(A) `{2, 3, 4, 1, 7, 6, 8}`
(B) `{2, 3, 4, 1, 7, 9, 11}`
(C) `{5, 6, 7, 4, 7, 9, 8}`
(D) `{5, 6, 7, 4, 7, 6, 8}`
(E) `{5, 6, 7, 4, 0, 9, 11}`

4. Consider the following interface used by a company to represent the items it has available for online purchase.

```
public interface OnlinePurchaseItem
{
    double getPrice();
    String getItem();
    String getMonth();
}
```

The company bills at the end of the quarter. Until then, it uses an `ArrayList` of `OnlinePurchaseItem` objects to track a customer's purchases.

```
private ArrayList<OnlinePurchaseItem> items;
```

The company decides to offer a 20 percent off promotion on all items purchased in September. Which of the following code segments properly calculates the correct total price at the end of the quarter?

I.
```
double total = 0.0;
for (int i = 0; i < items.size(); i++)
{
    if (items.get(i).getMonth().equals("September"))
        total += 0.80 * items.get(i).getPrice();
    else
        total += items.get(i).getPrice();
}
```

II.
```
double total = 0.0;
for (OnlinePurchaseItem purchase : items)
{
    if (purchase.get(i).getMonth().equals("September"))
        total = total + 0.80 * purchase.getPrice();
    else
        total += items.get(i).getPrice();

}
```

III.
```
double total = 0.0;
for (int i = items.size(); i >= 0; i--)
{
    if (items.get(i).getMonth().equals("September"))
        total = total + 0.80 * items.get(i).getPrice();
    else
        total += items.get(i).getPrice();
}
```

(A) I only
(B) II only
(C) I and II only
(D) II and III only
(E) I, II, and III

5. Consider the following method.

```
public int mystery(int n)
{
    if (n <= 1)
        return 1;
    return 2 + mystery(n - 1);
}
```

What value is returned by the call `mystery(5)`?

(A) `1`
(B) `7`
(C) `8`
(D) `9`
(E) `10`

6. Consider the following code segment.

```
String myString = "H";
int index = 0;
while (index < 4)
{
    for (int i = 0; i < index; i++)
        myString = "A" + myString + "A";
    index ++;
}
```

What is the value of `myString` after executing the code segment?

(A) `"H"`
(B) `"AHA"`
(C) `"AAAHAAA"`
(D) `"AAAAHAAAA"`
(E) `"AAAAAAHAAAAAA"`

7. Consider the following statement.

```
int var = (int)(Math.random() * 50) + 10;
```

What are the possible values of `var` after executing the statement?

(A) All integers from 1 to 59 (inclusive)
(B) All integers from 10 to 59 (inclusive)
(C) All integers from 10 to 60 (inclusive)
(D) All real numbers from 50 to 60 (not including 60)
(E) All real numbers from 10 to 60 (not including 60)

8. Consider the following statement.

```
System.out.print(13 + 6 + "APCSA" + (9 - 5) + 4);
```

What is printed as a result of executing the statement?

(A) `136APCSA(9 - 5)4`
(B) `19APCSA8`
(C) `19APCSA44`
(D) `136APCSA8`
(E) `136APCSA44`

9. Consider the following method.

```
public int guess(int num1, int num2)
{
    if (num1 % num2 == 0)
        return (num2 + num1) / 2;
    return guess(num2, num1 % num2) + (num1 % num2);
}
```

What is the value of `num` after executing the following code statement?

```
int num = guess(3, 17);
```

(A) `6`
(B) `7`
(C) `10`
(D) Nothing is returned. Modulus by 0 causes an `ArithmeticException`.
(E) Nothing is returned. Infinite recursion causes a stack overflow error.

10. Consider the following class.

```
public class Automobile
{
    private String make, model;
    public Automobile(String myMake, String myModel)
    {
        make = myMake;
        model = myModel;
    }

    public String getMake()
    {  return make;   }

    public String getModel()
    {  return model;   }

    /* Additional implementation not shown */
}
```

Consider the following code segment that appears in another class.

```
Automobile myCar = new Automobile("Ford", "Fusion");
Automobile companyCar = new Automobile("Toyota", "Corolla");
companyCar = myCar;
myCar = new Automobile("Kia", "Spectre");
System.out.println(companyCar.getMake() + " " + myCar.getModel());
```

What is printed as a result of executing the code segment?

(A) `Ford Spectre`
(B) `Ford Fusion`
(C) `Kia Spectre`
(D) `Toyota Corolla`
(E) `Toyota Spectre`

Questions 11–14 refer to the following class definition.

```
public class Depot
{
    private String city;
    private String state;
    private String country = "USA";
    private boolean active = true;

    public Depot(String myCity, String myState, String myCountry, boolean myActive)
    {
        city = myCity;
        state = myState;
        country = myCountry;
        active = myActive;
    }

    public Depot(String myState, String myCity)
    {
        state = myState;
        city = myCity;
    }

    public Depot(String myCity, String myState, boolean myActive)
    {
        city = myCity;
        state = myState;
        active = myActive;
    }

    public String toString()
    {
        return city + ", " + state + " " + country + " " + active;
    }

    /* Additional implementation not shown */
}
```

11. The `Depot` class has three constructors. Which of the following is the correct term for this practice?

(A) Overriding
(B) Procedural abstraction
(C) Encapsulation
(D) Polymorphism
(E) Overloading

12. Consider the following code segment in another class.

```
Depot station = new Depot("Oakland", "California");
System.out.println(station);
```

What is printed as a result of executing the code segment?

(A) `California, Oakland USA true`
(B) `California, Oakland USA active`
(C) `USA, California USA true`
(D) `Oakland, California USA active`
(E) `Oakland, California USA true`

GO ON TO THE NEXT PAGE

13. Consider the following class definition.

```
public class WhistleStop extends Depot
```

Which of the following constructors compiles without error?

I.
```
public WhistleStop()
{
    super();
}
```

II.
```
public WhistleStop()
{
    super("Waubaushene", "Ontario", "Canada", true);
}
```

III.
```
public WhistleStop(String city, String province)
{
    super(city, province);
}
```

(A) I only
(B) II only
(C) III only
(D) II and III only
(E) I, II, and III

14. Assume a correct no-argument constructor has been added to the `WhistleStop` class.

Which of the following code segments compiles without error?

I.
```
ArrayList<Depot> stations = new ArrayList<Depot>();
stations.add(new WhistleStop());
stations.add(new Depot("Mansonville", "Quebec", "Canada", false));
```

II.
```
ArrayList<WhistleStop> stations = new ArrayList<Depot>();
stations.add(new WhistleStop());
stations.add(new Depot("Orinda", "California", true));
```

III.
```
ArrayList<WhistleStop> stations = new ArrayList<WhistleStop>();
stations.add(new WhistleStop());
stations.add(new Depot("Needham", "Massachusetts", true));
```

(A) I only
(B) I and II only
(C) I and III only
(D) II and III only
(E) I, II, and III

15. Consider the following code segment.

```
String crazyString = "crazy";
ArrayList<String> crazyList = new ArrayList<String>();
crazyList.add("weird");
crazyList.add("enigma");
for (String s : crazyList)
{
    crazyString = s.substring(1, 3) + crazyString.substring(0, 4);
}
System.out.println(crazyString);
```

What is printed as a result of executing the code segment?

(A) `nicraz`
(B) `nieicr`
(C) `nieicraz`
(D) `einicraz`
(E) `weienicraz`

16. Consider the following method.

```
public int pathways(int n)
{
    int ans;
    if (n < -5)
        ans = 1;
    else if (n < 0)
        ans = 2;
    else if (n > 10)
        ans = 3;
    else
        ans = 4;
    return ans;
}
```

Which of the following sets of data tests every possible path through the code?

(A) `-6, -1, 15, 12`
(B) `-5, -3, 12, 15`
(C) `-8, -5, 8, 10`
(D) `-6, 0, 20, 7`
(E) `-10, -5, 10, 12`

Questions 17–18 refer to the following scenario.

A resort wants to recommend activities for its guests, based on the temperature (degrees F) as follows:

76º and above	go swimming
61º–75º	go hiking
46º–60º	go horseback riding
45º and below	play ping pong

17. Consider the following method.

```
public String chooseActivity(int temperature)
{
    String activity;
    if (temperature > 75)
        activity = "go swimming";
    if (temperature > 60)
        activity = "go hiking";
    if (temperature > 45)
        activity = "go horseback riding";
    else
        activity = "play ping pong";
    return activity;
}
```

Consider the following statement.

```
System.out.println("We recommend that you " + chooseActivity(temperature));
```

For which temperature range is the correct suggestion printed (as defined above)?

(A) `temperature > 75`
(B) `temperature > 60`
(C) `temperature > 45`
(D) `temperature <= 60`
(E) Never correct

18. After discovering that the method did not work correctly, it was rewritten as follows.

```
public String chooseActivity(int temperature)
{
    String activity;
    if (temperature > 45)
        activity = "go horseback riding";
    else if (temperature > 60)
        activity = "go hiking";
    else if (temperature > 75)
        activity = "go swimming";
    else
        activity = "play ping pong";
    return activity;
}
```

Consider the following statement.

```
System.out.println("We recommend that you " + chooseActivity(temperature));
```

What is the largest temperature range for which the correct suggestion is printed (as defined above)?

(A) Always correct
(B) Never correct
(C) `temperature <= 75`
(D) `temperature <= 60`
(E) `temperature <= 45`

19. Assume `int[] arr` has been correctly instantiated and is of sufficient size. Which of these code segments results in identical arrays?

I.
```
int i = 1;
while (i < 6)
{
    arr[i / 2] = i;
    i += 2;
}
```

II.
```
for (int i = 0; i < 3; i++)
{
    arr[i] = 2 * i + 1;
}
```

III.
```
for (int i = 6; i > 0; i = i - 2)
{
    arr[(6 - i) / 2] = 6 - i;
}
```

(A) I and II only
(B) II and III only
(C) I and III only
(D) I, II, and III
(E) All three outputs are different.

20. Consider the following code segment.

```
List<String> nations = new ArrayList<String>();
nations.add("Argentina");
nations.add("Canada");
nations.add("Australia");
nations.add("Cambodia");
nations.add("Russia");
nations.add("France");
for (int i = 0; i < nations.size(); i++)
{
    if (nations.get(i).length() >= 7)
        nations.remove(0);
}
System.out.println(nations);
```

What is printed as a result of executing the code segment?

(A) `[Canada, Russia, France]`
(B) `[Cambodia, Russia, France]`
(C) `[Australia, Cambodia, Russia, France]`
(D) `[Canada, Cambodia, Russia, France]`
(E) Nothing is printed. `IndexOutOfBoundsException`

21. Consider the following method.

```
public int doStuff(int[] numberArray)
{
    int index = 0;
    int maxCounter = 1;
    int counter = 1;
    for (int k = 1; k < numberArray.length; k++)
    {
        if (numberArray[k] == numberArray[k - 1])
        {
            if (counter <= maxCounter)
            {
                maxCounter = counter;
                index = k;
            }
            counter++;
        }
        else
        {
            counter = 1;
        }
    }
    return numberArray[index];
}
```

Consider the following code segment.

```
int[] intArray = {3, 3, 4, 4, 6};
System.out.println(doStuff(intArray));
```

What is printed as a result of executing the code segment?

(A) 1
(B) 3
(C) 4
(D) 5
(E) 6

Questions 22–23 refer to the following class.

```
public class Coordinates
{
    private int x, y;

    public Coordinates(int myX, int myY)
    {
        x = myX;
        y = myY;
    }

    public void setX(int myX)
    {  x = myX;  }

    public int getX()
    {  return x;  }

    public void setY(int myY)
    {  y = myY;  }

    public int getY()
    {  return y;  }
}
```

Consider the following code segment.

```
Coordinates c1 = new Coordinates(0, 10);
Coordinates c3 = c1;
Coordinates c2 = new Coordinates(20, 30);
c3.setX(c2.getY());
c3 = c2;
c3.setY(c2.getX());
c2.setX(c1.getX());
```

22. Which of the following calls returns the value 20?

(A) `c1.getX()`
(B) `c1.getY()`
(C) `c2.getX()`
(D) `c2.getY()`
(E) `c3.getX()`

23. Which of the following calls returns the value 30?

(A) `c1.getX()`
(B) `c2.getX()`
(C) `c3.getX()`
(D) All of the above
(E) None of the above

GO ON TO THE NEXT PAGE

Questions 24–25 refer to the following interface and class.

```
public interface Supernatural
{
   public String getType();
   public int getPower();
}

public class Werewolf implements Supernatural
{
   private String type;
   private int power;

   public Werewolf(String myType, int myPower)
   {
      type = myType;
      power = myPower;
   }

   public String getType()
   {  return type; }

   public int getPower()
   {  return power; }
}
```

24. Consider the following class declarations.

I. `public class BabyWolf extends Supernatural`
   `{  /* implementation not shown */  }`

II. `public class BabyWolf extends WereWolf`
   `{  /* implementation not shown */  }`

III. `public class BabyWolf implements Supernatural`
   `{  /* implementation not shown */  }`

IV. `public class BabyWolf implements WereWolf`
   `{  /* implementation not shown */  }`

Which of the declarations does not cause an error?

(A) I and II only
(B) II and III only
(C) III and IV only
(D) I and III only
(E) I and IV only

25. Consider the following interface.

```
public interface FictionalCharacter
{
   String getTitle();
   boolean isHuman();
}
```

Which of the following class declarations does not cause an error?

(A) `public class A extends Supernatural implements FictionalCharacter`
(B) `public class B extends BabyWolf, WereWolf implements FictionalCharacter`
(C) `public class C implements Supernatural, Werewolf`
(D) `public class D extends Werewolf implements Supernatural, FictionalCharacter`
(E) `public class E extends FictionalCharacter`

Questions 26–28 refer to the following classes.

```
public class Person
{
   private String firstName;
   private String lastName;

   public Person(String fName, String lName)
   {
      firstName = fName;
      lastName = lName;
   }

   public String getName()
   {
      return firstName + " " + lastName;
   }
}

public class Adult extends Person
{
   private String title;

   public Adult(String fname, String lName, String myTitle)
   {
      super(fname, lName);
      title = myTitle;
   }

   /*  toString method to be implemented in question 27  */
}

public class Child extends Person
{
   private int age;

   /*  Constructor to be implemented in question 26  */
}
```

26. Which of the following is a correct implementation of a `Child` class constructor?

(A)
```
public Child(String first, String last, String t, int a)
{
    super(first, last, t);
    age = a;
}
```
(B)
```
public Child()
{
    super();
}
```
(C)
```
public Child(String first, String last, int a)
{
    super(first, last);
    age = a;
}
```
(D)
```
public Child extends Adult (String first, String last, String t, int a)
{
    super(first, last, t);
    age = a;
}
```
(E)
```
public Child extends Person(String first, String last, int a)
{
    super(first, last);
    age = a;
}
```

27. Which of the following is a correct implementation of the `Adult` class `toString` method?

(A)
```
public String toString()
{
    System.out.println(title + " " + super.toString());
}
```
(B)
```
public String toString()
{
    return title + " " + super.toString();
}
```
(C)
```
public String toString()
{
    return title + " " + firstName + " " + lastName;
}
```
(D)
```
public void toString()
{
    System.out.println(title + " " + super.getName());
}
```
(E)
```
public String toString()
{
    return title + " " + super.getName();
}
```

28. Consider the following method declaration in a different class.

```
public void findSomeone(Adult someone)
```

Assume the following variables have been correctly instantiated and initialized with appropriate values.

```
Person p;
Adult a;
Child c;
```

Which of the following method calls compiles without error?

I. `findSomeone(p);`
II. `findSomeone(a);`
III. `findSomeone(c);`
IV. `findSomeone((Adult) p);`
V. `findSomeone((Adult) c);`

(A) II only
(B) I and II only
(C) II and IV only
(D) II, IV, and V only
(E) II, III, and IV only

29. Consider the following method. The method is intended to return `true` if the value `val` raised to the power `power` is within the tolerance `tolerance` of the target value `target`, and `false` otherwise.

```
public boolean similar(double val, double power, double target, double tolerance)
{
   /* missing code */
}
```

Which code segment below can replace `/* missing code */` to make the method work as intended?

(A)
```
double answer = Math.pow(val, power);
if (answer - tolerance >= target)
   return true;
return false;
```
(B)
```
return (Math.abs(Math.pow(val, power))) - target <= tolerance;
```
(C)
```
double answer = Math.pow(Math.abs(val), power);
return answer - target <= tolerance;
```
(D)
```
double answer = Math.pow(val, power) - target;
boolean within = answer - tolerance >= 0;
return within;
```
(E)
```
return (Math.abs(Math.pow(val, power) - target) <= tolerance);
```

30. Consider the following method.

```
public int changeEm(int num1, int num2, int[] values)
{
    num1 = values[num2];
    values[num1] = num2;
    num2++;
    return num2;
}
```

Consider the following code segment.

```
int[] values = {1, 2, 3, 4, 5};
int num1 = 2;
int num2 = 3;
num2 = changeEm(num1, num2, values);

System.out.println("num1 = " + num1 + "  values[" + num2 + "] = " + values[num2]);
```

What is printed as a result of executing the code segment?

(A) `num1 = 2  values[4] = 4`
(B) `num1 = 2  values[4] = 5`
(C) `num1 = 2  values[4] = 3`
(D) `num1 = 4  values[3] = 3`
(E) `num1 = 4  values[3] = 4`

Questions 31–32 refer to the following information.

In computer graphics, RGB colors can be specified several ways.

- Colors can be represented by ordered triples in which the first number represents red, the second green, and the third blue. The three numbers range in value from 0 to 255 (inclusive).
- Colors can be represented by a single six-digit hexadecimal number written with a # in front of it.

These two representations have a simple relationship. The leftmost two digits of the hexadecimal number are a direct conversion from decimal to hexadecimal of the red value in the ordered triple, the middle two digits are a direct conversion of the green value in the ordered triple, and the rightmost two digits are a direct conversion of the blue value in the ordered triple.

For example, the ordered triple (10, 16, 20) becomes #0A1014 in hexadecimal form since 10 ➔ 0A, 16 ➔ 10, and 20 ➔ 14, and the hexadecimal #110F03 becomes the ordered triple (17, 15, 3).

31. Convert the RGB ordered triple (125, 200, 78) to hexadecimal form.

(A) #12520078
(B) #7DC84E
(C) #713128414
(D) #202018
(E) #84D452

32. Convert the RGB color #EA27B5 in hexadecimal form to an ordered triple.

(A) (1410, 27, 115)
(B) (204, 40, 115)
(C) (234, 39, 181)
(D) (221, 39, 171)
(E) (24, 39, 16)

33. A programmer intends to apply the standard Binary Search algorithm on the following array of integers. The standard Binary Search algorithm returns the index of the search target if it is found and -1 if the target is not found. What is returned by the algorithm when a search for 50 is executed?

```
int[] array = {9, 100, 11, 45, 76, 100, 50, 1, 0, 55, 99};
```

(A) `-1`
(B) `0`
(C) `5`
(D) `6`
(E) `7`

34. Consider the following code segment.

```
int[][] nums = { {0, 1, 2}, {3, 4, 5}, {6, 7, 8} };

for (int x = 0; x < nums.length; x++)
{
    int temp = nums[x][0];
    nums[x][0] = nums[x][2];
    nums[x][2] = temp;
}
```

What are the values in array nums after the code segment is executed?
(A) `{ {0, 1, 0}, {3, 4, 3}, {6, 7, 6} }`
(B) `{ {0, 0, 0}, {3, 3, 3}, {6, 6, 6} }`
(C) `{ {2, 1, 0}, {5, 4, 3}, {8, 7, 6} }`
(D) `{ {0, 1, 2}, {3, 4, 5}, {6, 7, 8} }`
(E) `{ {8, 7, 6}, {5, 4, 3}, {2, 1, 0} }`

35. Consider the following code segment.

```
ArrayList<Integer> newList = new ArrayList<Integer>();
Integer[] list = {2, 4, 3, 3, 2, 5, 1};
newList.add(list[0]);

for (int i = 1; i < list.length; i++)
{
    if (list[i] > newList.get(newList.size() - 1))
        newList.add(list[i]);
}

System.out.println(newList);
```

What is printed as a result of executing the code segment?
(A) `[2, 4, 5]`
(B) `[2, 2, 1]`
(C) `[2, 4, 3, 5, 1]`
(D) `[1, 2, 3, 3, 4, 5]`
(E) `[2, 4, 3, 3, 5, 1]`

36. Consider the following code segment.

```
int[][] a = new int[5][5];

for (int r = 0; r < a.length; r++)
    for (int c = r + 1; c < a[0].length; c++)
        a[r][c] = 9;

for (int d = 0; d < a.length; d++)
    a[d][d] = d;

System.out.println(a[1][2] + " " + a[3][3] + " " + a[4][3]);
```

What is printed as a result of executing the code segment?

(A) `9 9 9`
(B) `0 0 0`
(C) `0 3 9`
(D) `9 3 9`
(E) `9 3 0`

37. Consider the following sorting method.

```
public void mysterySort(int[] arr)
{
    for (int j = 1; j < arr.length; j++)
    {
        int temp = arr[j];
        int index = j;
        while (index > 0 && temp < arr[index - 1])
        {
            arr[index] = arr[index - 1];
            index--;
        }
        arr[index] = temp;
    }
}
```

Consider the following code segment.

```
int[] values = {9, 1, 3, 0, 2};
mysterySort(values);
```

Which of the following shows the elements of the array in the correct order after the second pass through the outer loop of the sorting algorithm?

(A) `{0, 1, 2, 9, 3}`
(B) `{0, 1, 3, 9, 2}`
(C) `{0, 1, 9, 3, 2}`
(D) `{1, 3, 9, 0, 2}`
(E) `{1, 3, 0, 2, 9}`

38. Consider the following method.

```
public int puzzle(int m, int n)
{
    if (n == 1)
        return m;
    return m * puzzle(m, n - 1);
}
```

What is returned by the method call `puzzle(3, 4)`?

(A) 9
(B) 64
(C) 81
(D) 128
(E) Nothing is returned. Infinite recursion causes a stack overflow error.

39. Consider the following code segment.

```
int[][] table = new int[5][6];
int val = 0;
for (int k = 0; k < table[0].length; k++)
{
    for (int j = table.length - 1; j >= 0; j--)
    {
        table[j][k] = val;
        val = val + 2;
    }
}
```

What is the value of `table[3][4]` after executing the code segment?

(A) 22
(B) 30
(C) 34
(D) 42
(E) 46

40. Consider the following method.

```
public void selectionSort(String[] arr)
{
    for (int i = 0; i < arr.length; i++)
    {
        int small = i;
        for (int j = i; j < arr.length; j++)
        {
            /*  missing code */
            small = j;
        }
        String temp = arr[small];
        arr[small] = arr[i];
        arr[i] = temp;
    }
}
```

Assume `String[] ray` has been properly instantiated and populated with `String` objects.

Which line of code should replace `/* missing code */` so that the method call `selectionSort(ray)` results in an array whose elements are sorted from least to greatest?

(A) `if (arr[j].equals(arr[small]))`

(B) `if (arr[j] > arr[small])`

(C) `if (arr[j] < arr[small])`

(D) `if (arr[j].compareTo(arr[small]) > 0)`

(E) `if (arr[j].compareTo(arr[small]) < 0)`

STOP. End of Part I.

AP Computer Science A: Practice Exam 1

Part II (Free Response)

Time: 90 minutes
Number of questions: 4
Percent of total score: 50

Directions: Write all of your code in Java. Show all your work.

Notes:

- You may assume all imports have been made for you.
- You may assume that all preconditions are met when making calls to methods.
- You may assume that all parameters within method calls are not null.
- Be aware that you should, when possible, use methods that are defined in the classes provided as opposed to duplicating them by writing your own code.

1. A regular hexagon is a six-sided geometric figure in which all six sides are the same length and all the interior angles measure 120 degrees.

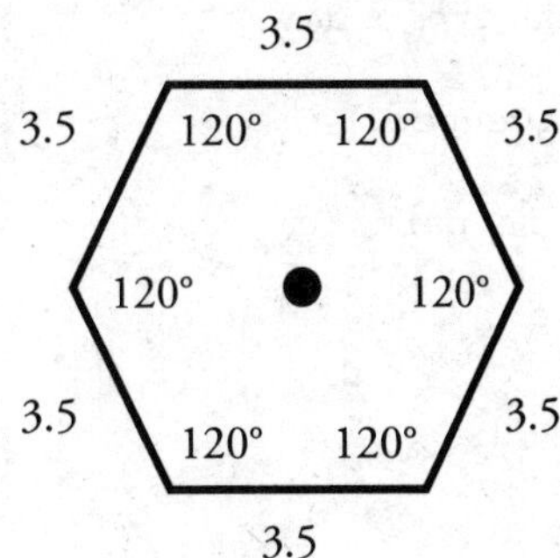

A regular hexagon is represented by the `RegularHexagon` class. The `RegularHexagon` class uses the `CoordinatePoint` class to specify its vertices (corners). The `CoordinatePoint` class represents a point on the coordinate plane.

```
public class RegularHexagon
{
   /** ArrayList points holds CoordinatePoint objects that represent
    *  the vertices of the RegularHexagon. The CoordinatePoint objects are in
    *  consecutive order. That is, if you connect the element at index 0 with
    *  the element at index 1 and so on to the element at index 5 and then
    *  back to the element at index 0, the hexagon will be formed correctly.
    */
   private ArrayList<CoordinatePoint> points;

   public RegularHexagon(ArrayList<CoordinatePoint> pts)
   {
      points = pts;
   }

   /** Returns the side length for a RegularHexagon
    *
    *  @return the length of each side of the RegularHexagon
    *  Precondition: The CoordinatePoint objects in points are in
    *                consecutive order.
    */
   public double getSideLength()
   {
      /*  to be implementedin part (a)   */
   }

   /** Returns the area of a RegularHexagon
    *
    *  @return the area of the RegularHexagon
    *  Precondition: The CoordinatePoint objects in points are in
    *                consecutive order.
    */
   public double getArea()
   {
      /*  to be implementedin part (b)   */
   }
```

```
    /** Returns the center of the RegularHexagon.
     *
     *  @return the CoordinatePoint that is the center of the RegularHexagon
     *  Precondition: The CoordinatePoint objects in points are in
     *                consecutive order.
     */
    public CoordinatePoint getCenter()
    {
       /*  to be implemented in part (c)  */
    }
}

public class CoordinatePoint
{
    private double x, y;

    public CoordinatePoint(double myX, double myY)
    {
       x = myX;
       y = myY;
    }

    public double getX()
    {  return x;  }

    public double getY()
    {  return y;  }
}
```

(a) Write the `getSideLength` method for the `RegularHexagon` class. The side length is found by calculating the distance between two consecutive coordinate points. The formula for computing distance is:

Example

Suppose two consecutive coordinate points of the regular hexagon are (1, 3) and (2, 4). The side length of the regular hexagon is:

$$distance = \sqrt{(x_2 - x_1)^2 + (y_2 - y_1)^2}$$

Complete the method `getSideLength`.

$$distance = \sqrt{(2-1)^2 + (4-3)^2} = \sqrt{2} = 1.414...$$

```
/** Returns the side length for a RegularHexagon
 *  @return the length of each side of the RegularHexagon
 */
public double getSideLength()
```

(b) Write the `getArea` method for the `RegularHexagon` class. The formula for computing the area of a regular hexagon is:

$$Area = \frac{(side\ length)^2 * 3\sqrt{3}}{2}$$

Example:
Suppose a regular hexagon has a side length of 5. Its area is:

$$Area = \frac{5^2 * 3\sqrt{3}}{2} = 64.9519\ldots$$

You may assume that the `getSideLength` method works as intended, regardless of what you wrote in part (a).

Complete the method `getArea`.

```
/** Returns the area of a RegularHexagon
 *   @return the area of the RegularHexagon
 */
public double getArea()
```

(c) Write the `getCenter` method for the `RegularHexagon` class. The center of a regular hexagon is located at the midpoint of any diagonal of a regular hexagon.

A diagonal of a regular hexagon is a line segment that connects two points that lie opposite each other on the hexagon.

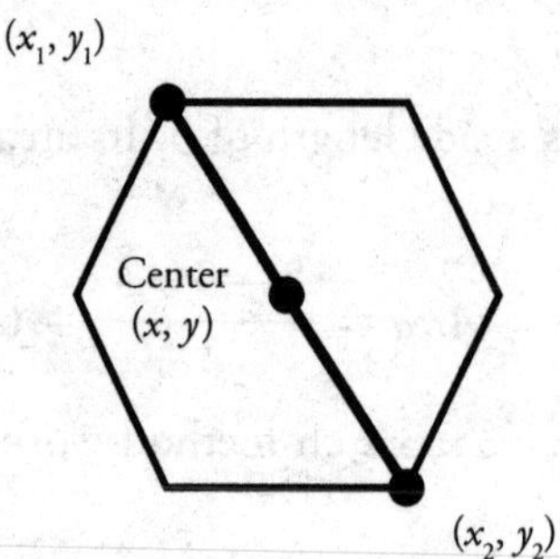

Point 0 and point 3 are endpoints of a diagonal, as are points 1 and 4, and points 2 and 5.

The midpoint formula is:

$$midpoint = \left(\frac{x_1 + x_2}{2}, \frac{y_1 + y_2}{2}\right)$$

Example

Given a RegularHexagon object with its first vertex at (2, 5) and its fourth vertex at (4, 3), the center of the hexagon is:

$$midpoint = \left(\frac{2 + 4}{2}, \frac{5 + 3}{2}\right) = (3, 4)$$

```
/** Returns the center of the RegularHexagon.
 *  @return the CoordinatePoint that is the center of the RegularHexagon
 */
public CoordinatePoint getCenter()
```

2. A train is composed of an engine and any number of train cars. The engine is rated to pull up to a maximum weight, based on its power. The weight of all the train cars combined, including the engine, must be below this maximum weight or the engine will not be able to move the train. The following complete `Engine` class is used to represent the train's engine.

Precondition: The maximum weight rating is greater than or equal to the weight of the `Engine` (every `Engine` object can pull itself).

```
/**
 *  Precondition: maximumWeight is greater than or
 *                equal to the weight of the engine.
 */
public class Engine
{
   private double weight, maximumWeight;

   public Engine(double wt, double maxWt)
   {
      weight = wt;
      maximumWeight = maxWt;
   }

   public double getWeight()
   {  return weight;  }

   public double getMaximumWeight()
   {  return maximumWeight;  }
}
```

Various train cars are represented by the `TrainCar` class. The weight of a `TrainCar` is its base weight plus the weight of its contents. The base weight of a `TrainCar` is the weight of the empty car. Many types of `TrainCar` objects exist, including passenger cars, boxcars, flatcars, cabooses, and so on. Each of these types has a different way of calculating the weight of its contents. Since each type of `TrainCar` does this differently, the `getTotalWeight` method is an abstract method.

```
public abstract class TrainCar
{
   private double baseWeight;

   public TrainCar(double baseWt)
   {  baseWeight = baseWt;  }

   public double getBaseWeight()
   {  return baseWeight;  }

   public abstract double getTotalWeight();
}
```

(a) `TrainCar` objects that are designed to transport people are called passenger cars and are represented by objects of the `PassengerCar` class. The weight of a `PassengerCar` is calculated by adding the base weight of the `PassengerCar` to the weight of the maximum number of people that the car can hold. The calculation assumes that the passenger car is filled to capacity and that each person, including luggage, weighs 300 pounds.

Write the complete `PassengerCar` class, including a constructor and any necessary instance variables and methods. The constructor takes as parameters the base weight of the car and the maximum number of passengers the car can hold. The class does not need to include a `toString` method or accessors and modifiers for any instance variables.

(b) Trains are represented by the `Train` class and contain an `Engine` followed by any number of `TrainCar` objects. You may assume that the various classes that extend `TrainCar` have been correctly implemented and are not shown.

The weight of a `Train` object is calculated by adding the weight of the `Engine` to the weight of each of the `TrainCar` objects in the `trainCars ArrayList`. The `Train` class contains a `removeExcessTrainCars` method that ensures the engine is able to pull the train (to be completed in part (b)).

```
public class Train
{
    private Engine engine;
    private ArrayList<TrainCar> trainCars;

    public Train(Engine e, ArrayList<TrainCar> tc)
    {
        engine = e;
        trainCars = tc;
    }

    /** Removes TrainCar objects from the end of the train
     *  until the train can be pulled by the Engine.
     *
     *  @return ArrayList<TrainCar> containing the removed cars in
     *          the order they were removed (the last car is
     *          item 0, etc.). If no cars are removed, the returned
     *          list will be empty.
     */
    public ArrayList<TrainCar> removeExcessTrainCars()
    {  /*  To be implemented in part (b)   */  }

    /* Additional implementation not shown */
}
```

Trains need to be checked to make sure that their weight can be pulled by their `Engine`. If the train is overweight, train operators must remove train cars from the end of the train until the train is within the acceptable weight range.

Write the `removeExcessTrainCars` method of the `Train` class that removes `TrainCar` objects one at a time from the end of the train until the train is less than or equal to the maximum weight allowed as given by the `getMaximumWeight` method of the `Engine` object. The removed train cars are added to the end of an `ArrayList` of `TrainCar` objects as they are removed. This `ArrayList` of removed `TrainCar` objects is returned by the method. If no `TrainCar` objects need to be removed, an empty `ArrayList` is returned.

Example:
A train is composed of the cars listed below. The engine has a maximum weight rating of 475,000 pounds.

Car	Weight
Engine	200,000
`trainCars[0]`	100,000
`trainCars[1]`	150,000
`trainCars[2]`	50,000
`trainCars[3]`	100,000
`trainCars[4]`	50,000

In the example, the train initially weighs 650,000 pounds. Train cars need to be removed one by one from the end of the train until the total weight is under the maximum allowed weight of 475,000.

- Removing the train car from index 4 lowers the weight to 600,000
- Removing the train car from index 3 lowers the weight to 500,000
- Removing the train car from index 2 lowers the weight to 450,000

At 450,000 pounds, the train is now in the acceptable weight range. The following `ArrayList` is returned by the method:

```
[trainCars[4], trainCars[3], trainCars[2]]
```

Complete the `removeExcessTrainCars` method.

```
/** Removes TrainCar objects from the end of the train
 *  until the train can be pulled by the Engine.
 *
 *  @return ArrayList<TrainCar> containing the removed cars in
 *          the order they were removed (the first car removed is
 *          item 0, etc.) If no cars are removed, the returned
 *          list will be empty.
 */
public ArrayList<TrainCar> removeExcessTrainCars()
```

3. The colors of the pixels on a TV or computer screen are made by combining different quantities of red, green, and blue light. Each of the three colors is represented by a value between 0 and 255, where 0 means none of that color light and 255 means as much of that color as there can be. This way of representing colors is referred to as RGB, for red-green-blue, and the color values are written in ordered triples like this (80, 144, 240). That particular combination means red at a value of 80, green at a value of 144, and blue at a value of 240. The resulting color is a kind of medium blue. The combination (0, 0, 0) produces black and (255, 255, 255) produces white.

Pixels can be modeled by the `Pixel` class shown below.

```
public class Pixel
{
    private int red;
    private int green;
    private int blue;

    public Pixel(int myRed, int myGreen, int myBlue)
    {
        red = myRed;
        green = myGreen;
        blue = myBlue;
    }
    public String toString()
    {
        return "(" + red + ", " + green + ", " + blue + ")";
    }
}
```

An artist wants to manipulate the pixels on a large computer display. She begins by creating three two-dimensional arrays the size of the screen, `rows` by `columns`. One array contains the red value of each pixel, one contains the blue value of each pixel, and one contains the green value of each pixel. She then calls methods of the `AlterImage` class to generate and then manipulate the data.

```
public class AlterImage
{
    /** Converts 3 arrays of color values into one array
     *  of Pixel objects
     *
     *  @param reds the red color values
     *  @param greens the green color values
     *  @param blues the blue color values
     *  @return the Pixel array generated with information
     *          in the red, green and blue arrays
     *
     *  Precondition:  The three color arrays are all the same
     *                 size and contain int values in the range 0-255.
     *  Postcondition: The returned array is the same size as the
     *                 3 color arrays.
     */
    public Pixel[][] generatePixelArray(int[][] reds, int[][] greens, int[][] blues)
    {  /* to be implemented in part (a) */  }
```

```
    /** Flips a 2D array of Pixel objects either
     *  horizontally or vertically
     *
     *  @param image the 2D array of Pixel objects to be processed
     *  @param horiz true: flip horizontally, false: flip vertically
     *  @return the processed image
     */
    public Pixel[][] flipImage(Pixel[][] image, boolean horiz)
    {  /* to be implemened in part (b) */  }

    /* Additional implementation not shown */

}
```

(a) Given the three arrays `reds`, `greens` and `blues`, and the `Pixel` and `AlterImage` classes, implement the `generatePixelArray` method that converts the raw information in the three color arrays into a `rows` by `columns` array of `Pixel` objects.

Precondition: The three color arrays are all the same size and contain `int` values in the range 0–255.
Postcondition: The returned array is the same size as the three color arrays.

```
/** Converts 3 arrays of color values into one array
 *  of Pixel objects
 *
 *  @param reds the red color values
 *  @param greens the green color values
 *  @param blues the blue color values
 *  @return the Pixel array generated with information
 *          from the red, green and blue arrays
 *
 *  Precondition:    The three color arrays are all the same
 *                   size and contain int values in the range 0-255.
 *  Postcondition:   The returned array is the same size as the
 *                   3 color arrays.
 */
public Pixel[][] generatePixelArray(int[][] reds, int[][] greens, int[][] blues)
```

(b) The artist uses various techniques to modify the image displayed on the screen. One of these techniques is to flip the image, either horizontally (along the *x*-axis, top-to-bottom) or vertically (along the *y*-axis, left-to-right), so that a mirror image is produced as shown below.

In this example, the data in the cells are the RGB values stored in the `Pixel` objects. To make the example clearer, red is set to the original row number, green is set to the original column number, and blue is constant at 100.

Original array:

0, 0, 100	0, 1, 100	0, 2, 100	0, 3, 100
1, 0, 100	1, 1, 100	1, 2, 100	1, 3, 100
2, 0, 100	2, 1, 100	2, 2, 100	2, 3, 100
3, 0, 100	3, 1, 100	3, 2, 100	3, 3, 100

Flipped horizontally:

3, 0, 100	3, 1, 100	3, 2, 100	3, 3, 100
2, 0, 100	2, 1, 100	2, 2, 100	2, 3, 100
1, 0, 100	1, 1, 100	1, 2, 100	1, 3, 100
0, 0, 100	0, 1, 100	0, 2, 100	0, 3, 100

Flipped vertically:

0, 3, 100	0, 2, 100	0, 1, 100	0, 0, 100
1, 3, 100	1, 2, 100	1, 1, 100	1, 0, 100
2, 3, 100	2, 2, 100	2, 1, 100	2, 0, 100
3, 3, 100	3, 2, 100	3, 1, 100	3, 0, 100

Write the `flipImage` method, which takes as parameters a 2-D array of `Pixel` objects and a `boolean` direction to flip, where true means horizontally and false means vertically. The method returns the flipped image as a 2-D array of `Pixel` objects.

```
/** Flips a 2D array of Pixel objects either
 *  horizontally or vertically
 *
 *  @param image the 2D array of Pixel objects to be processed
 *  @param horiz true: flip horizontally, false: flip vertically
 *  @return the processed image
 */
public Pixel[][] flipImage(Pixel[][] image, boolean horiz)
```

GO ON TO THE NEXT PAGE

4. The classes below represent families, the people in them and the neighborhoods they live in.

Every individual member of a family is represented by the `Person` class.

```
public class Person
{
    private String name;
    private int age;

    public Person(String myName, int myAge)
    {
        name = myName;
        age = myAge;
    }

    public String getName()
    {   return name;   }

    public int getAge()
    {   return age;   }

    public boolean matches(Person p)
    {   /* to be implemented in part (a) */   }
}
```

(a) Write the `Person` method `matches`. The method returns `true` if the values contained in the calling object's `name` and `age` instance variables match the values contained in the passed object's `name` and `age` instance variables.

```
/** Returns true if this object matches the parameter object
 *   @param person the Person object to be compared to this object
 *   @return true if matches, false if does not match
 */
public boolean matches(Person person)
```

(b) The `Family` class models a family that consists of one or more adults and zero or more children. The instance variables for the `Family` class must include:

- an `ArrayList` to hold members of the family who are adults, where adult is defined to be age 18 and over, and
- an `ArrayList` to hold members of the family who are children.

There are two methods in the `Family` class:

- the `add` method that takes one parameter of type `Person` and adds it to the appropriate `ArrayList`, and
- the `isInFamily` method that takes a `Person` object as a parameter and returns a `boolean` indicating whether that `Person` is a member of the object's family.

Write the complete `Family` class.

You may assume that the `matches` method of the `Person` class works as intended, regardless of what you wrote in part (a). You must use it appropriately in order to receive full credit for part (b).

(c) The `Neighborhood` class consists of a one or more `Family` objects, which are stored in an `ArrayList`.

```
public class Neighborhood
{
    ArrayList<Family> families = new ArrayList<Family>();

    public void add(Family newFamily)
    {
        families.add(newFamily);
    }

    public boolean isInNeighborhood(Person person)
    {
        /* to be implemented in part (c) */
    }
}
```

Implement the `isInNeighborhood` method of the `Neighborhood` class.

You may assume that the `Family` class and its `isInFamily` method work as intended, regardless of what you wrote in part (b). You must use them appropriately in order to receive full credit for part (c).

```
/** Determines whether a particular person is a member of
 *   the neighborhood.
 *   @param person the person in question
 *   @return true if a member of the neighborhood, otherwise false
 */
public boolean isInNeighborhood(Person person)
```

STOP. End of Part II.

Practice Exam 1 Answers and Explanations

Part I Answers and Explanations (Multiple Choice)

Bullets mark each step in the process of arriving at the correct solution.

1. The answer is E.

- Lines 1 and 2 give us: myValue = 17 and multiplier = 3
- Line 3: answer = 17 % 3 + 17 / 3
  - Using integer division: 17 / 3 = 5 remainder 2, giving us:
  - 17 / 3 = 5
  - 17 % 3 = 2
  - The expression simplifies to answer = 2 + 5 = 7 (remember order of operations)
- Line 4: answer = 7 * 3 = 21

2. The answer is B.

- Let's use DeMorgan's theorem to simplify the expression.

```
!(a < b || b <= c) && !(a < c || b >= a)
```

- Distribute the first !. Remember to change || to &&.

```
!(a < b) && !(b <= c) && !(a < c || b >= a)
```

- Distribute the second !

```
!(a < b) && !(b <= c) && !(a < c) && !(b >= a)
```

- It's easy to simplify all those !s. Remember that !< → >= and !<= → > (the same with > and >=).

```
(a >= b) && (b > c) && (a >= c) && (b < a)
```

- Since these are all &&s, all four conditions must be true. The first condition says a >= b and the fourth condition says b < a. No problem there. Options B and C both have a > b.
- The second condition says b > c. That's true in option B.
- The third condition says b < a. Still true in option B, so that's the answer.
- You could also solve this problem with guess and check by plugging in the given values, but all those ands and ors and nots tend to get pretty confusing. Simplifying at least a little will help, even if you are going to ultimately plug and play.

3. The answer is D.

- The while loop condition is (myArray[index] < 7). If you did not read carefully, you may have assumed that the condition was (index < 7), which, along with the index++ is the way you would write it if you wanted to access every element in the array.
- Because the condition is (myArray[index] < 7), we will start at element 0 and continue until we come to an element that is greater than or equal to 7. At that point we will exit the loop and no more elements will be processed.
- Every element before the 7 in the array will have 3 added to its value.

4. The answer is A.

- Option II uses a for-each loop. That is an excellent option for this problem, since we need to process each element in exactly the same way, but option II does not access the element correctly. Read the for-each loop like this: "For each OnlinePurchaseItem (which I will call purchase) in items. . . ." The variable purchase already holds the needed element. Using get(i) to get the element is unnecessary and there is no variable i.
- Option III uses a for loop and processes the elements in reverse order. That is acceptable. Since we need to process every item, it doesn't matter what order we do it in. However, option III starts at i = items.size(), which will cause an `IndexOutOfBoundsException`. Remember that the last element in an ArrayList is list.size() – 1 (just like a String or an array).
- Option I correctly accesses and processes every element in the ArrayList.

5. The answer is D.

- This is a recursive method. Let's trace the calls. The parts in *italics* were filled in on the way back up. That is, the calls in the plain type were written top to bottom until the base case returned a value. Then the answers were filled in *bottom to top.*
  - mystery(5) = 2 + mystery(4) = *2 + 7 =* ***9*** which gives us our final answer.
  - mystery(4) = 2 + mystery(3) = *2 + 5 = 7*
  - mystery(3) = 2 + mystery(2) = *2 + 3 = 5*
  - mystery(2) = 2 + mystery(1) = *2 + 1 = 3*
  - mystery(1) Base Case! return 1

6. The answer is E.

- The outer loop will execute three times, starting at index = 0 and continuing as long as index < 4, increasing by one each time through. So index will equal 0, 1, 2, 3 in successive iterations through the loop.
- The only statement inside the outer loop is the inner loop. Let's look at what the inner loop does in general terms.
- The inner loop executes from i = 0 to i < index, so it will execute index times.
- Each time through it puts an "A" in front of and after myString.
- Putting it all together:
  - The first iteration of the outer loop, index = 0, the inner loop executes 0 times, myString does not change.
  - The second iteration of the outer loop, index = 1, the inner loop executes one time, adding an "A" in front of and after myString. myString = "AHA"
  - The third iteration of the outer loop, index = 2, the inner loop executes two times adding two "A"s in front of and after myString. myString = "AAAHAAA"
  - The fourth (and last) iteration of the outer loop, index = 3, the inner loop executes three times, adding three more "A"s in front of and after myString. myString = "AAAAAAHAAAAAA"
- Putting it another way, when index = 1, we add one "A"; when index = 2, we add two "A"s; when index = 3, we add three "A"s. That's six "A"s all together added to the beginning and end of "H" → "AAAAAAHAAAAAA"

7. The answer is B.

- The general form for generating a random number between *high* and *low* is:

```
(int)(Math.random() * (high - low + 1)) + low
```

- high – low + 1 = 50, low = 10, so high = 59
- The correct answer is integers between 10 and 59 inclusive

8. The answer is C.

- This question requires that you understand the two uses of the + symbol.
- First, we execute what is in the parentheses. Now we have:

```
13 + 6 + "APCSA" + 4 + 4
```

- Now do the addition left to right. That's easy until we hit the string:

```
19 + "APCSA" + 4 + 4
```

- When you add a number and a string in any order, Java turns the number into a string and then concatenates the strings:

```
"19APCSA" + 4 + 4
```

- Now every operation we do will have a string and a number, so they all become string concatenations, not number additions.

```
"19APCSA44"
```

9. The answer is B.

- This is a recursive method. Let's trace the calls. The parts in *italics* were filled in on the way back up. That is, the calls in the plain type were written top to bottom until the base case returned a value. Then the answers were filled in *bottom to top*.
- Be very careful to get the order of num1 and num2 correct in the modulus operation and the parameters correct in the recursive call.
  - guess(3, 17) = guess(17, 3) + 3 = *4 + 3* = *7* which gives us our final answer.
  - guess(17, 3) = guess(3, 2) + 2 = *2 + 2* = *4*
  - guess(3, 2) = guess(2, 1) + 1 = *1 + 1* = *2*
  - guess(2,1) Base case! return 3/2 = 1 (integer division)

10. The answer is A.

- myCar is instantiated as a reference variable pointing to an Automobile object with instance variables make = "Ford" and model = "Fusion".

```
myCar → ["Ford" "Fusion"]
```

- companyCar is instantiated as a reference variable pointing to an Automobile object with instance variables make = "Toyota" and model = "Corolla".

```
myCar → ["Ford" "Fusion"]        companyCar → ["Toyota" "Corolla"]
```

- The statement companyCar = myCar reassigns the companyCar reference variable to point to the same object as the myCar variable. They are now aliases of each other. The original companyCar object is now inaccessible (unless yet another variable we don't know about points to it).

```
myCar → ["Ford" "Fusion"] ← companyCar        ["Toyota" "Corolla"]
```

- The next statement instantiates a brand-new Automobile object with make = "Kia" and model = "Spectre". It sets the myCar reference variable to point to this new object. This does not change the companyCar variable.
- The companyCar variable still points to the object with instance variables make = "Ford" and model = "Fusion".

```
myCar → ["Kia" "Spectre"]        ["Ford" "Fusion"] ← companyCar
```

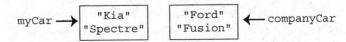

- So we print companyCar's make, which is "Ford" followed by myCar's model, which is "Spectre".

11. The answer is E.

- Methods or constructors with the same name (and return type, in the case of methods) may have different parameter lists. The parameters may differ in type or number. This is called *overloading*.

12. The answer is A.

- We can eliminate options B and D immediately by noticing that active is a boolean and will therefore print as either "true" or "false" and not as "active" even though that is more informative.
- We would expect option E to be correct by reading the code, but the constructor was called incorrectly. The overloaded constructor that is being called is the two-parameter constructor, and that constructor is going to assign its first parameter to state and its second parameter to city. Since we passed city, state, in that order, the city name and the state name have been assigned incorrectly and will therefore be printed incorrectly.
- Note that this is a confusing way to write an overloaded constructor! The order of the variables changes for no reason.

13. The answer is D.

- Option I is incorrect. Most classes have a no-argument constructor, either the default constructor provided automatically if no other constructor is provided, or one written by the programmer. However, the Depot class does not. The super call causes a compile time error.
- Option II is correct. It implements a no-argument constructor in the WhistleStop class, which passes the needed parameters to the Depot constructor. It's hard to imagine why you would want to default to constructing a station in the tiny town of Waubaushene, but if you want to do that, this code will do it for you.
- Option III is also correct. It takes two parameters and correctly passes them to the Depot constructor via the call to super. However, this code will probably not give you the result you are expecting. Just like in question 12, the order of the city and province has been switched.

14. The answer is A.

- Option I is correct. The ArrayList is correctly instantiated as an ArrayList of Depot objects. Since a WhistleStop object *is-a* Depot object, a WhistleStop object may be added to the ArrayList. WhistleStop contains a no-argument constructor, and the Depot constructor is called correctly as well.
- Option II is incorrect. A WhistleStop *is-a* Depot, but not the other way around. We cannot put Depot objects into an ArrayList that is declared to hold WhistleStop objects. The ArrayList declaration will generate a compile-time error.
- Option III is incorrect. The ArrayList is correctly instantiated as an ArrayList of WhistleStop objects. We can add WhistleStop objects to this ArrayList, but we cannot add Depot objects. The second add will generate a compile-time error.

15. The answer is B.

- When we start the for-each loop, the variable crazyString = "crazy" and the ArrayList crazyList has two elements ["weird," "enigma"].
- You can read the for-each like this: "For each String (which I am going to call s) in crazyList. . . ."
- So for each String s in crazyList we execute the assignment statement:

```
crazyString = s.substring(1, 3) + crazyString.substring(0, 4);
```

which takes two characters starting at index 1 from s and concatenates the first four characters of crazyString.

- The first time through the loop, s = "weird" and crazyString = "crazy" so the assignment becomes:

  ```
  crazyString = "ei" + "craz" = "eicraz"
  ```

  Remember that substring starts at the first parameter and ends *before* the second parameter, and remember that we start counting the first letter at 0.
- The second time through the loop, s = "enigma" and crazyString = "eicraz" so the assignment becomes:

  ```
  crazyString = "ni" + "eicr" = "nieicr"
  ```

16. The answer is E.

- There are four possible paths through the code. We need one data element that will pass through each path.
  - The first path is executed if n < -5.
  - The second path is executed if -5 ≤ n < 0 (because if n < -5, the first path is executed and we will not reach the second path).
  - The third path is executed if n > 10.
  - The fourth path is executed in all other cases.
- We need to be sure that we have at least one piece of data for all four of those cases.
- Option A does not test the fourth path, because 12 and 15 are both > 10.
- Option B does not test the first path, because it doesn't have a number that is < -5.
- Option C does not test the third path, because it doesn't have a number that is > 10.
- Option D does not test the second path, because 0 is not less than 0.
- Option E tests all four paths.

17. The answer is D.

- This statement should be written with cascading if-else clauses. When the clause associated with a true condition is executed, the rest of the statements should be skipped. Unfortunately, that is not how it is written. For example, if the parameter temp = 80, the first if is evaluated and found to be true and activity = "go swimming"; however, since there is no else, the second if is also evaluated and found to be true, so activity = "go hiking". Then the third if is evaluated and found to be true, so activity = "go horseback riding". The else clause is skipped and "go horseback riding" is returned (incorrectly).
- All temperatures that are true for multiple if statements will cause activity to be set incorrectly. Only the values of temp that will execute the last if or the else clause will return the correct answer.
- In order for this code to function as intended, all ifs except the first one should be preceded with *else*.

18. The answer is D.

- This time, there are if-else clauses, but they are in the wrong order. It is important to write the most restrictive case first. Let's test this code by using 80 degrees as an example. 80 > 75, and that is the condition we intend to execute, but unfortunately, it is also > 45, which is the first condition in the code, so that is the if clause that is executed and activity is set to "go horseback riding". Since the code is written as a cascading if-else, no more conditions are evaluated and flow of control jumps to the return statement.
- Two sets of values give the right answer:
  - Values that execute the first if clause correctly—that is, values that are greater than 45, but less than or equal to 60.
  - Values that fail all the if conditions and execute the else clause—that is, values that are less than or equal to 45.
- Those two conditions can be combined into temperatures <= 60.
- In order for this code to function as intended, the order of the conditions needs to be reversed.

19. The answer is A.

- Let's look at the three code segments one by one.
- Option I:
  - When we enter the loop, i = 1.
  - arr[1 / 2] = arr[0] = 1 (don't forget integer division)
  - i = i + 2 = 3, 3 < 6 so we enter the loop again.
  - arr [3 / 2] = arr[1] = 3
  - i = i + 2 = 5, 5 < 6 so we enter the loop again.
  - arr[5 / 2] = arr[2] = 5
  - i = i + 2 = 7, 7 is not < 6 so we exit the loop.
  - The array arr = [1, 3, 5]
- Option II:
  - When we enter the loop, i = 0.
  - arr[0] = 2 * 0 + 1 = 1
  - increment i to 1, 1 < 3 so we enter the loop again.
  - arr[1] = 2 * 1 + 1 = 3
  - increment i to 2, 2 < 3 so we enter the loop again.
  - arr[2] = 2 * 2 + 1 = 5
  - increment i to 3, 3 is not < 3 so we exit the loop.
  - The array arr = [1, 3, 5]
- Option III:
  - When we enter the loop, i = 6.
  - arr[(6 – 6) / 2] = arr[0] = 6 – 6 = 0
  - Since we know the first element should be 1, not 0, we can stop here and eliminate this option.
- Options I and II produce the same results.

20. The answer is C.

- Let's picture our ArrayList as a table. After the add statements, ArrayList nations looks like this:

Argentina	Canada	Australia	Cambodia	Russia	France

- It looks like perhaps the loop is intended to remove all elements whose length is greater than or equal to 7, but that is not what this loop does. Let's walk through it.
- i = 0. nations.size() = 6, 0 < 6 so we enter the loop.
  - nations.get(0).length() gives us the length of "Argentina" = 9.
  - 9 >= 7, so we remove the element at location 0. Now our ArrayList looks like this:

Canada	Australia	Cambodia	Russia	France

- increment i to 1. nations.size() = 5, 1 < 5, so we enter the loop.
  - nations.get(1).length() gives us the length of "Australia" = 9. (We skipped over "Canada" because it moved into position 0 when "Argentina" was removed.)
  - 9 >= 7, so we remove the element at location 0. Now our ArrayList looks like this:

Australia	Cambodia	Russia	France

- increment i to 2. nations.size() = 4, 2 < 4 so we enter the loop.
  - nations.get(2).length() gives us the length of "Russia" = 6.
  - 6 is not >= 7 so we skip the if clause.
- increment i to 3. nations.size() = 4, 3 < 4 so we enter the loop.
  - nations.get(3).length() gives us the length of "France" = 6.
  - 6 is not >= 7 so we skip the if clause.
- increment i to 4. nations.size() = 4, 4 is not < 4 so we exit the loop with the ArrayList:

Australia	Cambodia	Russia	France

- Note: You have to be really careful when you remove items from an ArrayList in the context of a loop. Keep in mind that when you remove an item, the items with higher indices don't stay put. They all shift over one, changing their indices. If you just keep counting up, you will skip items. Also, unlike an array, the size of the ArrayList will change as you delete items.

21. The answer is C.

- doStuff is going to loop through the array, checking to see if each item is the same as the item before it. If it is, the values of index, maxCounter, and counter all change. If it is not, only counter changes.
- Begin by setting index = 0; maxCounter = 1; counter = 1; and noting that numberArray.length() = 5.
- First iteration: k = 1, enter the loop with index = 0; maxCounter = 1; counter = 1;
  - Item 1 = item 0, execute the if clause
    - counter <= maxCounter, execute the if clause
      - maxCounter = 1; index = 1
    - counter = 2
- Second iteration: k = 2, enter the loop with index = 1; maxCounter = 1; counter = 2;
  - Item 2 != item 1, execute the else clause
    - counter = 1
- Third iteration: k = 3, enter the loop with index = 1; maxCounter = 1; counter = 1;
  - Item 3 = item 2, execute the if clause
    - counter <= maxCounter, execute the if clause
      - maxCounter = 1; index = 3
    - counter = 2
- Fourth iteration: k = 4, enter the loop with index = 3; maxCounter = 1; counter = 2;
  - Item 4 != item 3, execute the else clause
    - counter = 1
- Fifth iteration: k = 5, which fails the condition and the loop exits.
- index = 3, numberArray[3] = 4 is returned and printed.

22. The answer is D.

- This question is explained along with question 23 below.

23. The answer is D.

- We will solve questions 22 and 23 together by diagramming the objects and reference variables.
- The first three assignment statements give us:

```
c1 → [x = 0,  y = 10] ← c3      c2 → [x = 20, y = 30]
```

- `c3.setX(c2.getY());` results in:

```
c1 → [x = 30, y = 10] ← c3      c2 → [x = 20, y = 30]
```

- `c3 = c2;` results in:

```
c1 → [x = 30, y = 10]      c3 →, c2 → [x = 20, y = 30]
```

- `c3.setY(c2.getX());` results in:

```
c1 → [x = 30, y = 10]      c3 →, c2 → [x = 20, y = 20]
```

- `c2.setX(c1.getX());` results in:

```
c1 → [x = 30, y = 10]      c3 →, c2 → [x = 30, y = 20]
```

- Question 22: Both c2.getY() and c3.getY() return 20, but only c2.getY() is an option.
- Question 23: c1.getX(), c2.getX(), and c3.getX() all return 30.

24. The answer is B.

- You *implement* an interface and *extend* a class.
- Supernatural is an interface, so it must be implemented.
- Werewolf is a class, so it must be extended.

25. The answer is D.

- You may implement as many interfaces as you like, but you can extend only one class.

26. The answer is C.

- Option A is incorrect. If Child extended Adult, this would be the correct constructor, but Child extends Person. The Person constructor does not take a title, only a first name and last name.
- Option B is incorrect because Person does not have a no-argument constructor. The call to super() will cause a compile time error.
- Options D and E are incorrect because of the "extends" phrase. The keyword **extends** is used in the class declaration, not the constructor declaration.
- Option C is the only correctly written constructor.

27. The answer is E.

- Options A and D are incorrect because they print information to the console rather than returning it as a string. Option A also calls super.toString, which does not exist.
- Option B is incorrect because, like option A, it calls super.toString, which does not exist.
- Option C is incorrect because firstName and lastName are private instance variables of the Person class and cannot be accessed by the Adult class.
- Option E is a correctly written toString method.

28. The answer is C.

- Method findSomeone is expecting a parameter of type Adult. The question is, which of the options correctly provides a parameter of type Adult?
- Option I is incorrect. Here, the parameter is of type Person. Adult extends Person, not the other way around, so a Person is not an Adult.
- Option II is correct. The parameter is of type Adult.
- Option III is incorrect. Here, the parameter is of type Child. Child extends Person, not Adult, so a Child is not an Adult.
- Option IV is correct. A Person reference variable can be downcast to an Adult since Adult extends Person. At run-time, if the reference variable does not reference an Adult object as promised by the cast, the program will crash with a run-time error.
- Option V is incorrect. Child is not a superclass of Adult. Child cannot be downcast to Adult.

29. The answer is E.

- The basic formula for tolerance is:

```
Math.abs(a - b) <= tolerance;
```

- In our case, a is val^{power}, or Math.pow(val, power), and b is target, so the equation becomes:

```
Math.abs(Math.pow(val, power) - target) <= tolerance;
```

- That evaluates to a boolean. We could assign it to a boolean variable and return the variable, but we can also just return the expression as in option E.

30. The answer is C.

- This problem requires you to understand that primitives are passed by value and objects are passed by reference.
- When a primitive argument (or actual parameter) is passed to a method, its value is copied into the formal parameter. Changing the formal parameter inside the method has no effect on the value of the variable passed in. The caller's num1 will not be changed by the method.
- The caller's num2 is not changed when the num2++ is executed inside the method. That new value, however, is returned by the method, and the caller uses the returned value to reset num2. The value of num2 will change.
- When an object is passed to a method, its reference is copied into the formal parameter. The actual and formal parameters become aliases of each other; that is, they both point to the same object. Therefore, when the object is changed inside the method, those changes will be seen outside of the method. Changes to array values inside the method will be seen in array values outside of the method.
- num1 remains equal to 2, even though the method's num1 variable is changed to 4. num2 is reset to 4, because the method returns that value, and values[4] is set to 3.

31. The answer is B.

- To convert from decimal to hexadecimal (base 16), we need to remember the place values for base 16. A base 16 number has the 1's place on the right (like every base), preceded by the 16's place.

16	1
?	?

- Our first number is 125. How many times does 16 go into 125? 16 * 7 = 112. So we have a 7 in the 16's place. 125 – 112 = 13, so we have 13 left. We need to put 13 in the 1's place. That's D in hex. So our answer is 7D.
- Our second number is 200. Following the same procedure, 16 * 12 = 192. We want to put a 12 in the 16's place. That's C in hex. 200 – 192 = 8, so our number is C8.
- Our third number is 78. 16 * 4 = 64. Put a 4 in the 16's place. 78 – 64 = 14. That's E in hex. Our number is 4E.
- Putting the three hex values together in order, we get #7DC84E.

32. The answer is C.

- First split the hexadecimal number into its three components: EA, 27, B5.
- Consider EA. The E is in the 16's place. E is 14, so we have 14 * 16 = 224. A is 10, in the 1's place, so our first answer is 234.
- Consider 27. 2 * 16 = 32. 32 + 7 = 39.
- Last B5. B is 11. 11 * 16 = 176. 176 + 5 = 181
- Our ordered triple is (234, 39, 181).

33. The answer is A.

- The Binary Search algorithm should not be applied to this array! Binary Search only works on a sorted array. However, the algorithm will run; it just won't return reasonable results. Let's take a look at what it will do.
- First look at the middle element. That's 100. 50 < 100, so eliminate the second half of the array.

9	100	11	45	76	~~100~~	~~50~~	~~1~~	~~0~~	~~55~~	~~99~~

- Look at the middle element of the remaining part; 50 > 11, so eliminate the lower half.

~~9~~	~~100~~	~~11~~	45	76	~~100~~	~~50~~	~~1~~	~~0~~	~~55~~	~~99~~

- 50 does not appear in the remaining elements, return -1 (incorrectly, as it turns out).

34. The answer is C.

- After the array is instantiated, it looks like this:

0	1	2
3	4	5
6	7	8

- The for loop will execute three times: x = 0, x = 1, and x = 2.
- You should recognize that the code in the loop is a simple swap algorithm that swaps the value at nums[x][0] with the value at nums[x][2].
- The result is that in each row, element 0 and element 2 will be swapped.
- Our final answer is:

2	1	0
5	4	3
8	7	6

35. The answer is A.

- Let's trace the loop and see what happens. On entry we have:

  list (which will not change): newList:

2	4	3	3	2	5	1

2

- i = 1. In pseudocode, the if statement says "if (element 1 of list > the last element in newList) add it to newList" Since 4 > 2, add 4 to newList, which becomes:

2	4

- i = 2. "if (element 2 of list > the last element in newList) add it to newList" 3 is not > 4 so we do not add it.
- i = 3. "if (element 3 of list > the last element in newList) add it to newList" 3 is not > 4, we do not add it. (By this point, you may recognize what the code is doing and you may be able to complete newList without tracing the remaining iterations of the loop.)
- i = 4 "if (element 4 of list > the last element in newList) add it to newList" 2 is not > 4, we do not add it.
- i = 5. "if (element 5 of list > the last element in newList) add it to newList" 5 > 4, add 5 to newList which becomes:

2	4	5

- i = 6. "if (element 6 of list > the last element in newList) add it to newList" 1 is not > 5, so we do not add it.
- We have completed the loop; we exit and print newList.

36. The answer is E.

- After the array is instantiated, it is filled with 0s because that is the default value for an int, so anything we don't overwrite will be a 0.
- Let's look at the first loop. r goes from 0 to 4, so all rows are affected. c starts at r+1 and goes to 4, so not all columns are affected. When r = 0, c will = 1,2,3,4; when r = 1, c will = 2,3,4; when r = 2, c will = 3,4; when r = 3, c will = 4. That will assign 9 to the upper right corner of the grid like this.

0	9	9	9	9
0	0	9	9	9
0	0	0	9	9
0	0	0	0	9
0	0	0	0	0

- In the second loop, d goes from 0 to 4, so all rows are affected, and c always equals r, so the diagonals are filled in with their row/column number.

0	9	9	9	9
0	1	9	9	9
0	0	2	9	9
0	0	0	3	9
0	0	0	0	4

- Now we just have to look for the three cells specified in the problem.

37. The answer is D.

- The first thing to do is to identify the sorting algorithm implemented by mysterySort.
- We can tell it's not Merge Sort, because it is not recursive.
- There are several things we can look for to identify whether a sort is Insertion Sort or Selection Sort. Some easy things to look for:
  - In the inner loop, Insertion Sort shifts items over one slot.
  - After the loops are complete, Selection Sort swaps two elements.
- We can see items being shifted over and we can't see a swap, so this is Insertion Sort.
- Let's look at the state of the array in table form. Before the sort begins, we have this. The sort starts by saying that 9 (all by itself) is already sorted.

9	1	3	0	2

- The first iteration through the loop puts the first two elements in sorted order.

1	9	3	0	2

- The second iteration through the loop puts the first three elements in sorted order.

1	3	9	0	2

- Notice that in Insertion Sort, the elements at the end of the array are never touched until it is their turn to be "inserted." Selection Sort will change elements all through the array.

38. The answer is C.

- This is a recursive method. Let's trace the calls. The parts in *italics* were filled in on the way back up. That is, the calls in the plain type were written top to bottom until the base case returned a value. Then the answers were filled in *bottom to top.*
  - puzzle (3, 4) = 3 * puzzle (3, 3) = *3 * 27 =* ***81*** which gives us our final answer
  - puzzle (3, 3) = 3 * puzzle (3, 2) = *3 * 9 = 27*
  - puzzle (3, 2) = 3 * puzzle (3, 1) = *3 * 3 = 9*
  - puzzle (3, 1) Base Case! return 3
- It is interesting to notice that this recursive method finds the first parameter raised to the power of the second parameter.

39. The answer is D.

- This nested for loop is traversing the array in column major order. I can tell this is the case because the outer loop, the one that is changing more slowly, is controlling the column variable. For every time the column variable changes, the row variable goes through the entire row.
- In addition, although the outer loop is traversing the columns from least to greatest, the inner loop is working backward through the rows.
- Since we increase val by 2 each time, values are being filled in like this:

8	18	28	38	48	58
6	16	26	36	46	56
4	14	24	34	44	54
2	12	22	32	42	52
0	10	20	30	40	50

- You probably didn't need to fill in the whole table to figure out that table[3][4] = 42.

40. The answer is E.

- Since this is a Selection Sort algorithm, we know that the inner loop is looking for the smallest element.
- small is storing the index of the smallest element we have found so far. We need to find out if the current element being examined is smaller than the element at index small, or, in other words, if arr[j] is less than arr[small].
- Option C would be correct if we were sorting an array of ints, but < doesn't work with strings. We have to use the method compareTo. compareTo returns a negative value if the calling element is before the parameter value. We need:

```
if (arr[j].compareTo(arr[small] < 0))
```

Part II Answers and Explanations (Free Response)

Please keep in mind that there are multiple ways to write the solution to a free-response question, but the general and refined statements of the problem should be pretty much the same for everyone. Look at the algorithms and coded solutions, and determine if yours accomplishes the same task.

General Penalties (assessed only once per problem)**:**
-1 using a local variable without first declaring it
-1 returning a value from a void method or constructor
-1 accessing an array or ArrayList incorrectly
-1 overwriting information passed as a parameter
-1 including unnecessary code that causes a side effect such as a compile error or console output

1. Regular Hexagon

(a) **General Problem:** Write the getSideLength method for the RegularHexagon class.

Refined Problem: Use the distance formula to compute and return the length of one side of a regular hexagon. Use any two consecutive entries in ArrayList points. Access the x and y values by using the methods in the CoordinatePoint class.

Algorithm:

- Get the x coordinate of element 0 in the ArrayList points.
- Get the y coordinate of element 0 in the ArrayList points.
- Get the x coordinate of element 1 in the ArrayList points.
- Get the y coordinate of element 1 in the ArrayList points.
- Compute the distance between the two points, using the Math class methods pow and sqrt. Be careful to place parentheses accurately.
- Return the computed value.

Java Code:

```
public double getSideLength()
{
    double x1 = points.get(0).getX();
    double y1 = points.get(0).getY();
    double x2 = points.get(1).getX();
    double y2 = points.get(1).getY();

    return Math.sqrt(Math.pow(x2 - x1, 2) + Math.pow(y2 - y1, 2));
}
```

Common Errors:

- Be careful whenever you transcribe a formula. Check each parenthesis placement carefully. It's a good idea to count the open parentheses and the close parentheses to make sure you have the same number.

Java Code Alternate Solution #1:

You can skip storing the values in variables, though it makes the return statement pretty complex. You could also have stored the answer in a variable and returned that variable.

```
public double getSideLength()
{
    return Math.sqrt(Math.pow(points.get(1).getX() - points.get(0).getX(),2) +
            Math.pow(points.get(1).getY() - points.get(0).getY(), 2));
}
```

Java Code Alternate Solution #2:

You can use multiplication instead of Math.pow.

```
public double getSideLength()
{
    double x1 = points.get(0).getX();
    double y1 = points.get(0).getY();
    double x2 = points.get(1).getX();
    double y2 = points.get(1).getY();

    return Math.sqrt((x2 - x1) * (x2 - x1) + (y2 - y1) * (y2 - y1));
}
```

(b) **General Problem:** Write the getArea method for the RegularHexagon class.

Refined Problem: Use the area formula given in the problem to compute and return the area of the hexagon. Get the side length by using the method written in part (a).

Algorithm:

- Compute the area using the value returned by the sideLength method, and the Math class methods pow and sqrt. Be careful to place parentheses accurately.
- Return the computed value.

Java Code:

```
public double getArea()
{
    return (Math.pow(getSideLength(), 2) * 3 * Math.sqrt(3)) / 2;
}
```

(c) **General Problem:** Write the getCenter method for the RegularHexagon class.

Refined Problem: Use the accessor methods of the CoordinatePoint class to get the x and y values from elements that define a diagonal. Use the midpoint formula given in the problem to compute the point at the center of the hexagon. Create and return a CoordinatePoint object containing the center point.

Algorithm:
- Get the x coordinate of element 0 in the ArrayList points.
- Get the y coordinate of element 0 in the ArrayList points.
- Get the x coordinate of element 3 in the ArrayList points.
- Get the y coordinate of element 3 in the ArrayList points.
- Compute the x and y values of the midpoint of the two points.
- Create a new CoordinatePoint object with the calculated x and y points.
- Return the CoordinatePoint object.

Java Code:

```
public CoordinatePoint getCenter()
{
    double x1 = points.get(0).getX();
    double y1 = points.get(0).getY();
    double x2 = points.get(3).getX();
    double y2 = points.get(3).getY();

    double xCenter = (x1 + x2) / 2;
    double yCenter = (y1 + y2) / 2;
    CoordinatePoint center = new CoordinatePoint(xCenter, yCenter);

    return center;
}
```

Common Errors:
- Remember to put (x1 + x2) in parentheses. Otherwise, order of operations will yield x1 + (x2 / 2).

Java Code Alternate Solution:
- Like 1a, none of the variables are required. Sometimes a happy medium is a better solution than either of these examples.

```
public CoordinatePoint getCenter()
{
    return new CoordinatePoint(
        (points.get(0).getX() + points.get(3).getX()) / 2,
        (points.get(0).getY() + points.get(3).getY()) / 2);
}
```

Scoring Guidelines: Regular Hexagon

Part (a) **getSideLength** **3 points**
+1 Accesses the necessary values for any point in the RegularHexagon
+1 Accesses the necessary values for its consecutive point
+1 Calculates and returns the correct answer

Part (b) **getArea** **3 points**
+1 Calls the sideLength method and correctly uses the returned value
+1 Correctly calls a method from the Math class
+1 Calculates and returns the correct answer

Part (c) **getCenter** **3 points**
+1 Accesses the necessary values for any point in the RegularHexagon
+1 Accesses the necessary values for the point opposite that point
+1 Calculates and returns the correct answer

Sample Driver:

There are many ways to write these methods. Maybe yours is a bit different from our sample solutions and you are not sure if it works. Here is a sample driver program. Running it will let you see if your code works, and will help you debug it if it does not.

Copy RegularHexagonDriver into your IDE along with the complete RegularHexagon and CoordinatePoint classes (including your solutions). You will also need to add this import statement as the first line in your RegularHexagon class: `import java.util.ArrayList;`

```
public class RegularHexagonDriver {

   public static void main(String[] args) {

      ArrayList<CoordinatePoint> vertices = new ArrayList<CoordinatePoint>();
      vertices.add(new CoordinatePoint(100, 0));
      vertices.add(new CoordinatePoint(50, -87));
      vertices.add(new CoordinatePoint(-50, -87));
      vertices.add(new CoordinatePoint(-100, 0));
      vertices.add(new CoordinatePoint(-50, 87));
      vertices.add(new CoordinatePoint(50, 87));
      RegularHexagon hex = new RegularHexagon(vertices);

      System.out.println("Expected answers " + "(truncated to 2 decimal places):");
      System.out.println("Side Length: 100.34");
      System.out.println("Area: 26160.02");
      System.out.println("Get Center: (0.0, 0.0)\n");

      System.out.println("Your answers: ");
      System.out.println("Side Length: " + (int) (100 *hex.getSideLength()) / 100.0);
      System.out.println("Area: " + (int) (100 * hex.getArea()) / 100.0);
      System.out.println("Get Center: (" + hex.getCenter().getX() +
         ", "+ hex.getCenter().getY() + ")");
   }
}
```

2. Train

(a) **General Problem:** Complete the PassengerCar class, which extends the TrainCar abstract class.

Refined Problem: The PassengerCar class should include the necessary instance variables, a properly written constructor, and all the methods required by the TrainCar abstract class.

Algorithm:

- Use "extends TrainCar" in the class declaration.
- Create an instance variable to hold maxPassengers.
- Write a constructor that takes the car's baseWeight and its maxPassengers as parameters, calls super, passing the baseWeight, and sets the maxPassengers instance variable.
- Write the required getTotalWeight method, which returns the baseWeight (from the super class) + 300 * maxPassengers. Declare 300 as a constant so there are no magic numbers in the code (not required, but good programming style).

Java Code:

```
public class PassengerCar extends TrainCar
{
    private int maxPassengers;
    private final int PASSENGER_WEIGHT = 300;

    public PassengerCar(double baseWeight, int maxPgr)
    {
        super(baseWeight);
        maxPassengers = maxPgr;
    }

    public double getTotalWeight()
    {
        return super.getBaseWeight() + PASSENGER_WEIGHT * maxPassengers;
    }
}
```

Common Errors:

- The first statement in the constructor needs to be super(baseWeight). Remember that the call to super must come first.

(b) **General Problem:** Write the removeExcessTrainCars method of the Train class that removes cars from the train until the train is light enough to be pulled by the engine.

Refined Problem: Calculate the initial weight of the train. If the weight exceeds the weight limit, TrainCars are removed from the end of the train until the weight is in the acceptable range. All train cars that are removed are returned in an ArrayList in the order they are removed from the train.

Algorithm:

- Create an ArrayList of TrainCars to hold the removed cars.
- Create a variable to hold the total weight of the train.
- Calculate the weight of the train by adding the weight of the Engine with the weights of all the TrainCar objects in trainCars.
- While the total weight is greater than the Engine's maximum weight,
  - Remove a train car from the end of the train
  - Add it to the ArrayList to be returned
  - Subtract its weight from the total weight of the train
- Return the ArrayList of removed cars (which may be empty).

Java Code:

```
public ArrayList<TrainCar> removeExcessTrainCars()
{
    ArrayList<TrainCar> removed = new ArrayList<TrainCar>();
    double weight = engine.getWeight();

    for (TrainCar car : trainCars)
        weight += car.getTotalWeight();

    while (weight > engine.getMaximumWeight())
    {
        TrainCar removedCar = trainCars.remove(trainCars.size() - 1);
        weight -= removedCar.getTotalWeight();
        removed.add(removedCar);
    }
    return removed;
}
```

Common Errors:

- This method must be done in two consecutive loops. You have to know the total weight before you can start removing cars.
- Remember that remove also returns the removed element. You do not need a get followed by a remove.

Java Code Alternate Solution:

- The first loop that calculates the weight of the entire train can be written as a for loop.
- The second loop calculations can be written without a temp holding variable.

```
public ArrayList<TrainCar> removeExcessTrainCars()
{
    ArrayList<TrainCar> removed = new ArrayList<TrainCar>();
    double weight = engine.getWeight();

    for (int i  = 0; i < trainCars.size(); i++)
        weight += trainCars.get(i).getTotalWeight();

    while (weight > engine.getMaximumWeight())
    {
        weight -= trainCars.get(trainCars.size() - 1).getTotalWeight();
        removed.add(trainCars.remove(trainCars.size() - 1));
    }
    return removed;
}
```

Scoring Guidelines: Train

Part (a)	**PassengerCar class**	**3 points**

+1 Includes "extends TrainCar" in the class declaration

+1 Implements constructor including passing the correct parameter to super and initializing the correctly declared private instance variables

+1 Implements the getTotalWeight method; calls super to get the base weight, and returns the correctly calculated total weight

Part (b)	**removeExcessTrainCars**	**6 points**

+2 Calculates the initial weight of the train

- **+1** Accesses the weight of every element of the ArrayList; no bounds errors, no missed elements
- **+1** Calculates the total weight of the train correctly

+3 Removes necessary cars from the train

- **+1** Writes a loop with an appropriate terminating condition
- **+1** Removes a car from the end of the train and adjusts the total weight of the train
- **+1** Adds the car to the end of a correctly instantiated ArrayList

+1 Returns the correctly generated ArrayList

3. Pixels

(a) **General Problem:** Write the generatePixelArray method to convert three 2-D int arrays into a 2-D array of Pixel objects.

Refined Problem: Use a nested for loop to instantiate a Pixel object for each element of the Pixel object array. Use the corresponding values in the red, green, and blue arrays as parameters to the Pixel constructor. When the loops are complete, return the completed array.

Algorithm:

- Create a 2-D array of type Pixel to hold the new Pixel objects. The dimensions of this array are the same as any of the color arrays because there is a precondition saying all the arrays must be the same size.
- Write a for loop that goes through all the rows of the new array.
  - Nested in that for loop, write a for loop that goes through all the columns of the new array. (These can be switched. Row-major order is more common, but column-major order will also work here.)
    - Instantiate a new Pixel object, passing the values in the red, green, and blue arrays at the position given by the two loop counters, and assign it to the element of the Pixel array given by the two loop counters.
- When the loops are complete and every element has been processed, return the completed Pixel array.

Java Code:

```
public static Pixel[][] generatePixelArray(int[][] reds, int[][] greens, int[][] blues)
{
   Pixel[][] pix = new Pixel[reds.length][reds[0].length];
   for (int r = 0; r < pix.length; r++)
      for (int c = 0; c < pix[r].length; c++)
         pix[r][c] = new Pixel(reds[r][c], greens[r][c], blues[r][c]);
   return pix;
}
```

(b) **General Problem:** Write a flipImage method that takes a 2-D array of Pixel objects and flips it into a mirror image, either vertically or horizontally.

Refined Problem: Create a new array to hold the altered image. Determine whether to flip the image horizontally or vertically. Write a nested for loop to move all the Pixels to their mirror-image location in the new array, either horizontally or vertically. When the loops are complete, return the new array.

Algorithm:

- Instantiate a 2-D Pixel array that has the same dimensions as the array passed as a parameter. This array will hold the altered image.
- Create an if-else statement with one clause for a horizontal flip and one for a vertical flip.
- If the flip is horizontal, write a nested for loop that goes through the array in row-major order.
  - For each iteration of the loop, an entire row is moved into its new "flipped" place in the altered array.
- Otherwise, the flip is vertical. Write a nested for loop that goes through the array in column-major order.
  - For each iteration of the loop, an entire column is moved into its new "flipped" place in the altered array.
- Return the altered array.

Java Code:

```
public static Pixel[][] flipImage(Pixel[][] image, boolean horiz)
{
    Pixel[][] flipped = new Pixel[image.length][image[0].length];
    if (horiz)
        for (int r = 0; r < image.length; r++)
            for (int c = 0; c < image[0].length; c++)
                flipped[r][c] = image[image.length - 1 - r][c];
    else
        for (int c = 0; c < image[0].length; c++)
            for (int r = 0; r <image.length; r++)
                flipped[r][c] = image[r][image[0].length - 1 - c];
    return flipped;
}
```

Common Errors:

- Watch off-by-one errors. Always think: do I want length or length – 1? Using variables, like in the alternate solution, can help you be consistent.
- Be sure you understand the difference between row-major and column-major order.
- Check your answer with both even and odd numbers of rows and columns.
- Create a new array to hold the “flipped” version. Do not overwrite the array that is passed in. This is called *destruction of persistent data* and incurs a penalty.

Java Code Alternate Solution:

- This solution only loops halfway through the array. It flips a pair of lines on each iteration.

  This solution also uses a few extra variables to keep things easier to read. This is not necessary but it reduces the amount of typing and the chance of an off-by-one error with length, as opposed to length − 1.

```
public static Pixel[][] flipImage(Pixel[][] image, boolean horiz)
{
    Pixel[][] flipped = new Pixel[image.length][image[0].length];
    int maxR = image.length - 1;
    int maxC = image[0].length - 1;
    if (horiz)
    {
        for (int r = 0; r <= maxR / 2; r++)
        {
            for (int c = 0; c <= maxC; c++)
            {
                flipped[r][c] = image[maxR - r][c];
                flipped[maxR - r][c] = image[r][c];
            }
        }
    }
    else
    {
        for (int c = 0; c <= maxC / 2; c++)
        {
            for (int r = 0; r <= maxR; r++)
            {
                flipped[r][c] = image[r][maxC - c];
                flipped[r][maxC - c] = image[r][c];
            }
        }
    }
    return flipped;
}
```

Scoring Guidelines: Pixels

Part (a) **generatePixelArray** **3 points**

+1 Instantiates a 2-D array of Pixel objects with the correct dimensions

+1 Instantiates a new Pixel object with the corresponding elements of the red, green, and blue arrays for every element of the array; no bounds errors, no missed elements

+1 Returns the properly constructed and filled array

Part (b) **flipImage** **6 points**

+1 Correctly determines whether the image should be flipped vertically or horizontally

+2 Flips the array horizontally

- **+1** Correctly repositions at least one element
- **+1** Correctly repositions all elements

+2 Flips the array vertically

- **+1** Correctly repositions at least one element
- **+1** Correctly repositions all elements

+1 Returns the correctly generated flipped array

Sample Driver:

There are many ways to write these methods. Maybe yours is a bit different from our sample solutions and you are not sure if it works. Here is a sample driver program. Running it will let you see if your code works, and will help you debug it if it does not.

Copy PixelDriver and the complete Pixel class into your IDE. Add your generatePixelArray and flipImage methods to the bottom of the PixelDriver class.

```
public class PixelDriver {

   public static void main(String[] args) {
      int[][] red = { { 0, 0, 0, 0 }, { 1, 1, 1, 1 }, { 2, 2, 2, 2 },
                      { 3, 3, 3, 3 } };
      int[][] green = { { 0, 1, 2, 3 }, { 0, 1, 2, 3 }, { 0, 1, 2, 3 },
                        { 0, 1, 2, 3 } };
      int[][] blue = { { 100, 100, 100, 100 }, { 100, 100, 100, 100 },
                       { 100, 100, 100, 100 }, { 100, 100, 100, 100 } };
      System.out.println("generatePixelArray test");
      System.out.println("Expecting:");
      for (int r = 0; r < red.length; r++) {
         for (int c = 0; c < red[r].length; c++)
            System.out.print("(" + red[r][c] + ", " + green[r][c] +
                  ", " + blue[r][c] + ") ");
         System.out.println();
      }
      Pixel[][] pix = generatePixelArray(red, green, blue);
      System.out.println("\nYour Answer:");
      printIt(pix);
      final boolean HORIZ = true;
      final boolean VERT = !HORIZ;
      System.out.println("\nflipImage - Horizontal test");
      System.out.println("Expecting:");
      for (int r = red.length-1; r >=0; r--) {
         for (int c = 0; c < red[r].length; c++)
            System.out.print("(" + red[r][c] + ", " + green[r][c] +
                  ", " + blue[r][c] + ") ");
         System.out.println();
      }
      System.out.println("\nYour Answer:");
      Pixel[][] pix2 = flipImage(pix, HORIZ);
      printIt(pix2);

      System.out.println("\nflipImage - Vertical test");
      System.out.println("Expecting:");
      for (int r = 0; r < red.length; r++) {
         for (int c = red[r].length - 1; c >= 0; c--)
            System.out.print("(" + red[r][c] + ", " + green[r][c] +
                  ", " + blue[r][c] + ") ");
         System.out.println();
      }
      System.out.println("\nYour Answer:");
      Pixel[][] pix3 = flipImage(pix, VERT);
      printIt(pix3);
   }

   public static void printIt(Pixel[][] pix) {
      for (int r = 0; r < pix.length; r++) {
         for (int c = 0; c < pix[r].length; c++) {
            System.out.print(pix[r][c] + " ");
         }
         System.out.println();
      }
   }
   // Enter your generatePixelArray and flipImage methods here
}
```

4. Families

(a) **General Problem:** Write a matches method for the Person class.

Refined Problem: Write a matches method that compares the name and age of the Person object passed as a parameter with this object and returns true if they are the same and false if they are different.

Algorithm:

- Write an if statement with two conditions: this object's name is the same as parameter object's name and this object's age is the same as parameter object's age. Be sure to use == to compare int variables and the equals method to compare String variables.
  - If both conditions are true, return true.
- Otherwise, return false.

Java Code:

```
public boolean matches(Person person)
{
   return (name.equals(person.getName()) && age == person.getAge())
}
```

Java Code Alternate Solution:

- This solution uses methods instead of directly accessing the instance data. It also uses *this* to refer to the object itself. The use of *this* is optional. However, comparing this object to a parameter object can be confusing, and having an object name for both sides of the equals can help. Lastly, this solution uses an if-else statement rather than just returning the value of the boolean condition.

```
public boolean matches(Person person)
{
   if (this.getName().equals(person.getName()) &&
         this.getAge() == person.getAge())
      return true;
   else
      return false;
}
```

(b) **General Problem:** Write the entire Family class.

Refined Problem: Write the Family class. It will have two instance variables: an ArrayList of Person objects to hold the adults in the family and an ArrayList of Person objects to hold the children in the family. There will also be two methods: an add method that adds a new Person object to the correct ArrayList based on age, and an isInFamily method that determines if a Person passed as a parameter is in either of the ArrayLists, and is therefore a member of the family.

Algorithm:

- Write the class declaration.
- Instantiate the two ArrayLists of Person objects, one called adults and one called children.
- Write the add method. It must check the age of the Person object passed as a parameter. If it is greater than or equal to 18, add the Person object to the adults ArrayList; if it is less than 18, add the Person object to the children ArrayList.
- Write the isInFamily method. Traverse both ArrayLists looking for a Person object that equals the Person object passed as a parameter. If one is found, immediately return true. If no match is found after traversing both ArrayLists, return false. Be sure to use the matches method of the Person class written in part (a) to compare the Person objects.

Java Code:

```
public class Family
{
    ArrayList<Person> adults = new ArrayList<Person>();
    ArrayList<Person> children = new ArrayList<Person>();

    public void add(Person person)
    {
        if (person.getAge() >= 18)
            adults.add(person);
        else
            children.add(person);
    }

    public boolean isInFamily(Person person)
    {
        for (Person p : adults)
            if (p.matches(person))
                return true;
        for (Person p : children)
            if (p.matches(person))
                return true;
        return false;
    }
}
```

Common Errors:

There are several opportunities for error in the isInFamily method.

- Sometimes people begin with boolean found = false and then set it to true if the person is found. That's a perfectly fine approach provided no subsequent code will ever set it back to false. It is a common error to flip-flop that boolean depending on whether the current search finds the person, and obviously that does not work.
- The return false must be outside of the loops. You can return true immediately when the person is found, but you cannot return false until you have checked every person in every ArrayList.
- Make sure you use the matches method written in part (a). Do not rewrite the code here.

Java Code Alternate Solution:

The isInFamily method can be written with for loops instead of for-each loops.

```
public boolean isInFamily(Person person)
{
    for (int i = 0; i < adults.size(); i++)
        if (adults.get(i).matches(person))
            return true;
    for (int i = 0; i < children.size(); i++)
        if (children.get(i).matches(person))
            return true;
    return false;
}
```

(c) **General Problem:** Write the isInNeighborhood method of the Neighborhood class.

Refined Problem: Write method isInNeighborhood, which takes a Person object as a parameter and determines if that Person object appears in any of the families in the ArrayList of Family objects.

Algorithm:

- Traverse the ArrayList<Family> families.
- Use the isInFamily method of the Family class to determine if the Person object passed as a parameter is a member of the Family object. Be sure to use the method written in part (b).
- If the Person object is found in a Family, return true.
- If the traversal completes without locating the Person in a Family, return false.

Java Code:

```
public boolean isInNeighborhood(Person person)
{
    for (Family f : families)
    {
        if (f.isInFamily(person))
            return true;
    }
    return false;
}
```

Common Errors:

- Like in part (b), you can return true (or set a boolean) as soon as you find one true response, but you cannot return false until everything has been checked. The return false must be outside the loop.
- Be sure to use the isInFamily method.

Java Code Alternate Solution:

This solution uses a for loop instead of a for-each loop.

```
public boolean isInNeighborhood(Person person)
{
    for (int i = 0; i < families.size(); i++)
    {
        if (families.get(i).isInFamily(person))
            return true;
    }
    return false;
}
```

Scoring Guidelines: Families

Part (a)	**matches**	**2 points**
+1	Checks the conditions for equality by comparing this object's name with parameter object's name and this object's age with parameter object's age	
+1	Returns the correct value	

Part (b)	**Family class**	**5 points**
+1	Instantiates 2 ArrayList<Person>, one for adults, one for children	
+1	The add method checks if person.getAge() >= 18 and adds person to the correct ArrayList	
+3	The isInFamily method	
	+1 Traverses the ArrayList of adults and the ArrayList of children; no bounds errors, no missed elements	
	+1 For both lists, uses the matches method to determine if the Person object is present in the ArrayList	
	+1 Returns true if element is found and returns false only after checking every element of both ArrayLists	

Part (c)	**isInNeighborhood**	**2 points**
+1	Calls isInFamily(person) for each element of the ArrayList; no bounds errors, no missed elements	
+1	Returns true if element is found, and false only after checking every element of the ArrayList	

Scoring Worksheet

This worksheet will help you to approximate your performance on Practice Exam 1 in terms of an AP score of 1–5.

Part I (Multiple Choice)

Number right (out of 40 questions) = ________

Part II (Short Answer)

See the scoring guidelines included with the explanations for each of the questions and award yourself points based on those guidelines.

Question 1: Points Obtained (out of 9 possible) ________

Question 2: Points Obtained (out of 9 possible) ________

Question 3: Points Obtained (out of 9 possible) ________

Question 4: Points Obtained (out of 9 possible) ________

Add the points and multiply by 1.1111 ________ × 1.1111 = ________

Add your totals from both parts of the test (round to nearest whole number) Total Raw Score ________

Approximate conversion from raw score to AP score

Raw Score Range	AP Score	Interpretation
62–80	5	Extremely Well Qualified
44–61	4	Well Qualified
31–43	3	Qualified
25–30	2	Possibly Qualified
0–24	1	No Recommendation

AP Computer Science A: Practice Exam 2

Multiple-Choice Questions

ANSWER SHEET

1	A	B	C	D	E	21	A	B	C	D	E
2	A	B	C	D	E	22	A	B	C	D	E
3	A	B	C	D	E	23	A	B	C	D	E
4	A	B	C	D	E	24	A	B	C	D	E
5	A	B	C	D	E	25	A	B	C	D	E
6	A	B	C	D	E	26	A	B	C	D	E
7	A	B	C	D	E	27	A	B	C	D	E
8	A	B	C	D	E	28	A	B	C	D	E
9	A	B	C	D	E	29	A	B	C	D	E
10	A	B	C	D	E	30	A	B	C	D	E
11	A	B	C	D	E	31	A	B	C	D	E
12	A	B	C	D	E	32	A	B	C	D	E
13	A	B	C	D	E	33	A	B	C	D	E
14	A	B	C	D	E	34	A	B	C	D	E
15	A	B	C	D	E	35	A	B	C	D	E
16	A	B	C	D	E	36	A	B	C	D	E
17	A	B	C	D	E	37	A	B	C	D	E
18	A	B	C	D	E	38	A	B	C	D	E
19	A	B	C	D	E	39	A	B	C	D	E
20	A	B	C	D	E	40	A	B	C	D	E

AP Computer Science A: Practice Exam 2

Part I (Multiple Choice)

Time: 90 minutes
Number of questions: 40
Percent of total score: 50

Directions: Choose the best answer for each problem. Some problems take longer than others. Consider how much time you have left before spending too much time on any one problem.

Notes:

- You may assume all import statements have been included where they are needed.
- You may assume that the parameters in method calls are not null.
- You may assume that declarations of variables and methods appear within the context of an enclosing class.

1. Consider the following code segment.

```
for (int h = 5; h >= 0; h = h - 2)
{
    for (int k = 0; k < 3; k++)
    {
        if ((h + k) % 2 == 0)
            System.out.print((h + k) + "    ");
    }
}
```

What is printed as a result of executing the code segment?

(A) `5    3    1`
(B) `6    4    2`
(C) `51    31    11`
(D) `6    4    2    0`
(E) `51    31    11    00`

2. Consider the following code segment.

```
List<String> foods = new ArrayList<String>();

foods.add("Hummus");
foods.add("Soup");
foods.add("Sushi");
foods.set(1, "Empanadas");
foods.add(0, "Salad");
foods.remove(1);
foods.add("Curry");
System.out.println(foods);
```

What is printed as a result of executing the code segment?

(A) `[Salad, Empanadas, Sushi, Curry]`
(B) `[Hummus, Salad, Empanadas, Sushi, Curry]`
(C) `[Hummus, Soup, Sushi, Empanadas, Salad, Curry]`
(D) `[Soup, Empanadas, Sushi, Curry]`
(E) `[Hummus, Sushi, Salad, Curry]`

GO ON TO THE NEXT PAGE

3. Consider the following output.

```
A  A  A  A  P  P  P  P  C  C  C  C
```

Which of the following code segments will produce this output?

I.
```
String[] letters = {"A", "P", "C", "S"};
for (int i = letters.length; i > 0; i--)
{
    if (!letters[i].equals("S"))
    {
        for (int k = 0; k < 4; k++)
            System.out.print(letters[k] + "  ");
    }
}
```

II.
```
String letters = "APCS";
for (int i = 0; i < letters.length(); i++)
{
    String s = letters.substring(i, i + 1);
    for (int num = 0; num < 4; num++)
    {
        if (!s.equals("S"))
            System.out.print(s + "  ");
    }
}
```

III.
```
String[][] letters = { {"A", "P"}, {"C", "S"} };
for (String[] row : letters)
{
    for (String letter : row)
    {
        if (!letter.equals("S"))
            System.out.print(letter + "  ");
    }
}
```

(A) I only
(B) II only
(C) I and II only
(D) II and III only
(E) I, II, and III

Questions 4–5 refer to the following two classes.

```
public abstract class Vehicle
{
   private int fuel;

   public Vehicle(int fuelAmt)
   {
      fuel = fuelAmt;
   }

   public void start()
   {
      System.out.println("Vroom");
   }

   public void changeFuel(int change)
   {
      fuel = fuel + change;
   }

   public int getFuel()
   {
      return fuel;
   }

   public abstract void useFuel();
}

public class Boat extends Vehicle
{
   public Boat(int fuelAmount)
   {
      super(fuelAmount);
   }

   public void useFuel()
   {
      super.changeFuel(-2);
   }

   public void start()
   {
      super.start();
      useFuel();
      System.out.println("Remaining Fuel " + getFuel());
   }
}
```

GO ON TO THE NEXT PAGE

4. Assume the following declaration appears in a client program.

```
Boat yacht = new Boat(20);
```

What is printed as a result of executing the call `yacht.start()`?

(A) `Vroom`
(B) `Remaining Fuel 18`
(C) `Remaining Fuel 20`
(D) `Vroom`
`Remaining Fuel 18`
(E) `Vroom`
`Remaining Fuel 20`

5. Which of the following statements results in a compile-time error?

I. `Boat sailboat = new Boat(2);`
II. `Vehicle tanker = new Boat(20);`
III. `Boat tugboat = new Vehicle(10);`

(A) I only
(B) II only
(C) III only
(D) II and III only
(E) None of these options will result in a compile-time error.

6. Consider the following recursive method.

```
public int weird(int num)
{
    if (num <= 0)
        return num;
    return weird(num - 2) + weird(num - 1) + weird(num);
}
```

What value is returned as a result of the call `weird(3)`?

(A) `-2`
(B) `3`
(C) `4`
(D) `6`
(E) Nothing is returned. Infinite recursion causes a stack overflow error.

7. Consider the following method.

```
public boolean verifyValues(int[] values)
{
    for (int index = values.length - 1; index > 0; index--)
    {
        /* missing code */
            return false;
    }
    return true;
}
```

The method `verifyValues` is intended to return true if the array passed as a parameter is in ascending order (least to greatest), and `false` otherwise.

Which of the following lines of code could replace `/* missing code */` so the method works as intended?

(A) `if (values[index] <= values[index - 1])`
(B) `if (values[index + 1] < values[index])`
(C) `if (values[index] >= values[index - 1])`
(D) `if (values[index - 1] > values[index])`
(E) `if (values[index] < values[index + 1])`

8. Consider the following code segment.

```
int[][] matrix = new int[3][3];
int value = 0;

for (int row = 0; row < matrix.length; row++)
{
    for (int column = 0; column < matrix[row].length; column++)
    {
        matrix[row][column] = value;
        value++;
    }
}

int sum = 0;
for (int[] row : matrix)
{
    for (int number : row)
        sum = sum + number;
}
```

What is the value of `sum` after the code segment has been executed?

(A) `0`
(B) `9`
(C) `21`
(D) `36`
(E) `72`

9. Consider the following class used to represent a student.

```
public class Student
{
    private String name;
    private int year;

    public Student()
    {
        name = "name";
        year = 0;
    }

    public Student(String myName, int myYear)
    {
        name = myName;
        year = myYear;
    }
}
```

Consider the ExchangeStudent class that extends the Student class.

```
public class ExchangeStudent extends Student
{
    private String country;
    private String language;

    public ExchangeStudent(String myName, int myYear,
                           String myCountry, String myLanguage)
    {
        super(myName, myYear);
        country = myCountry;
        language = myLanguage;
    }
}
```

Which of the following constructors could also be included in the ExchangeStudent class without generating a compile-time error?

I.
```
public ExchangeStudent(String myName, int myYear, String myCountry)
{
    super(myName, myYear);
    country = myCountry;
    language = "English";
}
```

II.
```
public ExchangeStudent(String myCountry, String myLanguage)
{
    super("name", "2015");
    country = myCountry;
    language = myLanguage;
}
```

III.
```
public ExchangeStudent()
{
}
```

(A) I only
(B) II only
(C) III only
(D) I and III only
(E) I, II, and III

10. Consider the following code segment.

```
int number = Integer.MAX_VALUE;
while (number > 0)
{
    number = number / 2;
}

for (int i = number; i > 0; i++)
{
    System.out.print(i + "  ");
}
```

What is printed as a result of executing the code segment?

(A) `0`
(B) `1`
(C) `1073741823`
(D) Nothing will be printed. The code segment will terminate without error.
(E) Nothing will be printed. The first loop is an infinite loop.

11. Consider the following method.

```
/** Precondition: numbers.size() > 0
 */
public int totalValue(List<Integer> numbers)
{
    int total = 0;
    for (Integer val : numbers)
    {
        if (val > 1 && numbers.size() - 3 > val)
            total += val;
    }
    return total;
}
```

Assume that the `ArrayList` passed as a parameter contains the following `Integer` values.

`[0, 1, 2, 3, 4, 5, 6, 7, 8, 9]`

What value is returned by the call `totalValue`?

(A) `20`
(B) `21`
(C) `27`
(D) `44`
(E) `45`

Questions 12–13 refer to the following interface and class definitions, which are components of a video game.

```
public interface Ship
{
    // The move method changes the location of the Ship
    void move();
}
```

A `PlayerShip` object is controlled by the human player.

```
public class PlayerShip implements Ship
{
    public void move()
    {  /* implementation not shown  */  }

    // Returns the current y position of the PlayerShip
    public int getYPosition()
    {  /* implementation not shown  */  }

    /* Additional implementation not shown */
}
```

An `EnemyShip` object is controlled by the computer.

```
public class EnemyShip implements Ship
{
    public int xPosition;
    public int yPosition;

    public void move()
    {  /* missing code */  }

    /*  Additional implentation not shown */
}
```

12. An EnemyShip object has the following behavior:

- An EnemyShip always moves from right to left in the x direction by subtracting 1 from the xPosition for each move. When xPosition reaches 0, it is reset to 1000.
- An EnemyShip always moves toward a PlayerShip in the y direction, or stays in its current y position if the PlayerShip and the EnemyShip have the same yPosition (since the EnemyShip cannot get any closer).

Given PlayerShip player, which of the following is a correct implementation of the move method in the Enemyship class?

I.
```
public void move()
{
    if (xPosition == 0)
        xPosition = 1000;
    else
        xPosition = xPosition - 1;

    if (player.getYPosition() > 0)
        yPosition++;
}
```

II.
```
public void move()
{
    if (xPosition == 0)
        xPosition = 1000;
   else
      xPosition -= 1;

   if (player.getYPosition() > yPosition)
      yPostion++;
   else
      yPosition--;
}
```

III.
```
public void move()
{
    if (xPosition == 0)
        xPosition = 1000;
    else
        xPosition = xPosition - 1;

    if (player.getYPosition() > yPosition)
        yPosition++;
    else if (player.getYPosition() < yPosition)
        yPosition--;
}
```

(A) I only
(B) II only
(C) III only
(D) I and II only
(E) II and III only

GO ON TO THE NEXT PAGE

13. Which of the following classes correctly implements the `Ship` interface?

I.
```
public class BattleShip implements Ship
{
    private int x;
    private int y;

    public Ship(int myX, int myY)
    {
        x = myX;
        y = myY;
    }

    public boolean move()
    {
        return x + y > 0;
    }
}
```

II.
```
public abstract class Zoomer implements Ship
{
    public abstract void move();
}
```

III.
```
public class Sailboat extends Ship
{
    private int x;
    private int y;
    public Sailboat()
    {
        x = 0;
        y = 1000;
    }

    public void move()
    {
        x++;
        y--;
    }
}
```

(A) I only
(B) II only
(C) III only
(D) I and II only
(E) II and III only

14. A bank will approve a loan for any customer who fulfills one or more of the following requirements:

- Has a credit score ≥ 640
- Has a cosigner for the loan
- Has a credit score ≥ 590 and has collateral

Which of the following methods will properly evaluate the customer's eligibility for a loan?

I.
```
public boolean canGetLoan(int credit, boolean cosigner, boolean coll)
{
   if (credit >= 640 || cosigner || credit >= 590 && coll)
      return true;
   return false;
}
```

II.
```
public boolean canGetloan(int credit, boolean cosigner, boolean coll)
{
   if (!credit >= 640 || !cosigner || !(credit >= 590 && coll))
      return false;
   else
      return true;
}
```

III.
```
public boolean canGetLoan(int credit, boolean cosigner, boolean coll)
{
   if (coll && credit >= 590)
      return true;
   if (credit >= 640)
      return true;
   if (cosigner)
      return true;
   return false;
}
```

(A) I only
(B) II only
(C) III only
(D) I and III only
(E) II and III only

15. Consider the following code segment.

```
int value = 33;
boolean calculate = true;
while (value > 5 || calculate)
{
   if (value % 3 == 0)
      value = value - 2;
   if (value / 4 < 3)
      calculate = false;
}
System.out.print(value);
```

What is printed as a result of executing the code segment?

(A) `0`
(B) `1`
(C) `8`
(D) `33`
(E) Nothing will be printed. It is an infinite loop.

GO ON TO THE NEXT PAGE

16. Assume that planets has been correctly instantiated and initialized to contain the names of the planets. Which of the following code segments will reverse the order of the elements in `planets`?

(A)
```
for (int index = planets.length / 2; index >= 0; index--)
{
    String temp = planets[index];
    planets[index] = planets[index + 1];
    planets[index + 1] = temp;
}
```

(B)
```
int index = planets.length / 2;
while (index >= 0)
{
    String s = planets[index];
    planets[index] = planets[index + 1];
    index--;
}
```

(C)
```
String[] newPlanets = new String[planets.length];
for (int i = planets.length - 1; i >= 0; i--)
{
    newPlanets[i] = planets[i];
}
planets = newPlanets;
```

(D)
```
for (int index = 0; index < planets.length; index++)
{
    String temp = planets[index];
    planets[index] = planets[planets.length - 1 - index];
    planets[planets.length - 1 - index] = temp;
}
```

(E)
```
for (int index = 0; index < planets.length / 2; index++)
{
    String temp = planets[index];
    planets[index] = planets[planets.length - 1 - index];
    planets[planets.length - 1 - index] = temp;
}
```

17. When designing a class hierarchy, what should be true of an abstract class?

(A) An abstract class should contain the methods and class variables that all of its subclasses have in common, even if the correct implementations of the methods are unknown.
(B) An abstract class must be the largest most complicated superclass for which all other classes should derive.
(C) An abstract class should contain all final variables and static methods of the hierarchy.
(D) An abstract class should contain the specific implementations of every method inherited from an interface that the abstract class implements.
(E) An abstract class should contain basic implementation of all methods so that subclasses can override its implementation with more specific and correct methods.

18. Consider the following code segment.

```
int num = (int)(Math.random() * 30 + 20);
num += 5;
num = num / 5;
System.out.print(num);
```

What are the possible values that could be printed to the console?

(A) All real numbers from 4 to 10 (not including 10)
(B) All integers from 5 to 10 (inclusive)
(C) All integers from 20 to 49 (inclusive)
(D) All integers from 25 to 54 (inclusive)
(E) All real numbers from 30 to 50 (inclusive)

19. Consider the following code segment.

```
String exampleString = "computer";
List<String> words = new ArrayList<String>();
for (int k = 0; k < exampleString.length(); k++)
{
    words.add(exampleString.substring(k));
    k++;
}

System.out.println(words);
```

What is printed as a result of executing the code segment?

(A) `[computer, mputer, uter, er]`
(B) `[computer, comput, comp, co]`
(C) `[computer, computer, computer, computer]`
(D) `[computer, omputer, mputer, puter, uter, ter, er, r]`
(E) Nothing is printed. There is an `ArrayListIndexOutOfBoundsException`.

20. Consider the following incomplete method.

```
public boolean validation(int x, int y)
{
    /* missing code */
}
```

The following table shows several examples of the desired result of a call to `validation`.

x	y	Result
1	0	true
3	6	true
4	1	true
1	3	false
2	6	false
5	5	false

Which of the following code segments should replace `/* missing code */` to produce the desired return values?

(A) `return x > y;`
(B) `return (x % y) > 1`
(C) `return (x + y) % 2 == 0`
(D) `return (x + y) % x == y;`
(E) `return (x + y) % 2 > (y + y) % 2;`

GO ON TO THE NEXT PAGE

21. Consider the following method that is intended to remove all Strings from `words` that are less than six letters long.

```
                 private void letterCountCheck(ArrayList<String> words)
                 {
Line 1:              int numWord = 0;
Line 2:              while (numWord <= words.size())
                     {
Line 3:                  if (words.get(numWord).length() < 6)
                         {
Line 4:                      words.remove(numWord);
                         }
                         else
                         {
Line 5:                      numWord++;
                         }
                     }
                 }
```

Which line of code contains an error that prevents `letterCountCheck` from working as intended?

(A) `Line 1`

(B) `Line 2`

(C) `Line 3`

(D) `Line 4`

(E) `Line 5`

22. Consider the following recursive method.

```
public void wackyOutput(String wacky)
{
    if (wacky.length() < 1)
        return;
    wacky = wacky.substring(1, wacky.length());
    wackyOutput(wacky);
    System.out.print(wacky);
}
```

What is printed as a result of executing the call `wackyOutput("APCS")`?

(A) `PC`

(B) `SCPA`

(C) `SCSPCS`

(D) `PCSCSS`

(E) `SCSPCSAPCS`

23. Consider the following method.

```
public String calculate(int num1, int num2)
{
   if (num1 >= 0 && num2 >= 0)
      return "Numbers are valid";
   return "Numbers are not valid";
}
```

Which of the following methods will give the exact same results as `calculate`?

I.
```
public String calculate1(int num1, int num2)
{
   if (!(num1 < 0 || num2 < 0))
      return "Numbers are valid";
   else
      return "Numbers are not valid";
}
```

II.
```
public String calculate2(int num1, int num2)
{
   if (num1 < 0)
      return "Numbers are not valid";
   if (num2 < 0)
      return "Numbers are not valid";
   else
      return "Numbers are valid";
}
```

III.
```
public String calculate3(int num1, int num2)
{
   if (num1 + num2 < 0)
      return "Numbers are not valid";
   return "Numbers are valid";
}
```

(A) I only
(B) II only
(C) III only
(D) I and II only
(E) I, II, and III

GO ON TO THE NEXT PAGE

24. Consider the following class declaration.

```
public class TestClass
{
    private double value;

    public TestClass(double myValue)
    {
        value = myValue;
    }

    public void addValue(double add)
    {
        value += add;
    }

    public void reduceValue(double reduce)
    {
        value -= reduce;
    }

    public double getValue()
    {
        return value;
    }
}
```

The following code segment is executed in the main method.

```
TestClass test1 = new TestClass(9.0);
TestClass test2 = new TestClass(17.5);
test1.reduceValue(3.0);
test2.addValue(1.5);
test2 = test1;
test2.reduceValue(6.0);
System.out.print(test2.getValue() + test1.getValue());
```

What is printed to the console as a result of executing the code segment?

(A) `0.0`
(B) `3.0`
(C) `12.0`
(D) `19.0`
(E) `27.5`

25. Assume `truth1` and `truth2` are `boolean` variables that have been properly declared and initialized.

Consider this expression.

```
(truth1 && truth2) || ((!truth1) && (!truth2))
```

Which expression below is its logical equivalent?

(A) `truth1 != truth2`
(B) `truth1 || truth2`
(C) `truth1 && truth2`
(D) `!truth1 && !truth2`
(E) `truth1 == truth2`

26. The following insertion sort method sorts the array in increasing order. This method includes an additional `System.out.print(value + "  ")` statement that is not normally included in the insertion sort.

```
// postcondition: reals is sorted in increasing order
public void sortIt(double[] reals)
{
    for (int i = 1; i < reals.length; i++)
    {
        double value = reals[i];

        int j;
        for (j = i - 1; j >= 0 && reals[j] > value; j--)
            reals[j + 1] = reals[j];
        reals[j + 1] = value;

        System.out.print(value + "  ");    //Additional statement
    }
}
```

Consider the following array.

```
double[] costs = {2.6,  3.1,  7.8,  1.4,  6.6,  5.2,  9.1};
```

What is printed as a result of executing the call `sortIt(costs)`?

(A) `3.1  7.8  1.4  6.6  5.2`
(B) `3.1  7.8  1.4  6.6  5.2  9.1`
(C) `2.6  3.1  7.8  1.4  6.6  5.2  9.1`
(D) `1.4  2.6  3.1  5.2  6.6  7.8  9.1`
(E) `3.1  3.1  7.8  7.8  1.4  1.4  6.6  6.6  5.2  5.2  9.1  9.1`

GO ON TO THE NEXT PAGE

27. Consider the following interface.

```
public interface Polygon
{
   /** returns true if the ordered pair is inside the polygon
    *
    *         false otherwise
    */
   boolean contains(int x, int y);
}
```

Consider the following class.

```
public class Rectangle implements Polygon
{
   private int topLeftX, topLeftY, bottomRightX, bottomRightY;

   /**
    *  Precondition: topLeftX < bottomRightX
    *  Precondition: topLeftY < bottomRightY
    *                The y value increases as it moves down the screen.
    *  @param topLeftX The x value of the top left corner
    *  @param topLeftY The y value of the top left corner
    *  @param bottomRightX The x value of the bottom right corner
    *  @param bottomRightY The y value of the bottom right corner
    */
   public Rectangle(int topLX, int topLY, int bottomRX, int bottomRY)
   {
      topLeftX = topLX;
      topLeftY = topLY;
      bottomRightX = bottomRX;
      bottomRightY = bottomRY;
   }
}
```

Which of the following is a correct implementation of the `contains` method?

(A)
```
public boolean contains(int x, int y)
{
   return x > topLeftX && y > topLeftY &&
          x < bottomRightX && y < bottomRightY;
}
```

(B)
```
public boolean contains(int x, int y)
{
   if (topLeftX > x)
      return true;
   else if (topLeftY > y)
      return true;
   else if (bottomRightX < x)
      return true;
   else if (bottomRightY < y)
      return true;
   return false;
}
```

(C)
```
public boolean contains(int x, int y)
{  return super.contains(x, y);  }
```

(D)
```
public boolean contains(int x, int y)
{  return Polygon.contains(x, y);  }
```

(E)
```
public boolean contains(int x, int y)
{
   boolean result = false;
   if (topLeftX < x && topLeftY < y)
      result = true;
   if (bottomRightX > x && bottomRightY > y)
      result = true;
   return result;
}
```

28. Consider the following method.

```
public ArrayList<String> rearrange(ArrayList<String> myList)
{
    for (int i = myList.size() / 2; i >= 0; i--)
    {
        String wordA = myList.remove(i);
        myList.add(wordA);
    }
    return myList;
}
```

Assume that `ArrayList<String>` list has been correctly instantiated and populated with the following entries.

`["Nora", "Charles", "Madeline", "Nate", "Silja", "Garrett"]`

What are the values of the `ArrayList` returned by the call `rearrange(list)`?

(A) `["Silja", "Garrett", "Nate", "Madeline", "Charles", "Nora"]`
(B) `["Madeline", "Charles", "Nora", "Nate", "Silja", "Garrett"]`
(C) `["Nate", "Silja", "Garrett", "Nora", "Charles", "Madeline"]`
(D) `["Nora", "Charles", "Madeline", "Nate", "Silja", "Garrett"]`
(E) Nothing is returned. `ArrayListIndexOutOfBoundsException`

29. Consider the following code segment.

```
int varA = -30;
int varB = 30;
while (varA != 0 || varB > 0)
{
    varA = varA + 2;
    varB = Math.abs(varA + 2);
    varA++;
    varB = varB - 5;
}
System.out.println(varA + "    " + varB);
```

What will be printed as a result of executing the code segment?

(A) `-30    30`
(B) `-4    0`
(C) `0    -4`
(D) `0    0`
(E) Nothing will be printed. It is an infinite loop.

30. Consider the following code segment.

```
int[][] grid = { {0,  1,  2},
                 {3,  4,  5},
                 {6,  7,  8},
                 {9, 10, 11} };

int[][] newGrid = new int[grid[0].length][grid.length];

for (int row = 0; row < grid.length; row++)
{
    for (int col = 0; col < grid[row].length; col++)
    {
        newGrid[col][row] = grid[row][col];
    }
}
```

What is the value of `newGrid[2][1]` as a result of executing the code segment?

(A) `3`
(B) `4`
(C) `5`
(D) `7`
(E) Nothing is printed. There is an `ArrayIndexOutOfBoundsException`.

31. Consider the following code segment.

```
int[][] grid = new int[3][3];
for (int row = 0; row < grid.length; row++)
{
    for (int col = 0; col < grid[row].length; col++)
    {
        grid[row][col] = 1;
        grid[col][row] = 2;
        grid[row][row] = 3;
    }
}
```

Which of the following shows the values in grid after executing the code segment?

(A)
```
{ {1, 1, 1},
  {1, 1, 1},
  {1, 1, 1} };
```
(B)
```
{ {2, 2, 2},
  {2, 2, 2},
  {2, 2, 2} };
```
(C)
```
{ {3, 2, 2},
  {2, 3, 2},
  {2, 2, 3} };
```
(D)
```
{ {3, 1, 1},
  {1, 3, 1},
  {1, 1, 3} };
```
(E)
```
{ {3, 2, 2},
  {1, 3, 2},
  {1, 1, 3} };
```

32. What can the following method best be described as?

```
public int mystery(int[] array, int a)
{
  for (int i = 0; i < array.length; i++)
  {
    if (array[i] == a)
      return i;
  }
  return -1;
}
```

(A) Insertion sort
(B) Selection sort
(C) Binary search
(D) Merge sort
(E) Sequential search

Questions 33–36 refer to the following classes.

```
public abstract class Soda implements Beverage
{
   private String size;
   private double price;

   public Soda()
   {  /* implementation not shown  */  }

   public Soda(String size, double price)
   {  /* implementation not shown  */  }

   public Soda(String size)
   {  /* implementation not shown  */  }

   /* no other constructors */

   public abstract int getCalories(String size);

   public boolean isHot()
   {   return false;   }

   public abstract boolean hasCaffeine();

   /* Additional implementation not shown */
}

public class Coffee implements Beverage
{
   private double price;

   public Coffee()
   {  /* implementation not shown  */  }

   public Coffee(double price)
   {  /* implementation not shown  */  }

   /* no other constructors */

   public boolean isHot()
   {  return true;  }

   public boolean hasCaffeine()
   {  return true;  }

   /* no other methods */
}

public class RootBeer extends Soda
{
   public RootBeer(String size, double price)
   {
      super(size, price);
   }

   /* no other constructors */

   /* Additional implementation not shown */
}
```

33. Which of the following will compile without error?

I. `Beverage yum1 = new Coffee();`
II. `Soda yum2 = new RootBeer("large", 3.00);`
III. `Beverage yum3 = new Soda("small");`

(A) I only
(B) II only
(C) I and II only
(D) II and III only
(E) I, II, and III

34. Which of the following must be included in the implementation of the `RootBeer` class?

I. `public RootBeer()`
II. `public boolean hasCaffeine()`
III. `public boolean isHot()`
IV. `public int getCalories()`

(A) I only
(B) I and III only
(C) II and IV only
(D) II, III, and IV only
(E) I, II, III, and IV

35. Which of the following methods could be part of the `Beverage` interface?

I. `boolean hasCaffeine()`
II. `boolean isHot()`
III. `int getCalories()`

(A) III only
(B) I and II only
(C) I and III only
(D) II and III only
(E) I, II, and III

36. Which of the following code segments will compile without error?

(A)
```
ArrayList<Beverage> drinks = new ArrayList<Beverage>();
for (Beverage b : drinks)
    System.out.println(b.hasCaffeine());
```
(B)
```
ArrayList<RootBeer> drinks = new ArrayList<Soda>();
for (Beverage b : drinks)
    System.out.println(b.getCalories());
```
(C)
```
ArrayList<Soda> drinks = new ArrayList<Beverage>();
for (Beverage b : drinks)
    System.out.println(b.getCalories());
```
(D)
```
ArrayList<Beverage> drinks = new ArrayList<Coffee>();
for (Beverage b : drinks)
    System.out.println(b.getCalories());
```
(E)
```
ArrayList<Soda> drinks = new ArrayList<Coffee>();
for (Beverage b : drinks)
    System.out.println(b.getCalories());
```

37. Consider the following method.

```
public int countLetters(String word, String letter)
{
    int count = 0;
    for (int index = 0; index < word.length(); index++)
    {
        if ( /*  missing code  */ )
            count++;
    }
    return count;
}
```

What could replace `/* missing code */` to allow the method to return the number of times `letter` appears in `word`?

(A) `word.substring(index).equals(letter)`
(B) `word.substring(index, index + 1) == letter`
(C) `word.indexOf(letter) == letter.indexOf(letter)`
(D) `word.substring(index, index + 1).equals(letter)`
(E) `letter.equals(word.substring(index).indexOf(letter))`

38. Assume `obscureAnimals` is an `ArrayList<String>` that has been correctly constructed and populated with the following items.

```
["okapi", "aye-aye", "cassowary", "echidna", "sugar glider", "jerboa"]
```

Consider the following code segment.

```
for (int i = 0; i < obscureAnimals.size(); i++)
{
    if (obscureAnimals.get(i).compareTo("pink fairy armadillo") < 0)
        obscureAnimals.remove(i);
}
System.out.print(animals);
```

What will be printed as a result of executing the code segment?

(A) `[]`
(B) `[sugar glider]`
(C) `[aye-aye, echidna, sugar glider]`
(D) `[aye-aye, echidna, sugar glider, jerboa]`
(E) Nothing will be printed. There is an `ArrayListIndexOutOfBoundsException`.

Questions 39–40 refer to the following classes.

Consider the following class declarations.

```
public class Building
{
    int sqFeet;
    int numRooms;

    public Building(int ft, int rms)
    {
        sqFeet = ft;
        numRooms = rms;
    }

    public int getSqfeet()
    {  return sqFeet;  }

    public String getSize()
    {
        return numRooms + " rooms and " + sqFeet + " square feet";
    }

    /* Additional implementation not shown */
}

public class House extends Building
{
    int numBedRooms;
    public House(int ft, int rms, int bedrms)
    {
        super(ft, rms);
        numBedRooms = bedrms;
    }

    public String getSize()
    {
        return super.getSqFeet() + " square feet and " + numBedRooms + " bedrooms.";
    }
}
```

39. Assume that `ArrayList<Building>` list has been correctly instantiated and populated with `Building` objects.

Which of the following code segments will result in the square feet in each building being printed?

I.
```
for (int i = 0; i < list.size(); i++)
    System.out.println(list.getSqFeet(i));
```
II.
```
for (int i = 0; i < list.size(); i++)
    System.out.println(list[i].getSqFeet());
```
III.
```
for (int i = 0; i < list.size(); i++)
    System.out.println(list.get(i).getSqFeet());
```
IV.
```
for (Building b : list)
    System.out.println(list.getSqFeet());
```
V.
```
for (Building b : list)
    System.out.println(b.getSqFeet());
```

(A) I and IV only
(B) I and V only
(C) II and IV only
(D) III and IV only
(E) III and V only

40. Consider the following code segment.

```
Building b1 = new Building(2000, 3);
Building b2 = new House(2500, 8, 4);
Building b3 = b2;
Building[] buildings = new Building[3];
buildings[0] = b1;
buildings[1] = b2;
buildings[2] = b3;
```

What will be printed by the following code segment?

```
for (int i = 0; i < buildings.length; i++)
   System.out.println(buildings[i].getSize());
```

(A)
```
3 rooms and 2000 square feet
8 rooms and 2500 square feet
8 rooms and 2500 square feet
```
(B)
```
3 rooms and 2000 square feet
2500 square feet and 4 bedrooms
2500 square feet and 4 bedrooms
```
(C)
```
2000 square feet and 3 bedrooms
2500 square feet and 4 bedrooms
2500 square feet and 4 bedrooms
```
(D)
```
3 rooms and 2000 square feet
2500 square feet and 4 bedrooms
3 rooms and 2000 square feet
```
(E) There is an error. Nothing will be printed.

STOP. End of Part I.

AP Computer Science A: Practice Exam 2

Part II (Free Response)

Time: 90 minutes
Number of questions: 4
Percent of total score: 50

Directions: Write all of your code in Java. Show all your work.

Notes:

- You may assume all imports have been made for you.
- You may assume that all preconditions are met when making calls to methods.
- You may assume that all parameters within method calls are not null.
- Be aware that you should, when possible, use methods that are defined in the classes provided as opposed to duplicating them by writing your own code.

1. A hex grid is a type of two-dimensional (2-D) map used by many games to represent the area that tokens, ships, robots, or other types of game pieces can move around on.

The `GamePiece` class is intended to represent the various game pieces required to play different games on the `HexGrid`. The implementation of the `GamePiece` class is not shown. You may construct a `GamePiece` object with a no-argument constructor.

A hex grid is represented in the `HexGrid` class by a 2-D array of `GamePiece` objects. The coordinates of a `GamePiece` represent its position on the `HexGrid` as shown in the following diagram. Many of the hex cells in the `HexGrid` will not be occupied by any `GamePiece` and are therefore `null`.

The rows in a `HexGrid` are in a staggered horizontal pattern where hexes in even columns rise higher than hexes in odd-numbered columns. The columns in a `HexGrid` always appear vertically aligned regardless of the value of the row.

The following is a representation of a `HexGrid` object that contains several `GamePiece` objects at specified locations in the `HexGrid`. The row and column of each `GamePiece` are given.

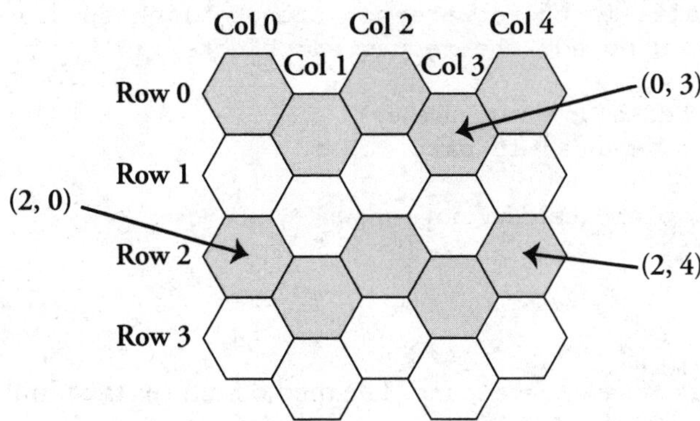


```
public class HexGrid
{
  /** A two-dimensional array of the GamePiece objects in the
   *  game. Each GamePiece is located in a specific hex cell
   *  as determined by its row and column number.
   */
  private GamePiece[][] grid;

  /** Returns the number of GamePiece objects
   *  currently occupying any hex cell in the grid.
   *
   *  @return the number of GamePiece objects currently in
   *          the grid.
   */
  public int getGamePieceCount()
  {  /*  to be implemented in part (a) */  }
```

```
    /** Returns an ArrayList of the GamePiece objects
     *  in the same column as and with a lower row number than
     *  the GamePiece at the location specified by the parameters.
     *  If there is no GamePiece at the specified location,
     *  the method returns null.
     *
     *  @param row the row of the GamePiece in question
     *  @param column the column of the GamePiece in question
     *  @return an ArrayList of the GamePiece objects "above" the
     *          indicated GamePiece, or null if no such object
     *          exists
     */
    public ArrayList<GamePiece> isAbove(int row, int col)
    {  /*  to be implemented in part (b) */  }

    /** Adds GamePiece objects randomly to the grid
     *  @ param the number of GamePiece objects to add
     *  @ return true if GamePieces were added successfully
     *           false if there were not enough blank spaces in the
     *           grid to add the requested objects
     */
    public boolean addRandom(int number)
    {  /*  to be implemented in part (c) */  }

    /* Additional implementation not shown */
}
```

(a) Write the method `getGamePieceCount`. The method returns the number of `GamePiece` objects in the hex grid.

```
/** Returns the number of GamePiece objects
 *  currently occupying any hex cell in the grid.
 *
 *  @return the number of GamePiece objects currently in
 *          the grid.
 */
public int getGamePieceCount()
```

(b) Write the method `isAbove`. The method returns an `ArrayList` of `GamePiece` objects that are located "above" the `GamePiece` in the `row` and `column` specified in the parameters. "Above" is defined as in the same column but with a lower row number than the specified location. In the diagram below, letters represent `GamePiece` objects. The objects may appear in any order in the `ArrayList`.

- If the method is called with the row and column of C, an `ArrayList` containing A and B will be returned.
- If the method is called with the row and column of F, an `ArrayList` containing E will be returned.
- If the method is called with the row and column of E, an empty `ArrayList` is returned.
- If the method is called with the row and column of an empty cell, `null` is returned.

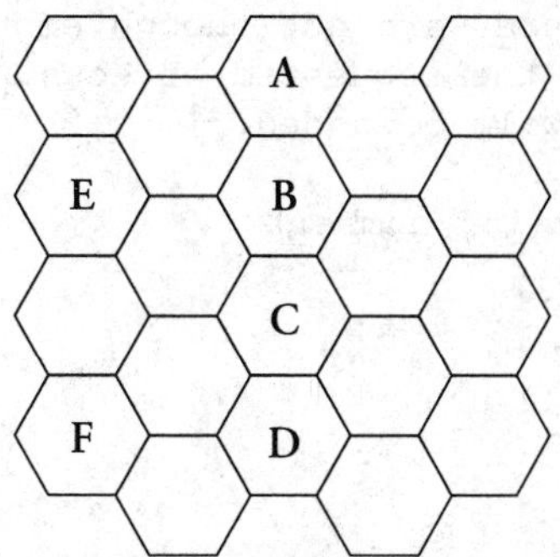


```
/** Returns an ArrayList of the GamePiece objects
 *  in the same column as and with a lower row number than
 *  the GamePiece at the location specified by the parameters.
 *  If there is no GamePiece at the specified location,
 *  the method returns null.
 *
 *  @param row the row of the GamePiece in question
 *  @param column the column of the GamePiece in question
 *  @return an ArrayList of the GamePiece objects "above" the
 *          indicated GamePiece, or null if no object exists
 *          at the specified location
 */
public ArrayList<GamePiece> isAbove (int row, int col)
```

(c) Write the addRandom method. This method takes a parameter that represents the number of GamePiece objects to be added to the grid at random locations. GamePiece objects can only be added to locations that are currently empty (null). If there are insufficient empty cells to complete the requested additions, no GamePiece objects will be added and the method will return false. If the additions are successful, the method will return true.

You may assume that getGamePieceCount works as intended, regardless of what you wrote in part (a). You must use getGamePieceCount appropriately in order to receive full credit for part (c).

```
/** Adds GamePiece objects randomly to the grid
  *  @param the number of GamePiece objects to add
  *  @return true if GamePieces were added successfully
  *          false if there were not enough empty cells in the
  *          grid to add the requested objects. In this case
  *          no objects will be added.
  */
public boolean addRandom(int number)
```

2. In mathematics, a monomial is a constant value multiplied by a variable raised to a power. The constant multiplier part of a monomial is called the coefficient and the value of the exponent is called the degree of the monomial.

Examples of Monomials

$3x^4$	coefficient = 3.0	degree = 4
$-7.2x^5$	coefficient = 7.2	degree = 5
$9.8x^2$	coefficient = 9.8	degree = 2

A monomial is represented by the following class.

```
public class Monomial
{
   private double coefficient;
   private int degree;

   public Monomial(double coeff, int deg)
   {
      coefficient = coeff;
      degree = deg;
   }

   public int getDegree()
   {  return degree;  }

   public double getCoefficient()
   {  return coefficient;  }
}
```

In mathematics, different numbers of monomials with different degrees can be combined to create polynomial expressions.

Since all polynomials of one variable have certain characteristics in common, they are represented by a `Polynomial` interface. The `Polynomial` interface requires implementing classes to provide the `getDegree`, `getNumTerms`, and `getName` methods, where degree, number of terms, and name are defined as follows:

- The *degree* is the value of the largest exponent on the variable.
- The *number of terms* is the number of monomials that combine to form the polynomial.
- The *name* of a polynomial is composed of two parts: the degree determines the first part, and the number of terms determines the second part as seen in the following charts and examples.

Degree	Name
0	Constant
1	Linear
2	Quadratic
3	Cubic
4 or more	Degree 4, etc.

Number of Terms	Name
1	Monomial
2	Binomial
3	Trinomial
4 or more	Polynomial

Examples

$3x^3 + 4x + 2$	degree = 3, number of terms = 3, name = "Cubic Trinomial"
$7.2x^5$	degree = 5, number of terms = 1, name = "Degree 5 Monomial"
$9.8x^2 + 3x - 7$	degree = 2, number of terms = 3, name = "Quadratic Trinomial"
$3x + 6$	degree = 1, number of terms = 2, name = "Linear Binomial"

The declaration for the `Polynomial` interface is shown below.

```
public interface Polynomial
{
    /** Returns the degree of the Polynomial
     *
     *  @return the degree of the polynomial
     */
    int getDegree();

    /** Returns the number of Monomials that combine
     *  to form the Polynomial.
     *
     *  @return the number of terms in the polynomial
     */
    int getNumTerms();

    /** Returns the name of the Polynomial including
     *  terminology for degree and number of terms.
     *
     *  @return the name of the Polynomial
     */
    String getName();
}
```

(a) A `Quadratic` is a degree 2 polynomial with one, two, or three monomial terms. The first monomial term must be of degree 2 for the object to be identified as `Quadratic`. The degree 1 and degree 0 terms are not required.

Examples of Quadratic Polynomials

(i) $3x^2 + 7x + 6$
(ii) $2.1x^2 + 8$
(iii) $4x^2$

Write the complete `Quadratic` class that properly implements the `Polynomial` interface. Include a constructor that takes an `ArrayList` of `Monomial` objects as a parameter, and any necessary instance variables and methods. It is not necessary to include additional methods like `toString`, or accessors and modifiers for the instance variables.

Precondition: The `Monomial` objects in the `ArrayList` passed to the constructor are in order of degree. The degree 2 monomial term is at index 0, the second highest degree at index 1 (if it exists), and the third highest degree at index 2 (if it exists).

The lists for the examples above would look like this:

(i)

coefficient = 3 degree = 2	coefficient = 7 degree = 1	coefficient = 6 degree = 0

(ii)

coefficient = 2.1 degree = 2	coefficient = 8 degree = 0

(iii)

coefficient = 4 degree = 2

GO ON TO THE NEXT PAGE

(b) A quadratic can be written using `a`, `b`, and `c` to represent the coefficients of the various monomial terms: `ax`^{2} `+ bx + c`. If a term does not exist, its coefficient is zero. Given this notation, the determinant is the name used for the expression `b`^{2} `- 4ac`. `a`, `b`, and `c` are passed to the method as parameters.

If the determinant is greater than or equal to zero, then real roots exist for this quadratic. If the determinant is negative, then there are no real roots. The `hasRoots()` method returns true if the determinant is greater than or equal to zero and false if the determinant is less than zero.

Write a `private boolean hasRoots(double a, double b, double c)` method to be included in the `Quadratic` class.

```
/** Returns true if the determinant is >= 0 and
 *          false if the determinant is negative.
 *
 *  @param a the coefficient of the degree 2 monomial
 *  @param b the coefficient of the degree 1 monomial
 *  @param c the coefficient of the degree 0 monomial
 *  @return true if the determinant is >= 0 and false otherwise
 */
private boolean hasRoots(double a, double b, double c)
```

(c) The solutions or *roots* of a `Quadratic` polynomial are the values of x that give a result of zero when the expression is evaluated. The roots can be determined by using the two-part quadratic formula (given below). The values of `a`, `b`, and `c` are defined as follows:

- `a` is the coefficient of the degree 2 monomial,
- `b` is the coefficient of the degree 1 monomial (`b = 0` if no such term exists), and
- `c` is the coefficient of the degree 0 monomial (`c = 0` if no such term exists).

$$\text{root } 1 = \frac{-b + \sqrt{b^2 - 4ac}}{2a} \quad \text{and root } 2 = \frac{-b - \sqrt{b^2 - 4ac}}{2a}$$

The two equations differ only in the operator before the square root. As there are two roots, they are returned in a double array of length 2.

If the determinant is negative, there are no real number solutions. This method must check the determinant before evaluating the quadratic equation and return `null` if no real roots exist. (Since the determinant is under a square root symbol, it is important not to evaluate the square root if the determinant is negative.)

You may assume that hasRoots works as intended, regardless of what you wrote in part (b). You must use hasRoots appropriately in order to receive full credit for part (c).

Complete the method `getRoots` to be added to the `Quadratic` class.

```
/** Returns the roots of the Quadratic as determined
 *  by the quadratic formula.  If the quadratic has
 *  no real roots, then null is returned.
 *
 *  @return the roots of the Quadratic or null
 */
public double[] getRoots()
```

3. A printing factory maintains many printing machines that hold rolls of paper.

```
public class Machine
{
   private PaperRoll paper;

   public Machine(PaperRoll roll)
   {  paper = roll;  }

   public PaperRoll getPaperRoll()
   {  return paper;  }

   /** Returns the current partial roll and replaces it with the new roll.
    *  @param pRoll a new full PaperRoll
    *  @return the old nearly empty PaperRoll
    */
   public PaperRoll replacePaper(PaperRoll pRoll)
   {
      /* to be implemented in part (a) */
   }

   /* Additional implementation not shown */
}

public class PaperRoll
{
   private double meters;

   public PaperRoll()
   {  meters = 1000; }

   public double getMeters()
   {  return meters;  }

   /* Additional implementation not shown */
}
```

The factory keeps track of its printing machines in array `machines`, and it keeps track of its paper supply in two lists: `newRolls` to hold fresh rolls and `usedRolls` to hold the remnants taken off the machines when they no longer hold enough paper to be usable. At the beginning of the day, `usedRolls` is emptied of all paper remnants and `newRolls` is refilled. When `newRolls` no longer has enough rolls available to refill all the machines, the factory must shut down for the day until its supplies are restocked. At that time, the amount of paper used for the day is calculated.

A partial `PrintingFactory` class is shown below.

```
public class PrintingFactory
{
   // All machines available in the company
   private Machine[] machines;

   // The available full paper rolls (1000 meters each)
   private ArrayList<PaperRoll> newRolls = new ArrayList<PaperRoll>();

   // The used paper roll remnants (less than 4.0 meters each)
   private ArrayList<PaperRoll> usedRolls = new ArrayList<PaperRoll>();

   public PrintingFactory(int numMachines)
   {   machines = new Machine[numMachines];   }

   /** Replaces the PaperRoll for any machine that has a
    *  PaperRoll with less than 4.0 meters of paper remaining.
    *  The used roll is added to usedRolls.
    *  A new roll is removed from newRolls.
    *  Precondition: newRolls is not empty.
    */
   public void replacePaperRolls(PaperRoll roll)
   {  /* to be implemented in part (b)   */  }

   /** Returns the total amount of paper that has been used.
    *  @return the total amount of paper that has been used from the PaperRolls
    *          in the usedRolls list plus the paper that has
    *          been used from the PaperRolls on the machines.
    */
   public double getPaperUsed()
   {  /* to be implemented in part (c)   */  }

   /* Additional implementation not shown */
}
```

(a) Write the `replacePaper` method of the `Machine` class. This method returns the nearly empty `PaperRoll` and replaces it with the `PaperRoll` passed as a parameter.

```
/** Returns the current partial roll and replaces it with the new roll.
 *  @param pRoll a new full PaperRoll
 *  @return the old nearly empty PaperRoll
 */
public PaperRoll replacePaper(PaperRoll pRoll)
```

(b) Write the `replacePaperRolls` method of the `PrintingFactory` class.

At the end of each job cycle, each `Machine` object in `machines` is examined to see if its `PaperRoll` needs replacing. A `PaperRoll` is replaced when it has less than 4 meters of paper remaining. The used roll is added to the `usedRolls` list, while a new roll is removed from the `newRolls` list.

```
/** Replaces the PaperRoll for any machine in array machines that has
 *  a PaperRoll with less than 4.0 meters of paper remaining.
 *  The used roll is added to usedRolls.
 *  A new roll is removed from newRolls.
 */
public void replacePaperRolls()
```

(c) Write the `getPaperUsed` method of the `PrintingFactory` class.

At the end of the day, the factory calculates how much paper has been used by adding up the amount of paper used from all the rolls in the `usedRolls` list and all the rolls still in the machines. A brand-new paper roll has 1000 meters of paper.

Example

The tables below show the status of the factory at the end of the day.

Paper in Machines	
Amount Left	**Amount Used**
900.0	100.0
250.0	750.0
150.0	850.0

Paper in `usedRolls` List	
Amount Left	**Amount Used**
3.5	996.5
2.0	998.0
1.5	998.5
3.0	997.0

Total used: 100.0 + 750.0 + 850.0 + 996.5 + 998.0 + 998.5 + 997.0 = 5690.0 meters

Complete the method `getPaperUsed`.

```
/** Returns the total amount of paper that has been used.
 *  @return the total amount of paper left on the PaperRolls
 *          in the usedRolls list plus the paper that has
 *          been used from the PaperRolls on the machines.
 */
public double getPaperUsed()
```

4. In the seventeenth century, Location Numerals were invented as a new way to represent numbers. Location Numerals have a lot in common with the binary number system we use today, except that in Location Numerals, successive letters of the alphabet are used to represent the powers of two, starting with $A = 2^0$ all the way to $Z = 2^{25}$.

To represent a given number as a Location Numeral (LN), the number is expressed as the sum of powers of two with each power of two replaced by its corresponding letter.

$A = 2^0$ $B = 2^1$ $C = 2^2$ $D = 2^3$ $E = 2^4$ and so on

Let's consider the decimal number 19.

- 19 can be expressed as a sum of powers of two like this: 16 + 2 + 1.
- So 19 would be written in binary like this: 10011.
- We can covert 19 to LN notation by choosing the letters that correspond with 1s: EBA.
- Since changing the order of the letters does not change their value, LN is usually written in alphabetical order: ABE.

Here are a few more examples.

Base 10	Sum of Powers	Binary	LN	Ordered
7	$4 + 2 + 1 = 2^2 + 2^1 + 2^0$	111	CBA	ABC
17	$16 + 1 = 2^4 + 2^0$	10001	EA	AE
12	$8 + 4 = 2^3 + 2^2$	1100	DC	CD

A partial `LocationNumeral` class is shown below.

```
public class LocationNumeral
{
    /** Returns the decimal value of a single LN letter
     *
     *  @param letter String containing a single LN letter
     *  @return the decimal value of that letter
     *  Precondition: the parameter contains a single uppercase letter
     */
    public int getLetterValue(String letter)
    {
        String letters = "ABCDEFGHIJKLMNOPQRSTUVWXYZ";
        /* to be implemented in part (a) */
    }

    /** Return the Location Numeral letter representation for a
     *  decimal value that is a power of two.
     *
     *  @param the decimal value to be converted to a Location Numeral letter
     *  @return the corresponding Location Numeral letter
     *  Precondition: The parameter must be an integer power of two between
     *                2 raised to the power 0 and 2 raised to the power 25.
     */
    public String getLetter(int value)
    {  /* implementation not shown */  }
```

GO ON TO THE NEXT PAGE

```
    /** Returns the decimal value of a simplified Location Numeral
     *
     *  @param numeral String representing a Location Numeral
     *  @return the decimal value of the Location Numeral.
     *  Precondition:  The characters in the parameter are
     *                 uppercase letters A-Z only.
     */
    public int getDecimalValue(String numeral)
    {  /* to be implemented in part (b) */  }

    /** Sorts the letters of a Location Numeral into alphabetical order
     *
     *  @param numeral String representing a Location Numeral
     *  @return the String in alphabetical order
     */
    public String sortLocationNumeral(String numeral)
    {  /* implementation not shown */  }

    /** Adds two Location Numerals and returns a simplified result
     *
     *  @param num1  String representing the first addend as a LN
     *  @param num2  String representing the second addend as a LN
     *  @return String which represents the sum as a simplified LN
     *
     *  Precondition:  The characters in the parameters are
     *                 uppercase letters A-Z only.
     *  Postcondition: The letters in the returned String are in
     *                 alphabetical order and no letter occurs more
     *                 than once.
     */
    public String add(String num1, String num2)
    {  /* to be implemented in part (c) */  }

    /* Additional implementation not shown */
}
```

(a) Write the method `getLetterValue` that returns the numerical value of a single LN letter.

The value of each LN letter is equal to 2 raised to the power of the letter's position in the alphabet string. Given `String letters = "ABCDEFGHIJKLMNOPQRSTUVWXYZ"`, return the decimal value of the LN letter passed as a parameter.

```
/** Returns the decimal value of a single LN letter
 *
 *  @param letter String containing a single LN letter
 *  @return the decimal value of that letter
 *  Precondition: The parameter contains a single uppercase letter
 */
public int getLetterValue(String letter)
{
   String letters = "ABCDEFGHIJKLMNOPQRSTUVWXYZ";
}
```

(b) Write the method `getDecimalValue` that takes a Letter Numeral and returns its decimal value.

For example, if passed the String "ACE", the method will add the decimal values of A, C, and E and return the result. In this case: 1 + 4 + 16 = 21.

You may assume that `getLetterValue` works as intended, regardless of what you wrote in part (a).

```
/** Returns the decimal value of a simplified Location Numeral
 *
 *  @param numeral String representing a Location Numeral
 *  @return the decimal value of the Location Numeral.
 *  Precondition:  The characters in the parameter are
 *                 uppercase letters A-Z only.
 */
public int getDecimalValue(String numeral)
```

(c) Because each letter in Location Numerals doubles the value of the previous letter, you can simplify a Location Numeral by replacing any two occurrences of the same letter with the next letter of higher value. For example:

AA → B (1 + 1 = 2)
CCDD → DE (4 + 4 + 8 + 8 = 8 + 16)
AACCE → BDE (1 + 1 + 4 + 4 + 16 = 2 + 8 + 16)
AAB → BB → C (1 + 1 + 2 = 2 + 2 = 4)
AABCD → BBCD → CCD → DD → E (1 + 1 + 2 + 4 + 8 = . . . = 16)

Notice that in the fourth and fifth examples, the process has to be done iteratively.

This makes adding Location Numerals very simple as shown by the following algorithm.

- Concatenate the two Location Numerals to be added.
- Sort the letters into alphabetical order and simplify the result.
    - BDE + AC = BDEAC = ABCDE
    - AB + ABC = ABABC = AABBC = BCC = BD

Write the method add that takes two Location Numerals and adds them together, returning the simplified `sum`.

LocationNumeral class contains two methods that are available for you to use: `sortLocationNumeral` and `getLetter`. Their implementations are not shown. Their documentation and method declarations are repeated below.

You may assume that getLetterValue works as intended, regardless of what you wrote in part (a).

```
/** Sorts the letters of a Location Numeral into alphabetical order
 *
 *  @param numeral String representing a Location Numeral
 *  @return the String in alphabetical order
 */
public String sortLocationNumeral(String numeral)
{  /* implementation not shown */  }

/** Return the Location Numeral letter representation for a decimal value
 *  that is a power of two.
 *  @param the decimal value to be converted to a Location Numeral letter
 *  @return the corresponding Location Numeral letter
 *  Precondition: The parameter must be an integer power of 2 between
 *                2 raised to the power 0 and 2 raised to the power 25.
 */
private String getLetter(int value )
{  /* implementation not shown */  }
```

Write the method `add`.

```
/** Adds two Location Numerals and returns a simplified result
 *
 *  @param num1  String representing the first addend as an LN
 *  @param num2  String representing the second addend as an LN
 *  @return A String which represents the sum as a simplified LN
 *
 *  Precondition:  The characters in the parameters are
 *                 uppercase letters A-Z only.
 *  Postcondition: The letters in the returned String are in
 *                 alphabetical order and no letter occurs more
 *                 than once.
 */
public String add(String num1, String num2)
```

STOP. End of Part II.

Practice Exam 2 Answers and Explanations

Part I Answers and Explanations (Multiple Choice)

Bullets mark each step in the process of arriving at the correct solution.

1. The answer is B.

- The outer loop will start at h = 5 and then, on successive iterations, h = 3 and h = 1.
- The inner loop will start at k = 0, then 1 and 2.
- The if statement will print h + k only if their sum is even (since % 2 = 0 is true only for even numbers).
- Since h is always odd, h + k will be even only when k is also odd. That only happens once per inner loop, when k = 1. Adding h and k, we get 5 + 1 = 6, 3 + 1 = 4, and 1 + 1 = 2, so 6 4 2 is printed.

2. The answer is A.

- Let's picture our ArrayList as a table. After the first three add statements we have:

Hummus	Soup	Sushi

- The set statement changes element 1:

Hummus	Empanadas	Sushi

- Add at index 0 gives us:

Salad	Hummus	Empanadas	Sushi

- Remove the element at index 1:

Salad	Empanadas	Sushi

- And finally, add "Curry" to the end of the list:

Salad	Empanadas	Sushi	Curry

3. The answer is B.

- Option I is incorrect. The outer loop begins at i = letters.length. Remember that the elements of an array start at index = 0 and end at index = length − 1. Starting the loop at i = letters.length will result in an `ArrayIndexOutOfBoundsException`.
- Option II is correct. The outer loop takes each letter from the string in turn, and the inner loop prints it four times, as long as it is not an "S".
- Option III is incorrect. It correctly traverses each element in the array, but then only prints each element once, not four times.

4. The answer is D.

- The constructor call creates a new Boat and the super call sets the fuel variable in the Vehicle class to 20.
- The call to the Boat class start method:
  - begins by calling the Vehicle class start method, which prints "Vroom",
  - then calls useFuel, which calls the Vehicle class changeFuel method, which subtracts 2 from the fuel variable, and
  - finally, the start method prints "Remaining fuel" followed by the value in the Vehicle class fuel variable, which is now 18.

5. The answer is C.

- The type on the left side of the assignment statement can be the same type or a super type of the object constructed on the right side of the assignment statement. The object constructed should have an *is-a* relationship with the declared type.
- Option I is correct. A Boat *is-a* Boat.
- Option II is correct. A Boat *is-a* Vehicle
- Option III is incorrect. Vehicle is a super class of Boat. A boat *is* a Vehicle, but a Vehicle doesn't have to be a Boat. The super class must be on the left side of the assignment statement.

In addition, we cannot construct an object of the Vehicle class, because Vehicle is an abstract class. Abstract classes can only appear on the left side of the assignment statement.

6. The answer is E.

- This is a recursive method. The first call results in:

  weird(3) = weird(3 – 2) + weird(3 – 1) + weird(3)
  = weird(1) + weird(2) + weird(3)

  A recursive method must approach the base case or it will infinitely recurse (until the Stack overflows). The call to weird(3) calls weird(3) again, which will call weird(3) again, which will call weird(3) again. . . .

7. The answer is D.

- It is important to notice that the for loop is traversing the array from the greatest index to the least. The array is correct if every element is greater than or equal to the one before it.
- We will return true if we traverse the entire array without finding an exception. If we find an exception, we return false, so we are looking for that exception, or values[index – 1] > values[index].
- Option A is incorrect because of the =.
- Note that since we must not use an index that is out of bounds, we can immediately eliminate any answer that accesses values[index + 1].

8. The answer is D.

- The first nested loop initializes all the values in the array in row major order. The first value entered is 0, then 1, 2, etc. After these loops terminate, the matrix looks like this:

0	1	2
3	4	5
6	7	8

- The second nested loop (for-each loops this time) traverses the array and adds all the values into the sum variable. 0 + 1 + 2 + . . . + 8 = 36.

9. The answer is D.

- Option I is correct. It calls the super constructor, correctly passing on 2 parameters, and then sets its own instance variables, one to the passed parameter, one to a default value.
- Option II is incorrect. It attempts to call the student class 2 parameter constructor, passing default values, but the year is being passed as a String, not as an int.
- Option III is correct. There will be an implicit call to Student's no-argument constructor.

10. The answer D.

- Integer.MAX_VALUE is a large number, but repeatedly dividing by 2 will eventually result in an answer of 0 and the loop will terminate. Remember that this is integer division. Each quotient will be truncated to an integer value. There will be no decimal part.
- In the second loop, the initial condition sets i = 0, which immediately fails the condition. The loop never executes, so nothing is printed, but the code segment terminates successfully.

11. The answer is A.

- The for-each loop looks at every item in the ArrayList and adds it to sum if:
  - the value of the item > 1 (all values except 0 and 1) AND
  - the size of the list − 3 > the value of the item. Since the list has 10 elements, we can rewrite this as the value of the item < 7, which is true for all elements 6 and below.
  - Both conditions are true for 2, 3, 4, 5, 6; so the sum is 20.

12. The answer is C.

- Movement in the x direction is correct for all three options.
- Option I is incorrect. It only moves in the positive y direction.
- Option II is incorrect. It does not deal correctly with the two ships having the same y position.
- Option III is correct. If the two ships have the same y position, the y position is not changed.

13. The answer is B.

- Option I is incorrect. The move method returns a boolean. The interface specifies a void method.
- Option II is correct. The implementation of the move method is abstract, but that's OK. Any class that extends Zoomer will have to provide an implementation.
- Option III is incorrect. It uses the keyword **extends** in its declaration, not the keyword **implements**. You extend a class and implement an interface.

14. The answer is D.

- Option I is correct. It might be clearer to put the && condition in parentheses, but since AND has a higher priority than OR, it will be executed first, even without the parentheses.
- Option II is incorrect. While it seems reasonable at first glance, DeMorgan's theorem tells us we can't just distribute the !. We also have to change OR to AND.
- Option III is correct. As soon as a condition evaluates to true, the method will return and no more statements will be executed. Therefore no else clauses are required. The method will return false only if none of the conditions are true.

15. The answer is E.

- The loop will continue to execute until either value <= 5 or calculate = false.
- When we enter the loop, value = 33, calculate = true.
  - Since 33 % 3 = 0 we execute the first if clause and set calculate = 31.
  - 31 / 4 = 7, which is > 5, so we do not execute the second if clause.
- The second iteration of the loop begins with value = 31 and calculate = true.
  - Since 31 % 3 != 0, we do not execute the first if clause.
  - Since 31 / 4 = 7, which is > 5, we do not execute the second if clause.
- At this point we notice that nothing is ever going to change. Value will always be equal to 31 and calculate will always be true. This is an infinite loop.

16. The answer is E.

- Remember, we only need to find one error to eliminate an option.
- Option A is incorrect. index starts in the middle of the array and goes down to zero. We swap the item at index with the item at index + 1. We are never going to address the upper half of the array.
- Option B is incorrect. The swap code is incomplete. We put planets[index] into s, but then never uses.
- Option C is incorrect. It's OK to use another array, but this code forgets to switch the order when it moves the elements into the new array. The new array is identical to the old one.
- Option D is incorrect. At first glance, there's no obvious error. The swap code looks correct. Although planets.length – 1 – index seems a bit complex, a few examples will show us that it does, indeed, give the element that should be swapped with the element at index: 0 ➔ 7 – 0 = 7, 1 ➔ 7 – 1 = 6, 2 ➔ 7 – 2 = 5, 3 ➔ 7 – 3 = 4, and so on. The trouble with this option is that "and so on" part. It swaps all the pairs and then continues on to swap them all again. After completing the four swaps listed above, we are done and we should stop.
- Option E is correct. The code is the same as option D, except it stops after swapping half the elements. The other half were the partners in those first swaps, so now everything is where it should be.

17. The answer is A.

- An abstract class contains method declarations for the methods that all of its subclasses have in common but may implement differently.

18. The answer is B.

- The general form for generating a random number between *high* and *low* is

```
(int)(Math.random() * (high - low + 1)) + low
```

- high – low + 1 = 30, low = 20, so high = 49
- After the first statement, num is an integer between 20 and 49 (inclusive).
- In the second statement, we add 5 to num. Now num is between 25 and 54 (inclusive).
- In the third statement we divide by 5, remembering integer division. Now num is between 5 and 10 (inclusive).

19. The answer is A.

- This loop takes pieces of the String "computer" and places them in an ArrayList of String objects.
- The loop starts at index k = 0, and adds the word.substring(0), or the entire word, to the ArrayList. The first element of the ArrayList is "computer".
- Then k is incremented by the k++ inside the loop, k = 1, and immediately incremented again by the instruction in the loop header, k = 2. It is bad programming style to change a loop index inside the loop, but we need to be able to read code, even if it is poorly written.
- The second iteration of the loop is k = 2, and we add word.substring(2) or "mputer" to the ArrayList.
- We increment k twice and add word.substring(4) or "uter".
- We increment k twice and add word.substring(6) or "er".
- We increment k twice and fail the loop condition. The loop terminates and the ArrayList is printed.

20. The answer is E.

- We need to plug the values of x and y into the given statements to see if they always generate the correct return value.
- Option A is incorrect. 3 > 6 is false, but should be true.
- Option B is incorrect. 1 % 0 will fail with a division by zero error.
- Option C is incorrect. (1 + 0) % 2 == 0 is false, but should be true.
- Option D is incorrect. (3 + 6) % 3 == 6 is false, but should be true.
- Option E works for all the given values.

21. The answer is B.

- The loop will continue until numWord = words.size(), which will cause an `ArrayIndexOutOfBoundsException.`
- Notice that numWord is only incremented if nothing is removed from the ArrayList. That is correct because, if something is removed, the indices of the elements after that element will change. If we increment numWord each time, we will skip elements.

22. The answer is C.

- This is an especially difficult example of recursion because the recursive call is not the last thing in the method. It's also not the first thing in the method. Notice that the print happens *after* the substring. So each iteration of the method will print one letter less than its parameter.
- Let's trace the code. The parts in *italics* were filled in on the way back up. That is, the calls in the plain type were written top to bottom until the base case returned a value. Then the answers were filled in *bottom to top.*

  wackyOutput("APCS") ➔ *print "PCS"* (printed fourth)
  wackyOutput("PCS") ➔ *print "CS"* (printed third)
  wackyOutput("CS") ➔ *print "S"* (printed second)
  wackyOutput("S") ➔ *print ""* (printed first)
  wackyOutput("") ➔ Base Case. Return without printing.

  Remember that the returns are read bottom to top, so the output is "SCSPCS" in that order.
- Note that "S".substring(1,1) is not an error. It returns the empty string.

23. The answer is D.

- Option I is correct. It uses DeMorgan's theorem to modify the boolean condition.

```
!(num1 >= 0 && num2 >= 0)
!(num1 >= 0 ) || !(num2 >= 0)
num1 < 0 || num2 < 0
```

- Option II is correct. It tests the two parts of the condition separately. If either fails, it returns "Numbers are not valid".
- Option III is incorrect. If num1 is a positive number and num2 is a negative number, but the absolute value of num2 < num1 (num1 = 5 and num2 = -2, for example), option III will return "Numbers are valid" although clearly both numbers are not >= 0.

24. The answer is A.

- test1 references a TestClass object with value = 17.5 and test2 references a TestClass object with value = 9.0.

test 1 → [value = 9.0] test 2 → [value = 17.5]

- After the addValue and reduceValue calls, our objects look like this:

test 1 → [value = 6.0] test 2 → [value = 19.0]

- The assignment statement changes test2 to reference the same object as test1.

test 1 → [value = 6.0] ← test 2 [value = 19.0]

- At this point, the object with value = 19.0 is inaccessible (unless another reference variable we don't know about is pointing to it). The next time Java does garbage collection, it will be deleted.
    - The next reduceValue call changes our object to this:

    test 1 → [value = 0.0] ← test 2

    - The getValue statements in the next line get the same value, as test1 and test2 are pointing to the same object. 0.0 + 0.0 = 0.0.

25. The answer is E.

- Let's consider what the expression is telling us.
  `(truth1 && truth2) || ((!truth1) && (!truth2))`
- The expression is true if both truth1 and truth2 are true OR if both truth1 and truth2 are false.
- Therefore, the expression is true as long as truth1 == truth2.

26. The answer is B.

- This method implements a standard insertion sort. In this sort, each item is taken in turn and placed in its correct position among the items to its left (with lower indices). The variable value holds the item whose turn it is to be inserted. Therefore, the values printed will simply be the values of the array, in the same order that they are processed. That's the same order in which they appear in the initial array, except that we skip element 0 (since one element, all alone, is already sorted).

27. The answer is A.

- In order for a rectangle to contain a point, the following conditions must be true:
  topleftX < x < bottomRightX and topLeftY < y < bottomRightY
- Option A is correct. These are the exact conditions used in option A, though they are in a different order.
- Option B is incorrect because it will return true if any one of the four conditions is true.
- Options C and D are incorrect because they attempt to make invalid calls to super or Polygon methods that do not exist.
- Option E is incorrect. It is similar to option B, except that it will return true if two of the required conditions are true.

28. The answer is A.

- Let's show the contents of the ArrayList graphically and trace the code. The method is called with this ArrayList.

Nora	Charles	Madeline	Nate	Silja	Garrett

- The first time through the loop, i = 3. The element at index 3 is removed and added to the end of the ArrayList.

Nora	Charles	Madeline	Silja	Garrett	Nate

- Decrement i, i = 2, which is >= 0, so execute the loop again. The element at index 2 is removed and added to the end.

Nora	Charles	Silja	Garrett	Nate	Madeline

- Decrement i, i = 1, which is >= 0, so execute the loop again. The element at index 1 is removed and added to the end.

Nora	Silja	Garrett	Nate	Madeline	Charles

- Decrement i, i = 0, which is >= 0, so execute the loop again. The element at index 0 is removed and added to the end.

Silja	Garrett	Nate	Madeline	Charles	Nora

- Decrement i, i = -1, which fails the loop condition. The loop terminates and the ArrayList is returned.

29. The answer is C.

- Let's trace the code.
- The loop will execute while varA != 0 or varB > 0. Another way to say that, using DeMorgan's theorem, is that the loop will stop executing when varA == 0 AND varB <= 0.
- varA starts at -30 and varB starts at 30. Here's a table of their values at the end of every iteration of the loop until the loop condition fails.

varA	varB
-27	21
-24	18
-21	15
-18	12
-15	9
-12	6
-9	3
-6	0
-3	-3
0	-4

30. The answer is C.

- The nested loops go through grid in row-major order, but assign into newGrid in column-major order. (Notice that newGrid is instantiated with its number of rows equal to the grid's number of columns and its number of columns equal to the grid's number of rows.) After executing the loops, newGrid looks like this:

```
{ {0, 3, 6,  9 },
  {1, 4, 7, 10 },
  {2, 5, 8, 11 } }
```

- newGrid [2][1] = 5

31. The answer is E.

- The nested loops traverse the entire array. Each time through, the three assignment statements are executed. The key to the problem is the order in which these statements are executed. We need to pay attention to which values will overwrite previously assigned values.
- On entering the loops, our array looks like this:

0	0	0
0	0	0
0	0	0

- We will execute the outer loop three times, and for each of those three times, we will execute the inner loop three times.
- The first iteration of both loops, we are changing element [0][0].
  - The first statement makes it a 1.
    - Note that this statement changes the array in row-major order.
  - The second statement makes it a 2.
    - Note that this statement changes the array in column-major order.
  - The third statement makes it a 3, so it will remain a 3.
    - Note that this statement only changes the diagonals.
- Completing the first cycle of the inner loop, the first statement assigns a 1 to [0][1] and [0][2] and the second statement assigns a 2 to [1][0] and [2][0]. The third statement repeatedly sets [0][0] to 3, and since that statement is last, it will remain a 3.

3	1	1
2	0	0
2	0	0

- The next cycle of the inner loop will assign a 1 to [1][0], [1][1], and[1][2]. A 2 will be assigned to [0][1], [1][1], and [2][1]. Then element [1][1] will be overwritten with a 3.

3	2	1
1	3	1
2	2	0

- The third (and last) cycle of the inner loop will assign a 1 to [2][0], [2][1], and [2][2]. A 2 will be assigned to [0][2], [1][2], and [2][2]. Then element [2][2] will be overwritten with a 3.

3	2	2
1	3	2
1	1	3

32. The answer is E.

- If this were a sorting algorithm, there would be only one parameter, the array to be sorted, and a sorted array would be returned, so we can eliminate the sorting algorithms. In addition, merge sort is recursive, and selection and insertion sorts require nested loops, neither of which appear in this code.
- Binary search repeatedly divides the remaining section of the array in half. There's no code that does anything like that. This is not a binary search.
- This code goes through the array one step at a time in order. This is a sequential search algorithm.

For questions 33–36, let's start by looking at the hierarchy described in the problem.

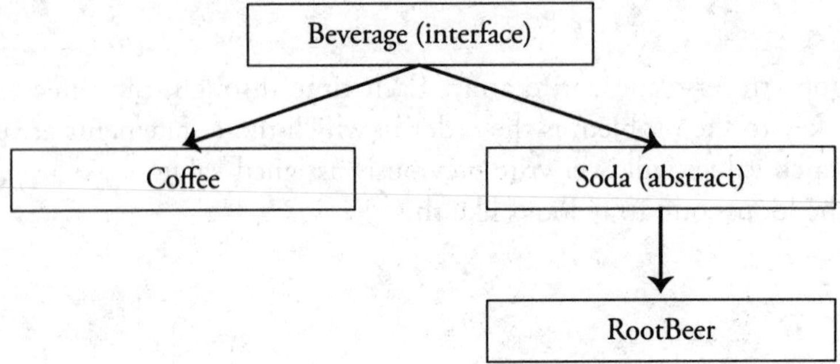

33. The answer is C.

- Option I is correct. Beverage is an interface. It cannot be instantiated, but it can be the type of a reference variable. Since Coffee *is-a* Beverage, and Coffee has a no-argument constructor, this statement is valid.
- Option II is correct. RootBeer *is-a* Soda and the parameters to the constructor are correct.
- Option III is incorrect. Soda is an abstract class and an object of type Soda cannot be instantiated.

34. The answer is C.

- The RootBeer class must implement all of the abstract methods in the Soda class. There are two: hasCaffeine and getCalories.

35. The answer is B.

- Since both the Soda class and the Coffee class implement Beverage, any methods that they have in common could be part of the Beverage interface, although they do not have to be.
- Both the Soda class and the Coffee class have implementations for isHot and hasCaffeine. Those methods could be part of the Beverage interface.
- Since the Coffee class does not have a getCalories method, getCalories cannot be part of the Beverage interface.

36. The answer is A.

- We need to be sure that the ArrayList type on the right side of the assignment statement has an *is-a* relationship with the ArrayList type on the left side of the assignment statement (is a sub-class, or the class itself).
  - Option B is incorrect because Soda is not a sub-class of RootBeer.
  - Option C is incorrect because Beverage is not a sub-class of Soda.
  - Option E is incorrect because Coffee is not a sub-class of Soda.
- We need to be sure that the method being called is a method appropriate to the declared type.
  - Option D is incorrect. Coffee does not have a getCalories method.

37. The answer is D.

- index starts at 0 and goes to the end of the String. We want a condition that will look at the letter in word at index, compare it to the letter parameter, and count it if it is the same.
- Option A is incorrect. The one-parameter substring includes everything from the starting index to the end of the word. We need one letter only.
- Option B is incorrect. We cannot compare Strings with ==.
- Option C is incorrect. indexOf will tell us whether a letter appears in a word and where it appears, but not how many times. It is possible to count the occurrences of letter using indexOf, but not this way. Option C does not change the if condition each time through the loop. It just asks the same question over and over.
- Option D is correct. It selects one letter at index and compares it to letter using the equals method, incrementing count if they match.
- Option E is incorrect. It will not compile. It tries to compare letter to the int returned by indexOf.

38. The answer is C.

- You might expect that the loop will remove any animal whose name comes before "sugar glider" lexicographically, but removing elements from an ArrayList in the context of a loop is tricky. Let's walk through it.
- We start with:

okapi	aye-aye	cassowary	echidna	sugar glider	jerboa

- The for loop begins, i = 0. Compare "okapi" with "pink fairy armadillo", and since "o" comes before "p", remove "okapi".

aye-aye	cassowary	echidna	sugar glider	jerboa

- The next iteration begins with i = 1. The remove operation has caused all the elements' index numbers to change. We skip "aye-aye", which is now element 0, and look at "cassowary", which we remove.

aye-aye	echidna	sugar glider	jerboa

- The next iteration begins with i = 2. Again the elements have shifted, so we skip "echidna", compare to "sugar glider" but that comes after "pink fairy armadillo" so it stays put leaving the ArrayList unchanged.

aye-aye	echidna	sugar glider	jerboa

- The next iteration begins with i = 3. Remove "jerboa".

aye-aye	echidna	sugar glider

- i = 4, which is not < animals.size(). We exit the loop and print the ArrayList.
- Remember that if you are going to remove elements from an ArrayList in a loop, you have to adjust the index when an element is removed so that no elements are skipped.
- Note that if we used the loop condition i < 6 (the size of the original ArrayList), then there would have been an IndexOutOfBoundsException, but because we used animals.size(), the value changed each time we removed an element.

39. The answer is E.

- Option I is incorrect. Among other errors, getNumRooms does not take a parameter.
- Option II is incorrect. Square brackets are used to access elements of arrays, but not of ArrayLists.
- Option III is correct.
- Option IV is incorrect. list is the name of the ArrayList, not the name being used for each element of the ArrayList.
- Option V is correct.

40. The answer is B.

- The array holds Building objects. As long as an object *is-a* Building, it can go into the array.
- The runtime environment will look at the type of the object and call the version of getSize written specifically for that object.
  - list[0] is a Building so the Building class getSize will be used to generate the String for list[0].
  - list[1] is a House so the House class getSize will be used to generate the String for list[1].
  - list[2] references the same object as list[1]. This will generate a duplicate of the list[1] response.

Part II Answers and Explanations (Free Response)

Please keep in mind that there are multiple ways to write the solution to a free-response question, but the general and refined statements of the problem should be pretty much the same for everyone. Look at the algorithms and coded solutions, and determine if yours accomplishes the same task.

General Penalties (assessed only once per problem):
-1 using a local variable without first declaring it
-1 returning a value from a void method or constructor
-1 accessing an array or ArrayList incorrectly
-1 overwriting information passed as a parameter
-1 including unnecessary code that causes a side effect such as a compile error or console output

1. Hex Grid

(a) **General Problem:** Write the getGamePieceCount method that counts how many GamePiece objects are located on the grid.

Refined Problem: Traverse the 2-D array of GamePiece objects and count the number of cells that are not null.

Algorithm:

- Create a variable to hold the count.
- Traverse the 2-D array of values using nested loops.
  - Inside the loops, evaluate each cell of the grid to see if it is null. When we find an element that is not null, increment the count variable.
- Once the entire array has been traversed, the value of the count variable is returned.

Java Code:

```
public int getGamePieceCount()
{
    int count = 0;
    for (int row = 0; row < grid.length; row++)
        for (int col = 0; col < grid[row].length; col++)
            if (grid[row][col] != null)
                count++;
    return count;
}
```

Common Errors:

- Remember that the number of rows is grid.length and the number of columns is grid[row].length or, since all columns are the same length, grid[0].length.
- The return statement has to be outside both loops.

Java Code Alternate Solution:

You can use for-each loops to traverse a 2-D array.

```
public int getGamePieceCount()
{
   int count = 0;
   for (GamePiece[] row : grid)
      for (GamePiece g : row)
         if (g != null)
            count++;
   return count;
}
```

(b) **General Problem:** Write the isAbove method that returns the GamePiece objects that are above an object at a given location on the grid.

Refined Problem: Create an ArrayList of GamePiece objects that are above the position passed in the parameter, where "above" is defined as having the same column number and a lower row number than another object. If there is no object at the location passed in, null is returned. If there are no objects above the parameter location, an empty ArrayList is returned.

Algorithm:

- Determine whether there is a GamePiece object at the location specified. If there is not, return null.
- Create an ArrayList to hold the GamePiece objects to be returned.
- Look in the specified column and the rows from 0 up to the specified row. If any of those cells contain GamePiece objects, add them to the ArrayList.
- Return the ArrayList.

Java Code:

```
public ArrayList<GamePiece> isAbove(int row, int col)
{
   if (grid[row][col] == null)
      return null;
   ArrayList<GamePiece> above = new ArrayList<GamePiece>();
   for (int r = 0; r < row; r++)
      if (grid[r][col] != null)
         above.add(grid[r][col]);
   return above;
}
```

Common Errors:

- Do not count the object itself. That means the loop must stop at r < row, not r <= row.

(c) **General Problem:** Write the addRandom method that adds a specified number of GamePiece objects to the grid in random locations.

Refined Problem: Check to see if there are enough blank cells to add the requested objects. If there are, use a random number generator to generate a row and column number. If the cell at that location is empty, add the object, otherwise, generate a new number. Repeat until all objects have been added.

Algorithm:

- Determine whether there are enough empty cells to hold the requested number of objects.
  - If there are not, return false.
- Loop until all objects have been added.
  - Generate a random int between 0 and number of rows.
  - Generate a random int between 0 and number of columns.
  - Check to see if that cell is free. If it is, add object and decrease number to be added by 1.
- Return true.

Note that this algorithm may take a very long time to run if the grid is large and mostly full. If that is the case, a more efficient algorithm is to put all available open spaces into a list and then choose a random element from that list.

Java Code:

```
public boolean addRandom(int number)
{
    if (getGamePieceCount() + number > grid.length * grid[0].length)
        return false;
    while (number > 0)
    {
        int row = (int)(Math.random() * grid.length);
        int col = (int)(Math.random() * grid[0].length);
        if(grid[row][col] == null)
        {
            grid[row][col] = new GamePiece();
            number--;
        }
    }
    return true;
}
```

Scoring Guidelines: Hex Grid

Part (a)	**getGamePieceCount**	**2 points**
+1	Compares every element to null; no bounds errors, no missed elements	
+1	Returns the correct count	

Part (b)	**isAbove**	**3 points**
+1	Returns null if the specified location is null	
+1	Compares to null all elements in the given column with a row number from 0 up to but not including the given row number; no bounds errors, no missed elements	
+1	Creates and returns ArrayList of all elements "above" the given location	

Part (c)	**addRandom**	**4 points**
+1	Returns false if there is not sufficient room to add requested elements; must use getGamePieceCount; returns true after requested number of elements have been added	
+1	Generates a random location; no bounds errors, no missed elements	
+1	Adds the element only if the generated location is null; continues to generate locations in the context of a loop until an element is successfully added (a non-null location is found)	
+1	Continues to add elements in the context of a loop until requested number of elements has been added	

2. Quadratic

(a) **General Problem:** Write the Quadratic class.

Refined Problem: Create an instance variable that is an ArrayList of Monomials. Write a one-parameter constructor that takes an ArrayList of Monomial objects and assigns it to the instance variable. The Quadratic class must implement the Polynomial interface. Implement the methods required by the interface, namely getDegree, getNumTerms, and getName.

Algorithm:

- Write a class declaration for Quadratic that includes "implements Polynomial."
- Declare an instance variable called `terms` that is an ArrayList of Monomial.
- Write a constructor with a parameter that is an ArrayList of Monomial and assign the parameter to the instance variable `terms`.
- Write method getDegree. Since this class is the Quadratic class, the degree will always be 2.
- Write method getNumTerms. The number of terms is the size of the ArrayList.
- Write method getName. The name is dependent on the number of elements in the ArrayList `terms` and the degree which we know is 2. Our choices are: *Quadratic Trinomial*, *Quadratic Binomial*, and *Quadratic Monomial*.

Java Code:

```
public class Quadratic implements Polynomial
{
    private ArrayList<Monomial> terms;

    public Quadratic(ArrayList<Monomial> quadTerms)
    {
        terms = quadTerms;
    }

    public int getDegree()
    {
        return 2;
    }

    public int getNumTerms()
    {
        return terms.size();
    }

    public String getName()
    {
        String[] names = {"Quadratic Monomial", "Quadratic Binomial",
                          "Quadratic Trinomial"};
        return names[getDegree() - 1];
    }
}
```

Common Errors:

- The ArrayList needs to be of type Monomial, not String or int.

Java Code Alternate Solution:

The name could be determined by using an if statement, rather than indexing into an array.

terms.size(), getDegree(), and this.getDegree() can all be used to provide the number of terms.

```
public String getName()
{
    if (terms.size() == 3)
        return "Quadratic Trinomial";
    else if (terms.size() == 2)
        return "Quadratic Binomial";
    return "Quadratic Monomial";
}
```

(b) **General Problem:** Write a hasRoots method that returns true if the quadratic has real roots and false if it does not.

Refined Problem: Use the coefficients passed to the method as parameters to evaluate $b^2 - 4ac$. Return true if the answer is greater than or equal to zero and false if it is less than zero.

Algorithm:
- Evaluate `Math.pow(b, 2) - 4 * a * c`
- If the answer is >= 0 return true.
- If the answer is < 0 return false.

Java Code:

```
public boolean hasRoots(double a, double b, double c)
{
   return Math.pow(b, 2) - 4 * a * c >= 0;
}
```

Common Errors:
- a, b, and c are doubles and Math.pow returns a double. Make sure that if you store the determinant in a variable, it is of type double, not int.
- Be sure to use greater than *or equal to* 0.

Java Code Alternate Solution:
- Use a variable to hold the determinant.
- Use multiplication rather than Math.pow.
- Use an if statement to determine the return.

```
public boolean hasRoots(double a, double b, double c)
{
   double answer = b * b - 4 * a * c;
   if (answer >= 0)
      return true;
   else
      return false;
}
```

(c) **General Problem:** Write a getRoots method for the Quadratic class that solves the quadratic equation.

Refined Problem: Find the values of all the coefficients. Call the hasRoots method to determine if real roots exist. If there are no real roots, return null. If there are real roots, solve the two versions of the quadratic formula and return the results in a two-element array of type double.

Algorithm:

- Instantiate and initialize doubles to hold the values of a, b, and c, and a two-element array of double to hold the roots.
- Use the methods getDegree and getCoefficient to assign values to a, b, and c.
  - The coefficient of the first term in the ArrayList will always be a.
  - If there are three terms in the ArrayList, the coefficient of the second term will be b and the coefficient of the third term will be c.
  - If there are two terms in the ArrayList, find the degree of the second term. If it is 1 assign the coefficient to b, if it is 0 assign the coefficient to c.
- Call hasRoots to determine if this quadratic has real roots.
  - If hasRoots returns false, return null.
- Find the two solutions, using the given equations.
- Store the solutions in the roots array and return it.

Java Code:

```
public double[] getRoots()
{
    double[] coeff = new double[3]; // stored in order a, b, c
    double[] roots = new double[2];

    for (int i = 0; i < terms.size(); i++)
    {
        int index = 2 - terms.get(i).getDegree();
        coeff[index] = terms.get(i).getCoefficient();
    }

    // This isn't necessary, but it makes the formula easier
    // to read and to enter correctly
    double a = coeff[0];
    double b = coeff[1];
    double c = coeff[2];

    if (!hasRoots(a, b, c))
       return null;

    roots[0] = (-b + Math.sqrt(Math.pow(b, 2) - 4 * a * c)) / (2 * a);
    roots[1] = (-b - Math.sqrt(Math.pow(b, 2) - 4 * a * c)) / (2 * a);

    return roots;
}
```

Common Errors:

- Remember that there doesn't have to be three terms in the ArrayList. You have to check before assigning to a, b, and c.
- If there are only two terms in the ArrayList, the first one will correspond to a, but the second could be either b or c.
- Be really careful when writing the quadratic formula. The entire numerator needs to be in parentheses, and the entire denominator needs to be in parentheses. Close the parentheses for Math.sqrt after the c. When you think you have it right, it's always a good idea to count the open parentheses and the close parentheses to make sure you have the same number.

Java Code Alternate Solution:

You do not have to store the coefficients in an array.

```
public double[] getRoots()
{
    double a = 0, b = 0, c = 0;
    double[] roots = new double[2];

    a = terms.get(0).getCoefficient();
    if (terms.size() == 3)
    {
        b = terms.get(1).getCoefficient();
        c = terms.get(2).getCoefficient();
    }
    else if (terms.size() == 2)
    {
        if (terms.get(1).getDegree() == 1)
            b = terms.get(1).getCoefficient();
        else
            c = terms.get(1).getCoefficient();
    }

    if (!hasRoots(a, b, c))
        return null;

    roots[0] = (-b + Math.sqrt(Math.pow(b, 2) - 4 * a * c)) / (2 * a);
    roots[1] = (-b - Math.sqrt(Math.pow(b, 2) - 4 * a * c)) / (2 * a);

    return roots;
}
```

Scoring Guidelines: Quadratic

Part (a)	**Quadratic class**	**4 points**

+1 Includes "implements Polynomial" in the class declaration

+1 Writes a correct 1 parameter constructor; assigns parameter to a correctly instantiated private ArrayList<Monomial>

+1 Implements getDegree and getNumTerms correctly

+1 Implements getName correctly

Part (b)	**hasRoots**	**1 point**

+1 Calculates the determinant correctly; returns false if negative, true otherwise

Part (c)	**getRoots**	**4 points**

+1 Correctly accesses and assigns coefficients to a, b, c (or equivalent)

+1 Calls method hasRoots passing the correct parameters; returns null if false

+1 Calculates roots

+1 Returns roots in a properly instantiated array

Sample Driver:

There are many ways to write these methods. Maybe yours is a bit different from our sample solutions and you are not sure if it works. Here is a sample driver program. Running it will let you see if your code works, and will help you debug it if it does not.

Copy QuadraticDriver into your IDE along with the complete Monomial class, the Polynomial interface, and Quadratic class (including your solutions).

You will need to add this import statement as the first line in your Polynomial class: `import java.util.ArrayList;`

```
import java.util.ArrayList;

public class QuadraticDriver {

   public static void main(String[] args) {
      ArrayList<Monomial> monoList = new ArrayList<Monomial>();
      monoList.add(new Monomial(2, 2));
      monoList.add(new Monomial(-8, 1));
      monoList.add(new Monomial(-10, 0));
      Quadratic quad1 = new Quadratic(monoList);
      System.out.print("Expecting:    Quadratic Trinomial   " +
            "Root 1: 5.0 Root 2: -1.0\nYour answers: ");
      testMethods(quad1);
      monoList.clear();
      monoList.add(new Monomial (1, 2));
      monoList.add(new Monomial (-4, 0));
      Quadratic quad2 = new Quadratic(monoList);
      System.out.print("\nExpecting:    Quadratic Binomial   " +
            "Root 1: 2.0 Root 2: -2.0\nYour answers: ");
      testMethods(quad2);
      monoList.clear();
      monoList.add(new Monomial (2, 2));
      monoList.add(new Monomial (6,1));
      Quadratic quad3 = new Quadratic(monoList);
      System.out.print("\nExpecting:    Quadratic Binomial   " +
            "Root 1: 0.0 Root 2: -3.0\nYour answers: ");
      testMethods(quad3);
      monoList.clear();
      monoList.add(new Monomial (1, 2));
      monoList.add(new Monomial (4,0));
      Quadratic quad4 = new Quadratic(monoList);
      System.out.print("\nExpecting:    Quadratic Binomial" +
            "  Imaginary Roots\nYour answers: ");
      testMethods(quad4);
      monoList.clear();
      monoList.add(new Monomial (-17, 2));
      Quadratic quad5 = new Quadratic(monoList);
      System.out.print("\nExpecting:    Quadratic Monomial   " +
            "Root 1: -0.0 Root 2: 0.0\nYour answers: ");
      testMethods(quad5);
   }

   private static void testMethods(Quadratic quad) {
      System.out.print(quad.getName());
      double[] roots = quad.getRoots();
      if (roots == null)
         System.out.println("  Imaginary roots");
      else
         System.out.println("  Root 1: " + roots[0] +
            " Root 2: " + roots[1]);
   }
}
```

3. Printing Factory

(a) **General Problem:** Write the replacePaper method of the Machine class.

Refined Problem: Take the new PaperRoll passed as a parameter and use it to replace the current roll. Return the used roll to the caller. Notice that this is a type of swapping and will require a temp variable.

Algorithm:

- Create a temp variable of type PaperRoll.
- Assign the machine's PaperRoll paper to temp.
- Assign the new PaperRoll passed as a parameter to the machine's PaperRoll.
- Return the PaperRoll in temp.

Java Code:

```
public PaperRoll replacePaper(PaperRoll pRoll)
{
    PaperRoll temp = paper;
    paper = pRoll;
    return temp;
}
```

Common Errors:

- It may seem like you want to do this:

```
return paper;
paper = pRoll;
```

But once the return statement is executed, flow of control passes back to the calling method. The second statement will never be executed.

(b) **General Problem:** Write the replacePaperRolls method of the PrintingFactory class.

Refined Problem: Traverse the machines ArrayList checking for Machine object with PaperRoll objects that contain less than 4.0 meters of paper. Any PaperRoll objects containing less than 4.0 meters of paper need to be replaced. Get a new PaperRoll object from the newRolls ArrayList, and pass it to the Machine class replacePaper method. Place the returned used PaperRoll object on the usedRolls ArrayList.

Algorithm:

- Write a for-each loop to traverse the machines array.
  - If the current Machine object's PaperRoll object contains < 4.0 m of paper:
    - Take a PaperRoll object off of the newRolls ArrayList.
    - Call replacePaper passing the new roll as a parameter.
    - replacePaper will return the old PaperRoll object; put it on the usedRolls ArrayList.

Java Code:

```
public void replacePaperRolls()
{
    for (Machine m : machines)
    {
        if (m.getPaperRoll().getMeters() < 4.0)
        {
            PaperRoll newRoll = newRolls.remove(0);
            PaperRoll oldRoll = m.replacePaper(newRoll);
            usedRolls.add(oldRoll);
        }
    }
}
```

Common Errors:

- The syntax for accessing the amount of paper remaining is complex, especially when using a for loop instead of a for-each loop. Remember that the getMeters method is not a method of the Machine class. We have to ask each Machine object for access to its PaperRoll object and then ask the PaperRoll object how much paper it has left.

Java Code Alternate Solution #1:

Use for loops instead of for-each loops.

Use just one variable for the old and new rolls.

```
public void replacePaperRolls()
{
    for (int i = 0; i < machines.length; i++)
    {
        if (machines[i].getPaperRoll().getMeters() < 4.0)
        {
            PaperRoll roll = newRolls.remove(0);
            roll = machines[i].replacePaper(roll);
            usedRolls.add(roll);
        }
    }
}
```

Java Code Alternate Solution #2:

Combine the statements and eliminate the variables altogether. This could also be done in the for loop version.

```
public void replacePaperRolls()
{
    for (Machine m : machines)
        if (m.getPaperRoll().getMeters() < 4.0)
            usedRolls.add(m.replacePaper(newRolls.remove(0)));
}
```

(c) **General Problem:** Write the getPaperUsed method of the PrintingFactory class.

Refined Problem: The amount of paper used on each roll is 1000 minus the amount of paper remaining. First traverse the usedRolls ArrayList and add up the paper used. Next traverse the machines array and add up the paper used. These two amounts represent the total paper used. Return the sum.

Algorithm:

- Write a for-each loop to traverse the usedRolls ArrayList.
  - Add 1000 minus paper remaining on the current roll to a running sum.
- Write a for-each loop to traverse the machines array.
  - Add 1000 minus paper remaining on the current machine's roll to the same running sum.
- Return the sum.

Java Code:

```
public double getPaperUsed()
{
    double sum = 0.0;
    for (PaperRoll p : usedRolls)
        sum = sum + (1000 - p.getMeters());

    for (Machine m : machines)
        sum = sum + (1000 - m.getPaperRoll().getMeters());

    return sum;
}
```

Java Code Alternate Solution #1:
Use for loops instead of for-each loops.
Replace sum = sum + … with sum += …

```
public double getPaperUsed()
{
    double sum = 0.0;
    for (int i = 0; i < usedRolls.size(); i++)
        sum += 1000 - usedRolls.get(i).getMeters();

    for (int i = 0; i < machines.length; i++)
        sum += 1000 - machines[i].getPaperRoll().getMeters();

    return sum;
}
```

Common Errors:

- Remember that ArrayLists and arrays use different syntax both to access elements and to find their total number of elements.

Java Code Alternate Solution #2:

There are other ways to compute the paper use. Here we add up all the leftover paper and then subtract the total from 1000 * the total number of rolls. This technique could be combined with the for loop version, as well as the for-each loop as shown here.

```
public double getPaperUsed()
{
    double totalLeft = 0.0;
    double totalUsed = 0.0;

    for (PaperRoll p : usedRolls)
       totalLeft += p.getMeters();

    for (Machine m : machines)
       totalLeft += m.getPaperRoll().getMeters();

    totalUsed = 1000 * (usedRolls.size() + machines.length) - totalLeft;

    return totalUsed;
}
```

Scoring Guidelines: Printing Factory

Part (a)	**replacePaper**	**2 points**

+1 Assigns the PaperRoll object passed as a parameter to the Machine object's PaperRoll variable; specifically: `paper = pRoll;`

+1 Returns the used PaperRoll object

Part (b)	**replacePaperRolls**	**3 points**

+1 Accesses all elements of machines; no bounds errors, no missed elements

+2 Processes low-paper rolls correctly

- +1 Removes a new roll from the newRolls ArrayList; calls the replacePaper method; passing the new roll as a parameter
- +1 Adds the used roll returned from the replacePaper method to the usedRolls ArrayList

Part (c)	**getPaperUsed**	**4 points**

+1 Accesses the amount of paper remaining for all elements of the usedRolls ArrayList; no bounds errors, no missed elements

+1 Accesses all elements of the machines array; no bounds errors, no missed elements

+1 Accesses the amount of paper remaining on a Machine object's PaperRoll

+1 Correctly computes and returns total paper used

4. Location Numerals

(a) **General Problem:** Write the getLetterValue method that returns the numerical value of the given letter.

Refined Problem: Find the parameter letter's position in the String letters. Return 2 raised to that power.

Algorithm:

- Use indexOf to find the position of letter in the String letters.
- Use Math.pow to raise 2 to the power of the position found.
- Return the result.

Java Code:

```
public int getLetterValue(String letter)
{
    String letters = "ABCDEFGHIJKLMNOPQRSTUVWXYZ";
    return (int) Math.pow(2, letters.indexOf(letter));
}
```

(b) **General Problem:** Write the getDecimalValue method that takes a Location Numeral and returns its decimal equivalent.

Refined Problem: Find the value of each letter in turn and add it to running total. Return the final answer.

Algorithm:

- Create a variable to hold the running total.
- Loop through the Location Numeral String from the letter at index 0 to the end of the String.
  - Isolate each letter, using substring.
  - Call the method getLetterValue, passing the isolated letter.
  - Add the result to the running total.
  - When the loop is complete, return the total.

Java Code:

```
public int getDecimalValue(String numeral)
{
    int total = 0;
    for (int i = 0; i < numeral.length(); i++)
    {
        String letter = numeral.substring(i, i + 1);
        total += getLetterValue(letter);
    }
    return total;
}
```

Common Errors:

- Don't end the loop at numeral.length() – 1. When you notice the expression `numeral.substring(i, i + 1)` you may think that the i + 1 will cause an out of bounds exception. It would if it were the first parameter in the substring, but remember that the substring will stop *before* the value of the second parameter, so it will not index out of bounds. If you terminate the loop at numeral.length() – 1, you will miss the last letter in the String.
- Be sure that you use the method you wrote in part (a). There is generally a penalty for duplication of code if you rewrite the function in another method. The problem usually has a hint when a previously written method is to be used. In this case, the problem says, "You may assume that getLetterValue works as intended, regardless of what you wrote in part (a)." That's a sure-fire indication that you'd better use the method to solve the current problem.

(c) General Problem: Write the add method to add two Location Numerals and return a simplified result.

Refined Problem: Given two Location Numerals to add together, concatenate them and then repeatedly sort and simplify until no more simplification can be done. Watch for places to use the existing methods sortLocationNumeral, getLetterValue, and getLetter.

Algorithm:

- Concatenate the two Strings.
- Sort the Location Numeral using the sortLocationNumeral method.
- As long as more simplification can be done:
  - Loop through the String from beginning to end.
    - Find and remove a pair of matching letters.
    - Use getLetterValue and getLetter to find the letter next highest in value.
    - Replace the removed pair with the new letter.
  - Sort the Location Numeral using the sortLocationNumeral method.
- Return the final result.

Java Code:

```
public String add(String num1, String num2)
{
    String sum = num1 + num2;
    boolean simplified = false;
    while (!simplified)
    {
        simplified = true;
        int i = 0;
        sum = sortLocationNumeral(sum);
        while (i < sum.length() - 1)
        {
            String currLetter = sum.substring(i, i + 1);
            if (currLetter.equals(sum.substring(i + 1, i + 2)))
            {
                String before = sum.substring(0, i);
                String after = sum.substring(i + 2);
                String newLetter = getLetter(2 * getLetterValue(currLetter));
                sum = before + newLetter + after;
                simplified = false;
            }
            else
            {
                i++;
            }
            sum = sortLocationNumeral(sum);
        }
    }
    return sum;
}
```

Common Errors:

- This is a tricky piece of code. There are many places where errors can sneak in. Remember that you can earn most of the points for a question even with errors or missing sections in your code.
- Make sure you simplify all the way. You can't just go through the String once. As the examples show, you need to keep simplifying until the simplification process doesn't make any changes. There are several ways to do this. The solution above uses a boolean flag that is set to true at the beginning of each iteration of the loop; it is then reset to false if any changes are made. The alternate solution compares the Strings at the beginning and end of the simplification loop to see if they are the same.
- The multiple substrings are tricky. Use a short example and work through it by hand to be sure you haven't made an off-by-one error.
- Using the getLetterValue and getLetter methods is an easy way to change one letter into the next, and since the directions remind us to use them, it is the way to go. However, if you couldn't figure out how to do it, there are other ways that it can be done, especially if you go outside of the Java subset. Remember that you may use any valid Java code; however, all questions can be solved within the Java subset, and it is recommended that you keep your code within the subset.

Java Code Alternate Solution:

The only difference between this code and the code above is the way in which it decides when the String is completely simplified.

```
public String add(String num1, String num2)
{
    String oldSum = num1 + num2;
    String sum = "";
    while (!sum.equals(oldSum))
    {
        int i = 0;
        sum = sortLocationNumeral(oldSum);
        while (i < sum.length() - 1)
        {
            String currLetter = sum.substring(i, i + 1);
            if (currLetter.equals(sum.substring(i + 1, i + 2)))
            {
                String before = sum.substring(0, i);
                String after = sum.substring(i + 2);
                String newLetter = getLetter(2 * getLetterValue(currLetter));
                sum = before + newLetter + after;
            }
            else
            {
                i++;
            }
            oldSum = sum;
            sum = sortLocationNumeral(oldSum);

        }
    }
    return sum;
}
```

Scoring Guidelines: Location Numerals

Part (a) **getLetterValue** **1 point**

+1 Returns the correct value

Part (b) **getDecimalValue** **3 points**

+1 Accesses every individual letter in numeral; no bounds errors, no missing values
+1 Calls getLetterValue for each letter
+1 Accumulates and returns the correct total value

Part (c) **add** **5 points**

+1 Concatenates num1 and num2 and calls sortLocationNumeral at least once
+2 Correctly completes one simplification cycle
  +1 Identifies a pair of duplicate values and identifies the letter to replace them
  +1 Constructs the new String with the duplicate replaced and sorts it
+1 Executes the simplification loop repeatedly until no more simplifications are made
+1 Returns the correct simplified sum

Sample Driver:
There are many ways to write these methods. Maybe yours is a bit different from our sample solutions and you are not sure if it works. Here is a sample driver program. Running it will let you see if your code works and will help you debug it if it does not.

Copy the LocationNumeralDriver into your IDE along with the LocationNumeral class (including your solutions). You will also need to:

- Add this import statement as the first line in your LocationNumeral class:

```
import java.util.Arrays;
```

- Add the completed getLetter method to the LocationNumeral class:

```
private String getLetter(int value)
{
    String letters = "ABCDEFGHIJKLMNOPQRSTUVWXYZ";
    int index = (int) (Math.log(value) / Math.log(2));
    return letters.substring(index, index + 1);
}
```

- Add the completed sortLocationNumeral method to the LocationNumeral class:

```
public String sortLocationNumeral(String numeral)
{
    char[] ar = numeral.toCharArray();
    Arrays.sort(ar);
    return String.valueOf(ar);
}
```

- Here's the LocationNumeralDriver:

```
public class LocationNumeralDriver {

   public static void main(String[] args) {
      LocationNumeral num = new LocationNumeral();

      System.out.println("getLetterValue test\nExpecting:" +
            "     A = 1 E = 16 M = 4096");
      System.out.println("Your answers: A = " +
         num.getLetterValue("A") + " E = " +
         num.getLetterValue("E") + " M = " +
         num.getLetterValue("M"));

      System.out.println("\ngetDecimalValue test\nExpecting:" +
            "    ABC = 7 EGM = 4176");
      System.out.println("Your answers: ABC = " +
           num.getDecimalValue("ABC") +
           " EGM = " + num.getDecimalValue("EGM"));

      System.out.println("\nadd test\nExpecting:" +
            "    BDE + AC = ABCDE    AB + ABC = BD");
      System.out.println("Your answers: BDE + AC = " +
            num.add("BDE", "AC") + "    AB + ABC = " +
            num.add("AB", "ABC"));
   }
}
```

Scoring Worksheet

This worksheet will help you to approximate your performance on Practice Exam 2 in terms of an AP score of 1–5.

Part I (Multiple Choice)

Number right (out of 40 questions) = ________

Part II (Short Answer)

See the scoring guidelines included with the explanations for each of the questions and award yourself points based on those guidelines.

Question 1: Points Obtained (out of 9 possible) ________

Question 2: Points Obtained (out of 9 possible) ________

Question 3: Points Obtained (out of 9 possible) ________

Question 4: Points Obtained (out of 9 possible) ________

Add the points and multiply by 1.1111 ________ × 1.1111 = ________

Add your totals from both parts of the test (round to nearest whole number) Total Raw Score ________

Approximate conversion from raw score to AP score

Raw Score Range	AP Score	Interpretation
62–80	5	Extremely Well Qualified
44–61	4	Well Qualified
31–43	3	Qualified
25–30	2	Possibly Qualified
0–24	1	No Recommendation

Appendix

Quick Reference Guide

Methods and fields from the Java library that may appear on the AP Computer Science A Exam:

```
class java.lang.Object
• boolean equals(Object other)          // compares this with other
• String toString()                     // returns a description of the object

class java.lang.Integer
• Integer(int value)                    // constructor for an Integer
• int intValue()                        // returns the value of the Integer
• Integer.MIN_VALUE                     // minimum value of an int or Integer
• Integer.MAX_VALUE                     // maximum value of an int or Integer

class java.lang.Double
• Double(double value)                  // constructor for a Double
• double doubleValue()                  // returns the value of the Double

class java.lang.String
• int length()                          // return the number of characters in String
• String substring(int from, int to)    // returns the substring beginning at from
                                        // and ending at to - 1
• String substring(int from)            // returns a String starting at from
• int indexOf(String str)               // returns index of the occurrence of str
                                        // returns -1 if str is not found
• int compareTo(String other)           // returns a negative # if this is < other
                                        // returns zero if this is equal to other
                                        // returns a positive # if this is > other

class java.lang.Math
• static int abs(int x)                 // returns the absolute value of x (int)
• static double abs(double x)           // returns the absolute value of x (double)
• static double pow(double base, double exponent)
                                        // returns the base raised to the exponent
• static double sqrt(double x)          // returns the square root of double
• static double random()                // returns a double in the range [0.0, 1.0)

interface java.util.List<E>
• int size()                            // returns the number of objects in the list
• boolean add(E obj)                    // appends obj to end of list; returns true
• void add(int index, E obj)            // inserts obj at position index [0, size]
• E get(int index)                      // returns the object at position index
• E set(int index, E obj)               // replaces element at index with obj
                                        // and returns element formerly at index
• E remove(int index)                   // removes element from position index
                                        // and returns the element formerly at index

class java.util.ArrayList<E> implements java.util.List<E>
```

Free-Response Scoring Guidelines

When grading your exam, the AP Computer Science A grader will compare your code for each question to the rubric for each question. After this process is complete, penalty points may be deducted from your score based on the list below. A penalty point can only be deducted once per question (even if you make the same mistake multiple times in the same question). You can't get a negative total. These are the most current set of guidelines as of the 2015 exam. Be aware that they are sometimes modified from year to year.

General Penalties (assessed only once per problem)

-1 using a local variable without first declaring it
-1 returning a value from a void method or constructor
-1 accessing an array or ArrayList incorrectly
-1 overwriting information passed as a parameter
-1 including unnecessary code that causes a side effect like a compile error or console output

No Penalty (*These may change from year to year!*)

- Extra code that doesn't cause any problems or doesn't print to the console
- Minor spelling mistakes such as "wile" for "while"
- Code that includes private or public on a local variable
- Forgetting to put public when writing a class or constructor header
- Accidently using the mathematical symbols for operators (×, •, +, ≤, ≥, <>, ≠)
- Accidently using "=" rather than "==" and vice versa
- Confusing length with size for either String, ArrayList, List, or array
- Accidently putting the size in array declaration such as int[25] myArray = new int[25];
- Missing a semicolon if you write the majority of them and you provide proper spacing
- Missing curly braces when your indenting is proper and you used them elsewhere
- Accidently forgetting to put parentheses on no-argument method or constructor calls
- Forgetting to put parentheses on if statements or while loops provided indenting is proper

Note: Suppose your code declares "Bunny bunny = new Bunny()" and then uses "Bunny.hop()" instead of "bunny.hop()". The mistake is not treated as a grammatical error. Instead, you will not earn points, because the grader is not allowed to assume that you know the difference between "Bunny" and "bunny" in the context of this problem.

List of Keywords in Java

The following words are keywords in Java and cannot be used as identifiers (variable names, class names, etc.).

abstract	assert	boolean	break	byte	case	catch	char
class	const	continue	default	do	double	else	enum
extends	final	finally	float	for	goto	if	implements
import	instanceof	int	interface	long	native	new	package
private	protected	public	return	short	static	strictfp	super
switch	synchronized	this	throw	throws	transient	try	void
volatile	while	true*	false*	null*	*Not keywords but can't be used!		

List of Required Runtime Exceptions

Runtime Exception	Possible Reason for Error
ArithmeticException	You divided by zero. You performed a mod by zero.
NullPointerException	You tried to make a null object reference perform a method. You tried to access or modify a field of a null object reference.
IndexOutOfBoundsException	This error occurs when doing ArrayList handling like add, get, remove and so on. The index for the ArrayList method is greater than or equal to the size of the ArrayList. The index for the ArrayList method is negative.
ArrayIndexOutOfBoundsException	You are attempting to use an index that is larger than (or equal to) the length of the array. You are attempting to use a negative array index
IllegalArgumentException	A method has been passed an inappropriate argument.

Common Syntax Errors for Beginning Java Programmers

<table>
<tr><th>Incorrect Code</th><th>Explanation</th><th>Correct Code</th></tr>
<tr><td><code>int count = 0</code></td><td>Forgot the semicolon</td><td><code>int count = 0;</code></td></tr>
<tr><td><code>system.out.println("smile);</code></td><td>Forgot the second double quote</td><td><code>System.out.println("smile");</code></td></tr>
<tr><td><code>double gpa = true;</code></td><td>Assigned a boolean to a double</td><td><code>double gpa = 3.75;</code></td></tr>
<tr><td><code>double a = 6.45;</code>
<code>int b = a;</code></td><td>Loss of precision since an integer will not accept the decimal portion</td><td><code>double a = 6.45;</code>
<code>double b = a;</code></td></tr>
<tr><td><code>system.out.println("OMG");</code></td><td>System starts with uppercase</td><td><code>System.out.println("OMG");</code></td></tr>
<tr><td><code>System.out.println('Yo');</code></td><td>Need double quotes rather than single quotes</td><td><code>System.out.println("Yo");</code></td></tr>
<tr><td><code>if (x > 5);</code>
<code>{</code>
<code>   // some random code</code>
<code>}</code></td><td>Semicolon before the curly brace</td><td><code>if (x > 5)</code>
<code>{</code>
<code>   // some random code</code>
<code>}</code></td></tr>
<tr><td><code>while(x > 5);</code>
<code>{</code>
<code>   // some random code</code>
<code>}</code></td><td>Semicolon before the curly brace</td><td><code>while(x > 5)</code>
<code>{</code>
<code>   // some random code</code>
<code>}</code></td></tr>
</table>

Online Resources

College Board AP Computer Science A Exam
https://apstudent.collegeboard.org/apcourse/ap-computer-science-a

Java API
http://docs.oracle.com/javase/8/docs/api/

If you have any questions, suggestions, or would like to report an error, we'd love to hear from you. Email us at 5stepstoa5apcsa@gmail.com.